Molecular Biology of G-Protein-Coupled Receptors

Mark R. Brann

Editor

Foreword by Leslie L. Iversen

61 Illustrations

Birkhäuser
Boston · Basel · Berlin

Mark R. Brann
Molecular Neuropharmacology Section
Department of Psychiatry
College of Medicine
University of Vermont
Burlington, VT 05405-0068
USA

Library of Congress Cataloging-in-Publication Data
Molecular biology of G-protein-coupled receptors / Mark R.
  Brann, editor : with a foreword by Leslie L. Iversen.
      p. cm. — (Applications of molecular genetics to
  pharmacology)
    Includes bibliographical references and index.
    ISBN 0-8176-3465-7 (H : alk. paper). — ISBN 3-7643-3465-7 (alk.
  paper)
    1. G proteins — Receptors.   2. G proteins — Physiological effect.
  3. Neurotransmitter receptors. I. Brann, Mark R., 1958–
  II. Series.
    [DNLM: 1. Guanine Nucleotide Regulatory Protein — physiology.
  2. Receptors, Endogenous Substances — physiology.   3. Receptors,
  Synaptic — physiology.   WL 102.8 M717]
  QP552.G16M65   1992
  612.8′22 — dc20
  DNLM/DLC
  for Library of Congress   91-33334
                                                                  CIP

Printed on acid-free paper.

ISBN 0-8176-3465-7
ISBN 3-7643-3465-7

Typeset by Lind Graphics, Inc., Upper Saddle River, NJ.
Printed and bound by Edwards Brothers, Inc., Ann Arbor, MI.
Printed in the United States of America.

9 8 7 6 5 4 3 2 1

Contents

Foreword vii
Leslie L. Iversen

Preface ix
Mark R. Brann

Contributors xiii

1. Rhodopsin 1
   Nicholas J. P. Ryba, Matthew D. Hall,
   and John B. C. Findlay

2. Structural, Functional, and Genetic Aspects of Receptors
   Coupled to G-Proteins 31
   Brian F. O'Dowd, Sheila Collins, Michel Bouvier,
   Marc G. Caron, and Robert J. Lefkowitz

3. Genetic Analysis of the β-Adrenergic Receptor 62
   Catherine D. Strader and Richard A. F. Dixon

4. The α-Adrenergic Receptors: New Subtypes, Pharmacology,
   and Coupling Mechanisms 76
   John W. Regan and Susanna Cotecchia

5. The 5-HT$_{1A}$ Receptor: From Molecular Characteristics
   to Clinical Correlates 113
   John R. Raymond, Salah El Mestikawy, and Annick Fargin

6. The Dopamine D1 Receptors 142
   Hyman B. Niznik, Roger K. Sunahara, Hubert H. M. Van Tol,
   Philip Seeman, David M. Weiner, Tom M. Stormann,
   Mark R. Brann, and Brian F. O'Dowd

7. The Dopamine D2 Receptor 160
   Olivier Civelli, James Bunzow, Paul Albert,
   Hubert H. M. Van Tol, and David Grandy

8. Muscarinic Acetylcholine Receptors 170
   S. V. Penelope Jones, Allan I. Levey, David M. Weiner,
   John Ellis, Elizabeth Novotny, Shua-Hua Yu, Frank Dorje,
   Jurgen Wess, and Mark R. Brann

9. Molecular Biology of Peptide and Glycoprotein Hormone
   Receptors 198
   David R. Poyner and Michael R. Hanley

10. Signal Transducing G-Proteins: α Subunits............................ 233
    Yoshito Kaziro

11. Immunological Probes of the Structure, Function, and
    Expression of Heterotrimeric GTP-Binding Proteins................. 270
    Allen M. Spiegel, Paul K. Goldsmith, William F. Simonds,
    Teresa Jones, Kevin Rossiter, and Cecilia G. Unson

Index 299

Foreword

LESLIE L. IVERSEN

The present series of volumes is well timed, as the impact of molecular genetics on pharmacology has been profound, and a comprehensive review of the rapid advances of the past decade is much needed. Since the pioneering work of Dale, Ariens, and others in the early years of this century, much of pharmacology has been founded on the concept of receptors. To begin with, the receptor was conceived of as a "black box," which recognized and transduced the biological effects of neurotransmitters, hormones, or other biological messengers—and which could also represent a target for man-made drugs.

It is only in the last two decades that "molecular pharmacology" has blossomed, first with the advent of radioligand binding techniques and second messenger studies which greatly facilitated the biochemical study of drug-receptor interactions, and latterly with increasing knowledge of the molecular architecture of the receptor proteins themselves. This started with the traditional biochemical approach of isolating and purifying the receptor molecules. This proved to be a task of immense technical difficulty because of the low density of receptors in most biological source tissues, although there were some notable successes, e.g., the purification of the nicotinic acetylcholine receptor from the electric organ of *Torpedo*. It was the application of molecular genetics technology during the 1980s, however, which really accelerated progress in this field. Following the successful cloning of the genes encoding the subunits of the nicotinic acetylcholine receptor, a flood of new information was unleashed. Indeed, by now a majority of all the classical drug-neurotransmitter receptors has been successfully cloned and sequenced, including more than 20 belonging to the family of G-protein-coupled receptors which form the focus for the first volume—*Molecular Biology of G-Protein-Coupled Receptors*—in this series.

What has this new information taught us about receptors? The amino acid sequences themselves have proved remarkably informative. In both the G-protein-coupled family and ion-channel receptors it has proved possible to deduce some of the key features of their molecular architecture from the primary sequence information. Thus, in all cases hydrophilic regions of amino acid sequence are interspersed with hydrophilic domains, which correspond to α-helical membrane-spanning regions of the molecule. The overall architecture has proved surprisingly similar in all members of a particular family of receptors, as with the G-protein-coupled series with their uniform sevenfold membrane-spanning domains. These key regions of the receptor molecule have also proved to be highly conserved through evolution.

Another big surprise to pharmacologists has been the remarkable diversity of receptor subtypes revealed by the molecular genetics studies. For example, pharmacologists had argued among themselves about the possible existence of more than one muscarinic receptor subtype. It came as a shock to learn some six years ago that no less than 5 genes encoding muscarinic receptor subtypes could be identified in mammalian tissues. Our ideas about the pharmacology of such well-studied classical receptors as the nicotinic and GABA-A have also changed radically with the discovery that multiple genes exist encoding different forms of the subunits of these receptors, offering a far richer diversity of subtypes than was ever thought possible—and in turn offering new challenges for the design of more selective agonist-antagonist drugs in the future.

It is the future possibilities opened by this new molecular era that are the most exciting. Apart from the diversity of receptor subtypes, which offers the possibility of more targeted drugs, we also have the opportunity of using human receptors for drug discovery. Through the use of immortalized cell lines expressing human receptors, or the development of transgenic animals expressing these receptors, it will become routine to use human receptors in primary drug discovery screens. The detailed knowledge of receptor structure will also eventually provide three-dimensional receptor models of sufficient precision and resolution to be used in the rational design of new drugs—in the same way as such information on enzyme structures is already being employed in the design of inhibitors.

I am looking forward to seeing the publication of the remaining volumes in this series, and believe that they will provide a landmark in the documentation of an important new departure for pharmacology.

Merck Sharp & Dohme Research Laboratories
Neuroscience Research Centre
Harlow, Essex CM20 2QR
United Kingdom

Preface

MARK R. BRANN

Molecular Biology of G-Protein-Coupled Receptors is the first volume in the series *Applications of Molecular Genetics to Pharmacology*. My motivation for starting this series came from two observations. First, with the molecular cloning of many families of drug receptors in the late 1980s, molecular genetics truly "came of age" in its application to mainstream pharmacology. Second, in spite of these landmark achievements, communication between pharmacologists and molecular biologists remains infrequent. In fact, now that many receptor clones have been available for several years, one can observe the development of a disturbing trend. From the viewpoint of classical pharmacologists, molecular biologists often report the characterization of receptors as if they were unaware of the fact that the molecules that they are studying have, in most cases, been the subject of intensive investigation for decades. Is it really necessary to repeat every observation with "cloned" receptors? On the other hand, why should years be spent on unraveling the details of the regulation of new genes before their pharmacological relevance has been documented? Perhaps, because of these views, the impact of molecular genetics on the classical pharmacological literature has been less than anticipated. Examples of citations such as "molecular cloning has supported the existence of receptor subtypes" or "receptor X has been cloned" are often found in the introduction sections of papers, but results from molecular genetic experiments are very rarely discussed in the context of the actual pharmacological data from these same papers. In fact, in the cases where clones have been available for several years (e.g., nicotinic, muscarinic, and β-adrenergic receptors), references to molecular biological experiments may have actually decreased in recent years.

Is this trend toward separate molecular genetic and pharmacological literatures based on a general lack of complementarity between the molecular genetic and classical pharmacological techniques and approaches? Or are these trends caused by difficulties in communication that are based on differences in the training and outlooks of pharmacologists and molecular biologists? Obviously, being the editor of this series, I am convinced that pharmacologists have a great deal to learn from molecular biologists and vice versa. The intent of *Applications of Molecular Genetics to Pharmacology* is to provide a transfer of information between these two divided camps. To achieve this goal I have recruited a series of contributors and editors who have pioneered the use of molecular genetic techniques for answering fundamental questions in pharmacology. The first volumes review the characterization of major classes of drug receptors (e.g., the G-protein-coupled receptors, and the ligand- and voltage-gated ion channels). Future volumes will review the structure-function relationships of receptors (e.g., crystallography, computational modeling, mutagenesis), and the identification and characterization of disease genes (e.g., molecular medicine, transgenic models of disease). The latter findings will be discussed in the context of rational drug design.

At the time the authors for *Molecular Biology of G-Protein-Coupled Receptors* were initially recruited, it was easy to cover the breadth of the G-protein-coupled receptors that had been subjected to analysis by molecular genetic techniques in one volume. In fact, I instructed the authors to avoid figures with sequence comparisons because I would prepare an exhaustive alignment figure for the preface. Don't turn the page, you won't find one. One will have to be content with the alignment figure from O'Dowd et al., which appears on pages 32–34 (19 sequences). While this alignment is far from exhaustive, more comparisons would be difficult to present using printed media. For a taste of how many sequences are available, consider the alignment figure of Ryba, Hall, and Findlay on pages 2 and 3 (20 sequences, and they only considered the visual pigments). Without even considering species homologs, more than 100 G-protein-coupled receptors have been cloned to date. These numbers do not include the "orphan receptors" (homologous clones with no known ligand or function) that are being cloned at an even more striking rate. For example, more than 100 genetically related clones have been identified in olfactory mucosa, and these may represent an even larger family of odorant receptors. Interestingly, many of these same receptors are expressed by sperm, from which dozens of other related genes have also been cloned. At a recent presentation at the University of Vermont, Dr. Randall Reed of Johns Hopkins University reported that approximately 1 out of every 300 clones in a typical genomic library has a G-protein-coupled receptor sequence from the family of olfactory receptors. Together, these data suggest that the G-protein-coupled receptors may represent the largest family of genes in the human genome.

In spite of the very rapid appearance of new receptor sequences, the last two years have seen the study of G-protein-coupled receptors reach a stage where consensus and common themes are being clearly established. Thus, the appearance of our book is well timed. The book starts with the chapter "Rhodopsin" because far more is known about the structure and function of the visual pigments than is known about any of the other G-protein-coupled receptors. Thus, Ryba, Hall, and Findlay present the volume's most complete model of the structure-function relationships of a G-protein-coupled receptor. Analogies to this model are drawn in all of the later chapters. In addition to being the flagship in terms of knowledge of structure-function relationships, the visual pigments were the first genes of the G-protein-coupled receptor family to be linked with human disease (e.g., altered visual pigment genes cause retinitis pigmentosa and color blindness).

In the chapter "Structural, Functional, and Genetic Aspects of Receptors Coupled to G-Proteins," O'Dowd et al. provide a rich source of information on the general features that are shared by all G-protein-coupled receptors. In addition, the authors present an elegant discussion of how the sensitivity of β-adrenergic receptors are controlled by various posttranslational modifications and by molecules such as β-ARK and β-arrestin. Strader and Dixon, in the next chapter, "Genetic Analysis of the β-Adrenergic Receptor," review the elegant work that proved the existence of a structure-function analogy between visual pigments and neurotransmitter receptors. They also demonstrate the power of site-directed mutagenesis when applied to the study of structure-function relationships.

Regan and Cotecchia in their chapter "The α-Adrenergic Receptors: New Subtypes, Pharmacology, and Coupling Mechanisms" review the striking diversity of α-adrenergic receptor subtype genes, and the significance of these genes in understanding adrenergic pharmacology and physiology. A similar theme of multiple receptor subtype genes and their physiological significance is explored for the serotonin receptor subtypes by Raymond, El Mestikawy, and Fargin in the chapter "The 5-HT$_{1A}$ Receptor: From Molecular Characteristics to Clinical Correlates." This article and those that follow describe the possible therapeutic utility of drugs that have subtype selectivity.

Niznik et al., in the chapter "The Dopamine D1 Receptors" and Civelli et al. in the chapter "The Dopamine D2 Receptor" describe the much-greater-than-expected heterogeneity that exists for dopamine receptor subtype genes. These authors also highlight the opportunities that now exist for developing new antipsychotic medications. Niznik et al. describe the curious existence of a family of "D1-like" un-translated pseudogenes, and Civelli et al. demonstrate just how far homology cloning can be extended, as the dopamine receptors were initially cloned from their distant cousin the β-adrenergic receptor. Jones et al. in "Muscarinic Acetylcholine Receptors" also deal with the issue of the genetic heterogeneity of receptor subtypes, and just how complex a task it is to relate genetically and pharmacologically derived data. These articles also describe the use of genetic techniques to define the pattern of expression of these newly discovered receptor subtypes.

Chapters 2 through 8 review information on receptors for monoaminergic ligands. These receptors share remarkable levels of sequence homology and have very similar structure-function relationships. For all of these receptors, the primary recognition of the ligand is by the hydrophobic core of the molecule. On the other hand, receptors for polypeptides have less sequence similarity and utilize hydrophilic epitopes in ligand binding. Poyner and Hanley, authors of the chapter "Molecular Biology of Peptide and Glycoprotein Hormone Receptors" review data from the very large family of receptors that bind polypeptides.

The final two chapters of this volume review features of the signal transducing G-proteins. In his chapter "Signal Transducing G-Proteins: α Subunits," Kaziro reviews the structure and function of the α subunits of the G-proteins. These are the subunits that define the selectivity of G-proteins for receptors and functional responses. As for the receptors, a central theme is the identification of an unexpected heterogeneity of genetic subtypes. Spiegel et al., the authors of the last chapter, "Immunological Probes of the Structure, Function, and Expression of Heterotrimeric GTP-Binding Proteins," review the use of immunological techniques to study G-proteins. These authors summarize the progress that has been made toward understanding the specificity of receptor/G-protein interaction and the functional significance of the newly discovered subunits.

Contributors

Paul Albert   Vollum Institute for Advanced Biomedical Research, Oregon Health Sciences University L-474, 3181 S.W. Sam Jackson Park Road, Portland, OR 97201

Michel Bouvier   Department of Biochemistry, University of Montreal, C.P. 6208 Succ. A, Montreal H3C 3T8, Canada

Mark R. Brann   Molecular Neuropharmacology Section, Department of Psychiatry, College of Medicine, University of Vermont, Burlington, VT 05405; *Former address:* Laboratory of Molecular Biology, National Institute of Neurological Disorder and Stroke, National Institutes of Health, Bethesda, MD 20892

James Bunzow   Vollum Institute for Advanced Biomedical Research, Oregon Health Sciences University L-474, 3181 S.W. Sam Jackson Park Road, Portland, OR 97201

Marc G. Caron   Howard Hughes Medical Institute, Departments of Medicine, Biochemistry, and Cell Biology, Duke University Medical Center, Box 3821 (468 CARL Building), Durham, NC 27710

Olivier Civelli   Vollum Institute for Advanced Biomedical Research, Oregon Health Sciences University L-474, 3181 S.W. Sam Jackson Park Road, Portland, OR 97201

Sheila Collins   Howard Hughes Medical Institute, Departments of Medicine, Biochemistry, and Cell Biology, Duke University Medical Center, Box 3821 (468 CARL Building), Durham, NC 27710

Susanna Cotecchia   Department of Medicine, Duke University Medical Center, Durham, NC 27710

Richard A. F. Dixon   Texas Biotechnology Corporation, Houston, TX 77030; *Former address:;* Departments of Biochemistry and Molecular Biology, Merck Sharp & Dohme Research Laboratories, Rahway, NJ 07065

Frank Dorje   Department of Pharmacology, University of Frankfurt, Theodo-Stem-ki, Gebäude 75A, W-6000 Frankfurt/Main, Germany; *Former address:* Laboratory of Molecular Biology, National Institute of Neurological Disorders and Stroke, National Institutes of Health, Bethesda, MD 20892

John Ellis   Molecular Neuropharmacology Section, Department of Psychiatry, College of Medicine, University of Vermont, Burlington, VT 05405

Salah El Mestikawy   Department of Cell Biology, Duke University Medical Center, Durham, NC 27710

Annick Fargin   Department of Pharmacology, University of Montreal, Montreal H3C 3T8, Canada

John B. C. Findlay   Department of Biochemistry and Molecular Biology, University of Leeds, Leeds LS2 9JT, United Kingdom

Paul K. Goldsmith   Molecular Pathophysiology Branch, National Institute of Diabetes, Digestive, and Kidney Disease, National Institutes of Health, Bethesda, MD 20892

David Grandy   Vollum Institute for Advanced Biomedical Research, Oregon Health Sciences University L-474, 3181 S.W. Sam Jackson Park Road, Portland, OR 97201

Matthew D. Hall   National Institute of Dental Research, National Institutes of Health, Bethesda, MD 20892; *Former address:* MRC Molecular Neurobiology Unit, University of Cambridge Medical School, Cambridge CB2 2QH, United Kingdom

Michael R. Hanley   Department of Biochemistry, School of Medicine, University of California, Davis, CA 95616-8635; *Former address:* MRC Molecular Neurobiology Unit, University of Cambridge Medical School, Cambridge CB2 2QH, United Kingdom

S. V. Penelope Jones   Molecular Neuropharmacology Section, Department of Psychiatry, College of Medicine, University of Vermont, Burlington, VT 05405

Teresa Jones   Molecular Pathophysiology Branch, National Institute of Diabetes, Digestive, and Kidney Disease, National Institutes of Health, Bethesda, MD 20892

Yoshito Kaziro Department of Molecular Biology, DNAX Research Institute of Molecular and Cellular Biology, Palo Alto, CA 94304; *Former address:* Institute of Medical Sciences, University of Tokyo, 4-6-1 Shirokenedai Minato-Ku, Tokyo 108, Japan

Robert J. Lefkowitz Howard Hughes Medical Institute, Departments of Medicine, Biochemistry, and Cell Biology, Duke University Medical Center, Box 3821 (468 CARL Building), Durham, NC 27710

Allan I. Levey Department of Neurology, College of Medicine, Emory University, Woodruff Memorial Research Bldg., Suite 6000, P.O. Drawer 5, Atlanta GA 30322-451; *Former address:* Laboratory of Molecular Biology, National Institute of Neurological Disorders and Stroke, National Institutes of Health, Bethesda, MD 20892

Hyman B. Niznik Departments of Pharmacology and Psychiatry, University of Toronto, Toronto, Ontario M5S 1A8; Laboratory of Molecular Neurobiology, Clarke Institute of Psychiatry, 250 College Street, Toronto, Ontario M5T 1R8, Canada

Elizabeth Novotny Nova Pharmaceutical Corporation, Baltimore, MD 21205; *Former address:* Laboratory of Molecular Biology, National Institute of Neurological Disorders and Stroke, National Institutes of Health, Bethesda, MD 20892

Brian F. O'Dowd Addiction Research Foundation, 33 Russell Street, Toronto, Ontario M5S 2S1, and Department of Pharmacology, Medical Sciences Building, University of Toronto, Toronto, Ontario M5S 1A8, Canada

David R. Poyner Department of Pharmaceutical Sciences, Aston University, Aston Triangle, Birmingham B4 7ET, United Kingdom; *Former address:* MRC Molecular Neurobiology Unit, University of Cambridge Medical School, Cambridge CB2 2QH, United Kingdom

John R. Raymond Department of Internal Medicine, Duke University Medical Center and Durham Veterans Administration Medical Center, Box 3459, Durham, NC 27710

John W. Regan Department of Pharmacology and Toxicology, College of Pharmacy, University of Arizona, Tucson, AZ 85721

Kevin Rossiter Molecular Pathophysiology Branch, National Institute of Diabetes, Digestive, and Kidney Disease, National Institutes of Health, Bethesda, MD 20892

Nicholas J. P. Ryba National Institute of Dental Research, National Institutes of Health, Bethesda, MD 20892; *Former address:* Department of Biochemistry and Molecular Biology, University of Leeds, Leeds LS2 9JT, United Kingdom

Philip Seeman   Departments of Pharmacology and Psychiatry, University of Toronto, Toronto, Ontario M5S 1A8, Canada

William F. Simonds   Molecular Pathophysiology Branch, National Institute of Diabetes, Digestive, and Kidney Disease, National Institutes of Health, Bethesda, MD 20892

Allen M. Spiegel   Molecular Pathophysiology Branch, National Institute of Diabetes, Digestive, and Kidney Disease, National Institutes of Health, Bethesda, MD 20892

Tom M. Stormann   Nova Pharmaceutical Corporation, Baltimore, MD 21205; *Former address:* Laboratory of Molecular Biology, National Institute of Neurological Disorders and Stroke, National Institutes of Health, Bethesda, MD 20892

Catherine D. Strader   Departments of Biochemistry and Molecular Biology, Merck Sharp & Dohme Research Laboratories, Rahway, NJ 07065

Roger K. Sunahara   Department of Pharmacology, University of Toronto, Toronto, Ontario M5S 1A8, Canada

Cecilia G. Unson   Department of Biochemistry, Rockefeller University, New York, NY 10021

Hubert H. M. Van Tol   Departments of Pharmacology and Psychiatry, University of Toronto, Toronto, Ontario M5S 1A8; *Former address:* Vollum Institute for Advanced Biomedical Research, Oregon Health Sciences University L-474, 3181 S.W. Sam Jackson Park Road, Portland, OR 97201

David M. Weiner   Department of Neurology, Columbia University School of Medicine, New York, NY 10032; *Former address:* Laboratory of Molecular Biology, National Institute of Neurological Disorders and Stroke, National Institutes of Health, Bethesda, MD 20892

Jurgen Wess   Laboratory of Molecular Biology, National Institute of Neurological Disorders and Stroke, National Institutes of Health, Bethesda, MD 20892

Shua-Hua Yu   Laboratory of Molecular Biology, National Institute of Neurological Disorders and Stroke, National Institutes of Health, Bethesda, MD 20892

Molecular Biology of
G-Protein-Coupled Receptors

1

Rhodopsin

Nicholas J. P. Ryba, Matthew D. Hall, and
John B. C. Findlay

1. Introduction

The initial determination of the primary structure of bovine (Ovchinnikov, 1982; Hargrave et al., 1983) and ovine (Findlay et al., 1981; Brett and Findlay, 1983) opsins and parts of the equine and porcine proteins (Pappin and Findlay, 1984) by the protein sequencing approach was rapidly followed by the derivation of the structure of a number of other species from their nucleotide sequences (for the latest compilation, see Figure 1.1). Among G-protein receptors, the rhodopsins have been well characterized by physical techniques; the biochemistry related to their function has also been documented more completely than for the others. This chapter will describe the amalgamation of the information obtained so far into a representation of the structure and its implications for the function of the visual pigment. In the absence of a crystallographically derived structure, the crude structural model we describe can be used predictively and to analyze results of both physical and mutagenic studies of rhodopsin structure and function. Moreover, it has relevance for the whole superfamily of G-protein-linked receptors, as it is likely that the structural framework will be well preserved. The final goal in all this work is a detailed understanding of the structure of rhodopsin and of the molecular mechanisms through which its functional attributes are expressed. Its significance extends not only to G-protein-linked receptors as a superfamily (Findlay and Eliopoulos, 1991) but to integral membrane proteins in general.

2. Structural Representation

In some respects integral membrane proteins may be easier to model at a relatively crude level than their water-soluble relatives. It seems likely, for

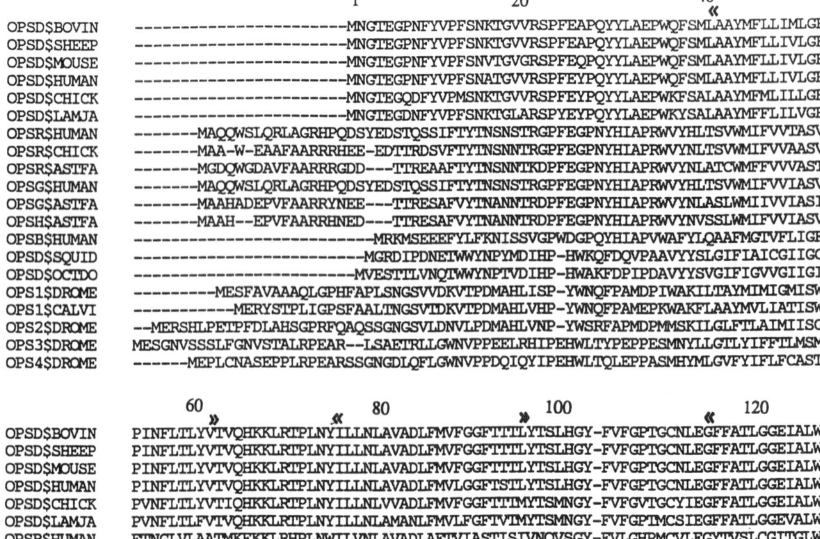

FIGURE 1.1. Alignment of opsin sequences. Alignment of published sequences of visual pigments obtained by application of the SOMAP program (Parry-Smith and Attwood, 1991). Boundaries of the intramembranous regions are delineated by arrowheads. Numbers correspond with residues of bovine rhodopsin.

The sequences were compiled from the following sources: **1**, bovine rhodopsin (Ovchinnikov, 1982; Hargrave et al., 1983); **2**, ovine rhodopsin (Pappin et al., 1984); **3**, mouse rhodopsin (Baehr et al., 1988); **4**, human rhodopsin (Nathans & Hogness, 1984); **5**, chicken rhodopsin (Takao et al., 1988); **6**, lamprey rhodopsin (Hisatomi et al., 1991); **7**, human red color pigment (Nathans et al., 1986); **8**, chicken red color pigment (Tokunaga et al., 1990); **9**, blind cave fish *(Astyanax fasciatus)* red color pigment (note: incomplete sequence at C-terminus) (Yokoyama

```
                200       «                220          »                240                    «
OPSD$BOVIN      NESFVIYMFVVHFIIPLIVIFFCYGQLVFTVKEAAA---------QQQESATTQ-----KAE-KEVTRMVIIMVI
OPSD$SHEEP      NESFVIYMFVVHFSIPLIVIFFCYGQLVFTVKEAAA---------QQQESATTQ-----KAE-KEVTRMVIIMVI
OPSD$MOUSE      NESFVIYMFVVHFTIPMIVIFFCYGQLVFTVKEAAA---------QQQESATTQ-----KAE-KEVTRMVIIMVI
OPSD$HUMAN      NESFVIYMFVVHFTIPMIIIFFCYGQLVFTVKEAAA---------QQQESATTQ-----KAE-KEVTRMVIIMVI
OPSD$CHICK      NESFVIYMFVVHFMIPLAVIFFCYGNLVCTVKEAAA---------QQQESATTQ-----KAE-KEVTRMVIIMVI
OPSD$LAMJA      NESYVVYMFVVHFLVPFVIIFFCYGRLLCTVKEAAA---------AQQESASTQ-----KAE-KEVTRMVVLMVI
OPSR$HUMAN      VQSYMIVLMVTCCIIPLAIIMLCYLQVWLAIRAVAK---------QQKESESTQ-----KAE-KEVTRMVVVMIF
OPSR$CHICK      VQSYMVVLMVTCCFFPLAIIILCYLQVWLAIRAVAA---------QQKESESTQ-----KAE-KEVSRMVVVMIV
OPSR$ASTFA      VQSYMIVLMITCCFIPLGIIILCYIAVWWAIRTVAQ---------QQKDSESTQ-----KAE-KEVSRMVVVMIM
OPSG$HUMAN      VQSYMIVLMVTCCITPLSIIVLCYLQVWLAIRAVAK---------QQKESESTQ-----KAE-KEVTRMVVVMVL
OPSG$ASTFA      VASYMVTLLLTCCILPLSVIIICYIFVWNAIHQVAQ---------QQKDSESTQ-----KAE-KEVSRMVVVMIL
OPSH$ASTFA      VASYMITLMLTCCILPLSIIIICYIFVWSAIHQVAQ---------QQKDSESTQ-----KAE-KEVSRMVVVMIL
OPSB$HUMAN      SESYTWFLFIFCFIVPLSLICFSYTQLLRALKAVAA---------QQQESATTQ-----KAE-REVSRMVVVMVG
OPSD$SQUID      S--NILCMYIFAFMCPIVVIFFCYFNIVMSVSNHEKEMAAMAKRLNAKELRKAQ--AGANAEMK-LAKISIVIVT
OPSD$OCTDO      S--FILCMYFCGFMLPIIIIAFCYFNIVMSVSNHEKEMAAMAKRLNAKELRKAQ--AGASAEMK-LAKISMVIIT
OPS1$DROME      S--YLIFYSIFVYYIPLFLICYSYWFIIAAVSAHEKAMREQAKKMNVKSLRSSED-AEKSAEGK-LAKVALVTIT
OPS1$CALVI      S--YLIFYSIFVYYLPLFLICYSYWFIIAAVSAHEKAMREQAKKMNVKSLRSSED-ADKSAEGK-LAKVALVTIS
OPS2$DROME      S--YLITYSLFVYYTPLFLICYSYWFIIAAVAAHEKAMREQAKKMNVKSLRSSED-CDKSAEGK-LAKVALTTIS
OPS3$DROME      L--FVACIFFFSFVCPTIMITYYYSQIVGHVFSHEKALRDQAKKMNVESLRSNVDKNKETAEIR-IAKAAITICF
OPS4$DROME      L--FVGTIFFFSFVCPTLMILYYYSQIVGHVFSHEKALREQAKKMNVESLRSNVDKSKETAEIR-IAKAAITICF


                260        »    280         «              300            »            320
OPSD$BOVIN      AFLICWLPYAGVAFYIFTHQGSDFGPIFMTIPAFFAKTSAVYNPVIYIMMNKQFRNCMVTTLCCGKNPLGDDEAS
OPSD$SHEEP      AFLICWLPYAGVAFYIFTHQGSDFGPIFMTIPAFFAKSSSVYNPVIYIMMNKQFRNCMLTTLCCGKNPLGDDEAS
OPSD$MOUSE      FFLICWLPYASVAFYIFTHQGSNFGPIFMTLPAFFAKSSSIYNPVIYIMLNKQFRNCMLTTLCCGKNPLGDDDAS
OPSD$HUMAN      AFLICWVPYASVAFYIFTHQGSNFGPIFMTIPAFFAKSAAIYNPVIYIMMNKQFRNCMLTTICCGKNPLGDDEAS
OPSD$CHICK      AFLICWVPYASVAFYIFTNQGSDFGPIFMTIPAFFAKSSAIYNPVIYIVMNKQFRNCMITTLCCGKNPLGDEDTS
OPSD$LAMJA      GFLVCWVPYASVAFYIFTHQGSDFGATFMTLPAFFAKSSALYNPVIYILMNKQFRNCMITTLCCGKNPLGDDESG
OPSR$HUMAN      AYCVCWGPYTFFACFAAANPGYAFHPLMAALPAYFAKSATTYNPVIYVFMNRQFRNCILQLFGKKVDDGSE---L
OPSR$CHICK      AYCFCWGPYTFFACFAAANPGYAFHPLAAALPAYFAKSATTYNPIIYVFMNRQFRNCILQLFGKKVDDGSE---V
OPSR$ASTFA      AYCFCWGPYTFFACFAAANPGYAFHPLAAAMPAYFAKSATTYNPVIYVFMNRQFRVCIMQLFGKKVDDGS
OPSG$HUMAN      AFCFCWGPYAFFACFAAANPGYPFHPLMAALPAYFAKSATTYNPVIYVFMNRQFRNCILQLFGKKVDDGSE---L
OPSG$ASTFA      AFILCWGPYASFATFSALNPGYAWHPLAAALPAYFAKSATTYNPIIYVFMNRQFRSCIMQLFGKKVEDASE----
OPSH$ASTFA      AFIVCWGPYASFATFSAVNPGYAWHPLAAAMPAYFAKSATTYNPIIYVFMNRQFRSCIMQLFGKKVEDASE----
OPSB$HUMAN      SFCVCYVPYAAFAMYMVNNRNHGLDLRLVTIPSFFSKSACIYNPIIYCFMNKQFQACIMKMVCGKAMTDESD--T
OPSD$SQUID      QFLLSWSPYAVVALLAQFGPIEWVTPYAAQLPVMFAKASAIHNPMIYSVSHPKFRERIASNFPWILTCCQYDEKE
OPSD$OCTDO      QFMLSWSPYAIIALLAQFGPAEWVTPYAAELPVLFAKASAIHNPIVYSVSHPKFREAIQTTFPWLLTCCQFDEKE
OPS1$DROME      LWFMAWTPYLVINCMGLFKF-EGLTPLNTIWGACFAKSAACYNPIVYGISHPKYRLALKEKCPCCVFGKVDDGKS
OPS1$CALVI      LWFMAWTPYTIINTLGLFKY-EGLTPLNTIWGACFAKSAACYNPIVYGISHPKYGIALKEKCPCCVFGKVDDGKA
OPS2$DROME      LWFMAWTPYLVICYFGLFKI-DGLTPLTTIWGATFAKTSAVYNPIVYGISHPKYRIVLKEKCPMCVFGNIDEPKP
OPS3$DROME      LFFCSWTPYGVMSLIGAFGDKTLLTPGATMIPACACKMVACIDPFVYAISHPRYRMELQKRCPWLALNEKAPE-S
OPS4$DROME      LFFVSWTPYGVMSLIGAFGDKSLLTQGATMIPACTCKLVACIDPFVYAISHPRYRLELQKRCPWLGVNEKSGEIS


                340
OPSD$BOVIN      TTVSKTETSQ-----VAPA
OPSD$SHEEP      TTVSKTETSQ-----VAPA
OPSD$MOUSE      ATASKTETSQ-----VAPA
OPSD$HUMAN      ATVSKTETSQ-----VAPA
OPSD$CHICK      A--GKTETSSVSTSQVSPA
OPSD$LAMJA      ASTSKTEVSSVSTSPVSPA
OPSR$HUMAN      SSASKTEVSSVSS--VSPA
OPSR$CHICK      ST-SRTEVSSVSNSSVSPA
OPSR$ASTFA
OPSG$HUMAN      SSASKTEVSSV--SSVSPA
OPSG$ASTFA      VSGSTTEVSTAS
OPSH$ASTFA      VSGSTTEVSTAS                         360                  380                 400
OPSB$HUMAN      CSSQKTEVSTVSSTQVGPN
OPSD$SQUID      IEDDKDAEAEIPAGEQSGGETADAAQMKEMMAMMQKMQAQQQQQPAY----PPQGYPPPQGYPPPPPQG-YPP-
OPSD$OCTDO      CEDANDAEEEVVASER-GGESRDAAQMKEMMAMMQKMQAQQ---AAYQPPPPPQGYPPQGYPP---QGAYPPP
OPS1$DROME      SD-AQSQATASEAESKA
OPS1$CALVI      SD-ATSQATNNESETKA       QGYPPQGYPPQGYPPPPQGPPPQGPPPQ--A---APPQGVDNQAYQA
OPS2$DROME      DAPASDTETTSEADSKA       QGYPPQGYPPQGYPP--QGYPPQGAPPQVEAPQGAPPQGVDNQAYQA
OPS3$DROME      SAVASTSTTQEPQQTTAA
OPS4$DROME      SA--QSTTTQEQQQTTAA                        420                 440
```

and Yokoyama, 1990b); **10,** human green color pigment (Nathans et al., 1986);
11, blind cave fish *(Astyanax fasciatus)* green color pigment (Yokoyama and
Yokoyama, 1990a); **12,** blind cave fish *(Astyanax fasciatus)* green color pigment
(Yokoyama and Yokoyama, 1990b); **13,** human blue color pigment (Nathans et al.,
1986); **14,** squid *(Loligo forbesi)* rhodopsin (Hall et al., 1991); **15,** octopus *(Octopus
dofleini)* rhodopsin (Ovchinnikov et al., 1988b); **16,** *Drosophila melanogaster* RH1
opsin (O'Tousa et al., 1985; Zuker et al., 1985); **17,** *Calliphora vicini* RH1 opsin
(Huber et al., 1990); **18,** *Drosophila melanogaster* RH2 opsin (Cowman et al.,
1986); **19,** *Drosophila melanogaster* RH3 opsin (Fryxell and Meyerowitz, 1987;
Zuker et al., 1987); **20,** *Drosophila melanogaster* RH4 opsin (Montell et al.,
1987).

3

example, that the intramembranous regions of most of these proteins will consist largely of regular helices oriented roughly perpendicular to the plane of the bilayer. (Some of the more complex ion channels may contain β-sheet elements.) Therefore, the two most vital pieces of information are the topography of the polypeptide (i.e., the identity and boundaries of the transmembrane segments) and the packing arrangement of these segments in the bilayer. However, in practice it has, so far, proved very difficult to satisfy these requirements for the vast majority of membrane proteins. Fortunately, rhodopsin is one of the few exceptions.

2.1. Disposition

Because of the amount of effort which was invested into generating and isolating the constituent peptides of opsin, it was possible to identify the attachment points for carbohydrate, phosphate, palmitate (Ovchinnikov et al., 1988a), and the prosthetic group 11-*cis*-retinal. Allied to this was the determination of side chains subject to modification with hydrophilic sulfhydryl reagents, succinic anhydride and diazodiiodosulfanilic acid, together with the peptide bonds subject to cleavage by a variety of proteases. Since all these experiments were carried out on the native protein in the membrane, the information provided a good description of the disposition of the polypeptide chain in the bilayer (see Figure 1.2). Moreover, there was surprisingly little discrepancy in the data (for reviews of these data, see Findlay and Pappin, 1986; Findlay et al., 1988). The outcome was the seven-transmembrane delineation for the superfamily with the N-terminal region in the extracellular phase and the C-terminus in the intracellular compartment.

The protein did not possess a classic signal peptide. Thus integration into the bilayer is likely to be largely entropically driven, although there have been suggestions that several of the hydrophobic segments could be associated with insertion signals (Friedlander and Blobel, 1985). The extracellular domain does have some hydrophobic character, indicating that it may have the potential to move through the bilayer relatively easily and that stabilization in an extracellular orientation may be, in part, reliant on glycosylation. One type of retinitis pigmentosa results from replacement of an extracellular, conserved proline (Pro[21]) by a histidine (Dryja et al., 1990). Therefore, the N-terminal region is thought likely to possess important structural features that are required for correct folding of rhodopsin. Finally, it is worth noting that the intracellular boundaries of the transmembrane helices contain the positively charged side chains often diagnostic of such orientation (Dalbey, 1990).

2.2. 3-D Modeling

No diffraction data of sufficiently high resolution to reveal the protein backbone has yet been reported. However, low-angle X-ray and neutron

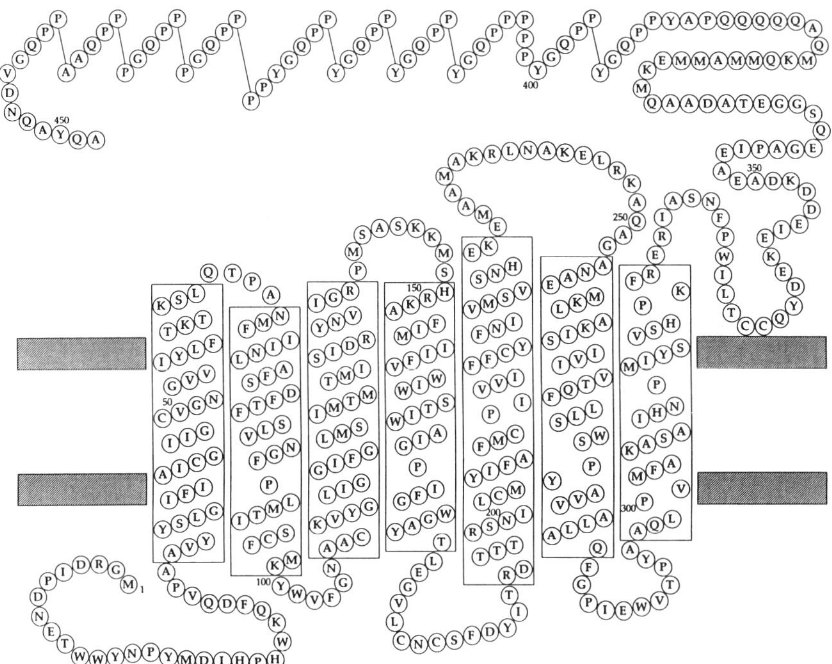

FIGURE 1.2. Membrane topology of squid rhodopsin. Disposition of squid *(Loligo forbesi)* rhodopsin in the bilayer based on that of ovine rhodopsin established by chemical labeling, proteolytic, and amino acid sequencing studies (for reviews, see Findlay and Pappin, 1986; Findlay et al., 1988).

diffraction studies have indicated that rhodopsin is an elongated globular protein with its long axis normal to the lipid bilayer. About one-half of the protein mass is embedded in the bilayer and there are approximately equal amounts exposed on each side of the membrane (see Dratz and Hargrave, 1983, for a review). Projection maps derived from two-dimensional crystallization of bovine (Dratz et al., 1985) and of frog rhodopsin (Corless et al., 1982) have confirmed that the cross-sectional area is very similar to that of bacteriorhodopsin. Fortunately, the packing arrangement for the intramembranous helices of this retinal-containing H^+-pump has been derived from electron diffraction (Henderson et al., 1990). Thus, we used the relative positions of the transmembrane segments of bacteriorhodopsin as the basis for our model building. Infrared dichroism and diamagnetic anisotropy suggested that the helices were largely oriented perpendicular to the bilayer surface, similar to bacteriorhodopsin.

Several spectroscopic studies had indicated that 50–60% of the mass of opsin was in the helical form (Stubbs et al., 1976). From the viewpoint of protein stability most of these helical regions are likely to correspond to the segments of the chain which cross the lipid bilayer. Extensive structure prediction not only supported this assumption but also suggested where the interhelical loops and regions of more irregular structure may lie (Pappin et al., 1984). A very large number (5040) of different permutations of these helices are possible, however, and when coming to our preferred arrangement, we were guided by several considerations. In the first instance, the two principal ones were a wish to avoid any "crossing-over" of the extramembranous regions of the protein and a prejudice toward packing helices in an antiparallel fashion. Both of these guidelines arose from considerations of possible mechanisms for the insertion and folding of the polypeptide chain in the bilayer during biosynthesis but there is no direct experimental support for imposing such restraints. A similar approach was used to fit the sequence of bacteriorhodopsin to its electron density (Engelman et al., 1980). However, the structures established to date and the increasing consistency of model predictions and experimental results does suggest that these may turn out to be fairly general features of integral membrane proteins.

The final major piece of evidence required for meaningful modeling was some assessment of the positions of amino acid side chains relative to the hydrophobic region of the bilayer. One way of obtaining information was to use photosensitive probes. Since extensive labeling of all the lipid-exposed side chains such as might occur with carbene-generating reagents could have rendered sensible interpretation of subsequent protein sequencing studies almost impossible, these studies were carried out with azido-iodobenzene. Great care was taken to guard against diffusion and prolonged reactivity of the photoactivated probe. Differential labeling of several reactive amino acids (e.g., Cys, Tyr, Trp, Lys, and His) indicated which were accessible to the probe and therefore mostly exposed to the lipid phase. In addition to assessing the disposition of the helices with respect to the bilayer surfaces, these data suggested that several residues had locations within the protein structure that rendered them inaccessible to the labeling reagent (Davison and Findlay, 1986).

A low-resolution 3-D representation of the polypeptide was generated based on the above considerations and incorporating the chemical modification data. The helices were positioned within 40° of the normal to the plane of the membrane to be consistent with the evidence from bacteriorhodopsin and linear dichroism studies on both proteins (Dratz and Hargrave, 1983). The only major deviations from the regularity of the α-helices introduced were at intramembranous proline residues and a distortion in the seventh helix [transmembrane (TM) 7] immediately C-terminal to the retinal attachment point (Lys^{296}). In ovine and mouse rhodopsins, the sequence Ser-Ser-Ser, which adopts a looplike conformation in many protein structures, is present here. Moreover, a distortion of TM 7 was consistent with the labeling data and with the position of charged residues in the superfamily as a whole. A regular helix and intact structural framework throughout the superfamily would not allow all of these charges to be sequestered away from the lipid phase.

Finally, since cross-linking studies indicated that rhodopsin was monomeric both in the dark and when activated by light (Brett and Findlay, 1979), no account was taken of subunit interactions.

2.3. General Features

The data suggest that rhodopsin is a cylindrical molecule flattened at the equivalent to the external surface but projecting into the cytosol primarily due to the extended helices predicted for TM 5 and 6. This is consistent both with the sensitivity of this region to proteolytic cleavage and to its variability in size among the G-protein-linked receptor family. Despite earlier suggestions, the labeling and sequence data indicate that the protein is not held together by a network of disulfide bonds. The only such bond predicted both by failure to modify its constituent sulfhydryls and by the modeling results is situated at the external surface linking the start of TM 3 (Cys^{110}) to the surface region between TM 4 and TM 5 (Cys^{187}). These suggestions have been reinforced by the conservation of these two cysteine residues in the superfamily, by protein modification studies (Al-Saleh et al., 1987) and by protein engineering (Karnik et al., 1988). This latter investigation supported the labeling studies in that it indicated that all but two of the cysteines could be replaced without altering the ligand binding properties of rhodopsin. Mutating either of the two disulfide-bonded residues, however, was totally destructive, suggestive of an important role in the folding and/or stability of the proteins. Moreover, after modification of all other cysteines by site-directed mutagenesis, CN-labeling indicated that the protein contained a buried disulfide bond (Karnik and Khorana, 1990). Interestingly, the two cysteines thought to be esterified with palmitate (Ovchinnikov et al., 1988a) could be substituted without apparent detriment to functional behavior (Karnik et al., 1988), although the ability to activate the G-protein was only studied in detergent where the structural constraints imposed by palmitoylation would be expected to be minimized.

The model suggests that most of the polar residues within the bilayer are oriented toward the center of the protein (i.e., are not exposed to the lipid milieu). This organization would facilitate a pattern of hydrogen-bonds involving side-chain moieties which, together with helix dipole interactions, would contribute significantly to the kinetic stability of the protein. However, thermodynamic stability does not require all polar (as distinct from charged) residues to be sequestered away from the hydrophobic environment any more than hydrophobics (leucine, valine, etc.) need to be in exclusively hydrophobic locations. Moreover, serine and threonine can hydrogen-bond back to the main-chain peptide bond, thereby conferring an element of stability if the regularity of the helix is disturbed in any way.

However, such exposed residues are likely to be relatively few as substantial regions of polar character could open the possibility of aggregation. As yet no lipid-exposed charged residues have been identified in any membrane protein. Indeed, introduction of a charged amino acid can markedly affect the organization of such proteins (Cosson et al., 1991). However, in some, a dicyclohexylcarbodiimide-sensitive negative charge does seem to be located in a relatively hydrophobic environment (Finbow et al., 1991).

The thermodynamic stability of rhodopsin also results from interaction of the protein with the 11-cis chromophore. On bleaching, the denaturation temperature drops from $\sim 72°$ to $\sim 55°C$ (Miljanich et al., 1985). Though the interactions between the protein and chromophore are the prime determinants of this change, long-range alterations of the protein structure on bleaching probably, at least partly, determine its magnitude. The interhelical loops appear to contribute relatively little to the stability of the protein fold (Albert and Litman, 1978).

It is also interesting to note that the general features of the structure are apparently well conserved across the whole superfamily of G-linked receptors. For example, discriminators can be constructed which describe the character of the amino acids permitted at every position in the seven transmembrane segments (Attwood et al., 1991). These seven discriminators can then be used to identify precisely the corresponding segments not only in other rhodopsins but also in other receptors in the family where the degree of residue identity may be as low as 15%. Little or no cross-reactivity occurs with other integral membrane proteins, indicating that hydrophobicity is not an overwhelming influence. The observations that each helix has its own characteristic structure and that this character is maintained across the family tend to indicate a high degree of structural conservation (Figure 1.3).

3. Functional Aspects

With so many sequences for the light-sensitive visual pigment now available, it should be possible to extract some useful inferences from a comparative analysis of these data. Table 1.1 lists those residues which are conserved across the whole family of opsins and it is not unreasonable to suggest that they represent critical aspects of the structure and/or function of the protein. The interpretation of these conservations is further facilitated by comparisons with other receptors in the wider family (Table 1.1A), and by the functional consequences of site-directed mutagenesis.

3.1. Ligand Binding Site

3.1.1. Spectroscopic Characteristics

In the early 1960s, linear dichroism demonstrated that retinal was rigidly held within the protein and that the major axis of absorption (the

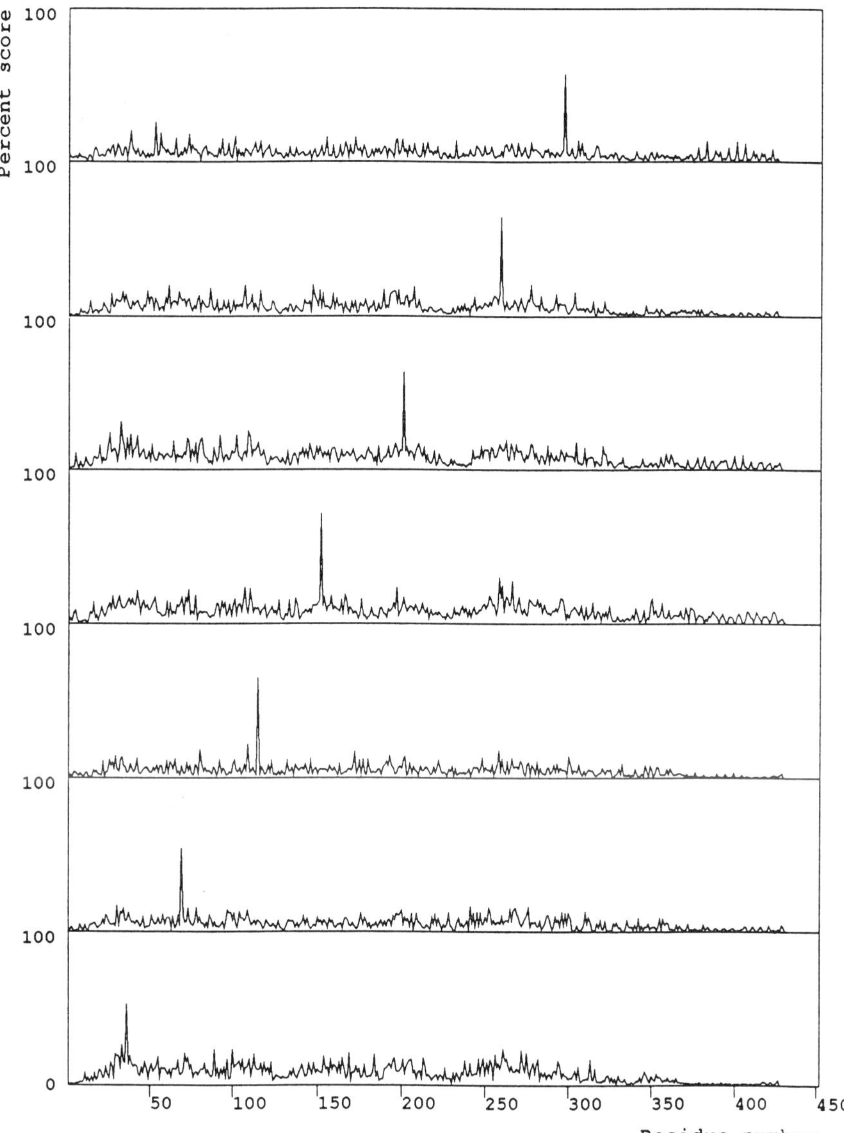

FIGURE 1.3. Discriminator scan of squid rhodopsin. The amino acid sequence of squid *(Loligo forbesi)* rhodopsin was scanned using helix discriminators based on a compilation of G-protein-linked receptor sequences (Attwood et al., 1991). The horizontal axis represents the sequence; the vertical axis represents the percentage score of each discriminator; the peak is a residue-by-residue match in the sequence, its leading edge marking the first position of the match.

TABLE 1.1. Sequence conservation among the opsins and G-protein-linked receptors

A. Conserved residues (identity) in opsins and throughout the superfamily			B. Conserved residues in opsins (near identity)			C. Conserved residues in opsins (character)	
Location[a]	Residue (opsins)	Residue (universal)	Location[a]	Residue	Comments	Location[a]	Residues
N-term	Pro^{23}	–	i1	Arg^{69}	(Gln in cephalopods)	TM 1	Leu^{50}, Leu^{57}
TM 1	Asn^{55}	Asn	TM 2	Asp^{83}	(Gly in blue, Asn in lamprey)	TM 1	Leu^{59}, Val^{61}
TM 1/i1	Lys^{66}	–				i1/TM 2	Leu^{76}, Leu^{77}
i1	Leu^{68}	–	oi/TM 3	Glu^{113}	(Tyr/Phe in all invertebrates)	TM 2	Leu^{79}
i1	Pro^{71}	–				TM 3	Ser^{127}, Leu^{131}
i1	Asn^{73}	–	TM 3	Gly^{114}	(Asn in RH1, RH2, RH4)	TM 3/i2	Glu^{134}, Tyr^{136}
i1/TM 2	Asn^{78}	–				TM 3/i2	Val^{139}, Lys^{141}
o1	Gly^{106}	–	i2/TM 4	Ala^{153}	(Ser in RH2)	i2	Pro^{142}, Arg^{147}
o1/TM3	Cys^{110}	Cys	TM 4	Trp^{161}	(Tyr in RH3)	TM 4	Ala^{158}
TM 3	Gly^{121}	–	TM 4	Pro^{171}	(Cys in RH3)	TM4/o2	Tyr^{178}
TM 3/i2	Arg^{135}	Arg	TM 4/o2	Gly^{174}	(Asn in octopus, Glu in RH3, Gln in RH4)	o2	Tyr^{192}, Thr^{193}

TM 3/i2	Val^{138}	—				TM 5	Val^{218}, Leu^{226}
TM 4/o2	Trp^{175}	Trp				TM 5/i3	Val^{230}
o2	Cys^{187}	Cys	o2	Cys^{185}	(Thr in all red/greens, octopus and insects)	i3/TM 6	Lys^{248}, Val^{250}
TM 5	Pro^{215}	Pro				i3/TM 6	Arg^{252}
TM 5	Ile^{219}	—	o2	Asp^{190}	(Leu in ovine)	TM 6	Met^{253}, Ile^{255}
TM 5	Tyr^{223}	Tyr	o3	Pro^{285}	(Ala in lamprey, Leu in blue, Gln in RH4)	TM 6	Phe^{261}
i3/TM 6	Ala^{246}	—				TM 7	Pro^{291}
i3/TM 6	Glu^{247}	—				TM 7/C-term	Phe^{313}
i3/TM 6	Glu^{249}	—	TM 7	Phe^{294}	(Ala in RH3, Thr in RH4)		
TM 7	Pro^{267}	Pro					
TM 7	Tyr^{268}	—	TM 7	Asn^{302}	(Asp in RH3, RH4)		
TM 7	Lys^{296}	—	TM 7/C-term	Arg^{314}	(Gln in blue, Gly in *Calliphora*)		
TM 7	Pro^{303}	Pro					
TM 7	Tyr^{306}	Tyr					

[a] Key to location: N-term, N-terminal to TM 1, outer (extracellular); TM 1–TM 7, transmembrane segments 1–7; i1-i3, inner (cytoplasmic) loops 1–3; o1-o3, outer (extracellular) loops 1–3; C-term, C-terminal to TM 7.

conjugated region of retinal) lies approximately parallel to the membrane surface. Since rhodopsin functions in the rod outer segment (a structure formed from a stack of planar disks containing rhodopsin in their membranes), and because the long axis of the outer segments (and rhodopsin) are arranged parallel to impinging light (Wald, 1968), this leads to about a threefold increase in light absorption over that of rhodopsin free in solution. The functional adaptation required to allow this chromophore binding arrangement is likely to have induced features unique to rhodopsin. Clearly all the other G-protein receptor subfamilies will also have their own idiosyncracies.

The absorption of light by the retinal chromophore immediately reveals unusual aspects of its environment. With the exception of the UV pigments, the absorbance of the protein–retinal complex is shifted to longer wavelengths (the opsin shift) than observed for retinal derivatives in solution. Photoisomerization of retinal proceeds with an unusually high quantum yield and whereas excitation of retinal results in a variety of photoproducts, single-photon absorption by rhodopsin yields exclusively the all-trans isomer (Wald, 1968). These observations suggest that 11-*cis*-retinal is distorted from its solution structure by its interactions with rhodopsin and that there are probably many van der Waals contacts between chromophore and protein. The bleaching of rhodopsin involves a general decrease in the interaction between chromophore and protein, probably resulting as much from changes in the structure of the binding site as from changes in the conformation of the bound retinal; it is important to note, however, that simple readdition of the 11-*cis*-retinal chromophore is sufficient to reform the dark-adapted protein (Wald and Brown, 1950), even while still phosphorylated (see Section 3.4).

A variety of spectroscopic methods have been applied to study the interaction between rhodopsin and its chromophore. These include the use of chromophore analogues (Renk and Crouch, 1989; Fukada et al., 1990; Koutalos et al., 1989), two-photon absorption using locked 11-cis derivatives (Birge et al., 1985), resonance Raman (Eyring et al., 1980), and Fourier transform infrared spectroscopy (Bagley et al., 1989; Ganter et al., 1989). Recently the powerful solid-state nuclear magnetic resonance (NMR) technique of low-temperature magic angle sample spinning (MASS-NMR) of rhodopsin containing isotopically enriched 11-*cis*-retinal has added considerable weight to earlier conclusions (Smith et al., 1987, 1990; Mollevanger et al., 1987). Thus, in bovine rhodopsin, the C6–7 bond has been shown to be in a twisted s-cis conformation by solid-state NMR. This is in contrast to the configuration found in bacteriorhodopsin (De Groot et al., 1989), and since it breaks the π-electron conjugation with the C5–6 double bond, it blue shifts the absorption of this rhodopsin relative to that of bacteriorhodopsin. The break in conjugation, which this twist introduces, decreases the dependence of the absorption of rhodopsin on the exact angle of the bond. Solid-state NMR has also confirmed the conclusions of resonance Raman experiments that retinal is linked to this protein via a protonated Schiff base (Smith et al., 1987). Moreover, the $C=N$ was shown to be in the anti-configuration, agreeing with both resonance Raman and Fourier transform infrared data. ^{13}C-NMR has also demonstrated that the Schiff base forms a strong hydrogen bond and that the protein induces perturbations in the C10–13 region of the polyene chain (Smith et al., 1990). The latter observation has been interpreted as resulting either from a charged amino acid close to the retinal C13 or from steric constraints of the binding pocket which distort the solution structure of the retinal C12–13 bond.

Whereas the membrane-spanning helices of the vertebrate rhodopsin are very similar, those of invertebrate opsins and the color pigments differ significantly. Therefore, it is difficult to speculate as to which of the constraints on chromophore binding are modified to fine-tune the absorption spectrum of, for example, the red and green color pigments (see next section). However, in the case of the two *Drosophila* UV photoreceptors, it is likely that at physiological pH the Schiff base is not protonated. The opsin shift of all these is, however, strongly dependent on three main parameters: the degree to which electron delocalization extends into the β-ionone ring (dependent on the configuration of the C6–7 single bond), the protonation state and hydrogen bond strength of the Schiff base nitrogen, and the electronic interaction of the chromophore with neighboring amino acids. These, in turn, are likely to depend on the nature and conformation of the amino acids at a relatively small number of positions on several of the membrane-spanning helices.

3.1.2. STRUCTURAL CONSIDERATIONS

The binding site for the chromophore probably corresponds fairly directly with the binding site of many receptors which utilize small ligands. (The polypeptide hormone subgroup possesses an extended N-terminal region which appears to be required for ligand binding.) However, whereas most of the receptors are activated or inhibited by noncovalently bound ligands, 11-*cis*-retinal is covalently linked through a lysyl-Schiff base to a residue situated toward the extracellular side of the C-terminal transmembrane helix (Findlay et al., 1981; Thomas and Stryer, 1982). Despite this difference, it is probably constructive to consider rhodopsin as a typical member of this protein superfamily. In the dark-adapted state, an antagonist (11-*cis*-retinal) is covalently bound, and the action of light can be viewed as a means of converting it to an agonist. Recent mutagenic experiments where the lysine was replaced with a glycine or an alanine, and the 11-*cis*-retinal by positively charged derivatives (Zhukovsky et al., 1991), demonstrate that the covalent link is not a prerequisite for bleaching or for activation of transducin.

Among the obvious conserved features of the membrane-spanning helices across the whole superfamily are 3 to 4 prolines. In the model, they appear to surround the putative ligand binding site, perhaps distorting the helices to provide an accommodating pocket. It will take some time to establish their structural role and whether these prolines are in the cis or trans conformation, but it may be relevant that the *Drosophila* mutant (nina-A) has a defect in a protein which it is thought may be a membrane-associated prolyl cis-trans isomerase (Schneuwly et al., 1989; Stamnes et al., 1991).

Apart from the UV-absorbing pigments, the wavelength shift induced by the protein is bathochromic and a "point-charge" hypothesis has been advanced to account for this phenomenon (Honig et al., 1979). It involves two negative charges about 3 Å away, one from the protonated Schiff base involving Lys^{296}, and the other from approximately the middle of the polyene chain. Being apparently isolated in a hydrophobic environment, Asp^{83} was originally put forward as a strong candidate for the counterion to the Schiff base. The model building results suggested at an early stage that this was unlikely to be the case (Findlay and Pappin, 1986). Since then, the absence of the residue in the human blue and lamprey pigments and its presence in all the other G-protein-coupled receptors reinforce this conclusion. The possible role of this residue as the second negative charge has been investigated by site-directed mutagenesis in a number of laboratories. No major change in absorbance maximum occurred on alterations of Asp^{83}, indicating that it does not interact closely with the chromophore (Sakmar et al., 1989; Zhukovsky and Oprian, 1989; Janssen et al., 1990; Nathans, 1990a). Possible functional roles for Asp^{83} are discussed in Section 3.2.

From the model, we indicated some time ago that if a negative charge was critical, the best candidate for the counterion appeared to be Glu^{113}, although it did not seem to be ideally placed for a direct, close interaction (Findlay and Pappin, 1986). Khorana and colleagues, on the basis of site-directed mutagenesis, have advanced the suggestion that this is indeed the elusive counterion (Zhukovsky and Oprian, 1989; Nathans, 1990b; Sakmar et al., 1991). Similar work with the β-adrenergic receptor suggested that an aspartic acid acted as the counterion to the catecholamine ligand (Strader et al., 1989), but it is important to note that the residues in the two proteins neither align with one another nor are in the same position in the model. Both these criteria place the aspartic acid approximately one and a half helical turns further into the bilayer than Glu^{113}.

The view of Glu^{113} acting as a simple counterion has been questioned by other observations. De Grip's group has also substituted Glu^{113} without significant effect on the absorbance characteristics of rhodopsin when the spectrum was measured at normal ionic strength (Janssen et al., 1990). More recently, Nathans has shown that mutation of this residue results in a pigment that is sensitive to the anionic composition of the measuring solution (Nathans, 1990b). Comparison of the sequence of vertebrate opsins with sequences of their invertebrate counterparts demonstrates that the corresponding residue is always aromatic in the latter group. However, the replacement of retinal with dihydroretinal derivatives indicate considerable similarity between vertebrate and invertebrate binding sites (Koutalos et al., 1989). The invertebrate pigments possess no suitably placed residue that could substitute for Glu^{113} (in squid, Glu^{290} is likely to be too far into the hydrophilic phase). Therefore, the universal role of Glu^{113} as a straightforward negative counterion still remains to be confirmed. Perhaps the data could be consistent with the view that the region of the protein which contains this residue is essential for maintaining the integrity of the retinal binding site and that the Schiff base is stabilized either by indirect interaction with charged residues, or a group of dipoles, or possibly even the end of a helix. The effects of mutation of this residue on the pH dependence of absorption and the stability of the photoproducts (Sakmar et al., 1989; Hubbard and St. George, 1958) and the relative stabilities of the retinal attachment in vertebrate and invertebrate pigments suggests that Glu^{113} may be involved in the catalysis of Schiff base hydrolysis.

And what of the second negative charge that the point-charge hypothesis puts forward? Because of its near ubiquitous occurrence in the receptor superfamily, Asp^{83} would seem to occupy a more critical common function in all the proteins. Its involvement in wavelength regulation appears less likely and mutation to Asn, for example, does not affect the opsin shift (Janssen et al., 1990). The only other likely candidate is Glu^{122}. Although this residue is at the correct height and orientation in the model to support this suggestion, there seems to be an easier affiliation with His^{211}, perhaps to form a salt bridge. This possibility is supported by the presence or absence of both residues in the various pigments — neither is ever present on its own. A lack of any central role in wavelength regulation is also indicated by the presence of His/Glu in only a few mammalian rhodopsins (see Figure 1.1). Against this is set the observation that a diazirine analogue of retinal covalently binds to Glu^{122} on photoactivation (Nakayama and Khorana, 1990). However, substitution of this residue by charged or polar residues produced only slight blue shifts in the absorption spectrum (Sakmar et al., 1989; Nathans, 1990a). Mutation of His^{211} to Cys or Phe did not affect the spectrum. At present, these data are best accommodated by the proposal that both Glu^{122} and His^{211} are in the vicinity of the retinal binding pocket thereby contributing to the environment that produces the opsin shift. In such a scenario, the "point-charge" hypothesis will require modification. One final point is that Glu^{113} and Glu^{122} would be on almost diametrically opposed positions in a regular α-helix. Thus if they were both in the vicinity of the retinal binding pocket, the regularity of the helix would have to break down, most probably in the region of Glu^{113}.

Since the ligand binding pocket is likely to be an interesting and complex series of interactions, are there any other biological systems that can provide didactic comparisons? A similar bathochromic shift occurs when ligands are bound to some carotenoid binding proteins, for example astaxanthin to crustacyanin (Zagalsky et al., 1990a,b; Keen et al., 1991).

The protein consists of two homologous subunits, and models can be constructed based on the crystal structures of the related ligand binding proteins (North, 1991). No charged residues are present in the binding pocket. Therefore, the aromatic and nonpolar residues must provide the environment that increases delocalization of the conjugated polyene-chain chromophore and thereby shifts the absorbance maximum (Zagalsky et al., 1991). A similar system is apparent in the case of bacteriorhodopsin (Henderson et al., 1990) where aromatic residues have been implicated in chromophore–protein interaction. Another example is the photoactive yellow pigment from *Ectothiorhodospira halophila,* for which a crystal structure has been determined (McRee et al., 1989). Although the identity of the chromophore is unknown, it appears that it is linked by a Schiff base to a lysine and that this is stabilized by interaction with a buried glutamate and also interacts strongly with a tyrosine residue. Labeling of bovine rhodopsin with photoactivated retinal pointed to the involvement of Trp^{126} with the binding site (Nakayama and Khorana, 1990). Mutagenesis to Ala or Leu, but not Phe, slightly blue shifted the absorbance maximum and reduced the bleaching rate (Nakayama and Khorana, 1991). Interestingly, this residue in ovine rhodopsin was inaccessible to labeling by the hydrophobic membrane probe, suggesting an internal location in the protein (Davison and Findlay, 1986). Other residues which have been implicated in the binding site include the conserved Trp and Tyr at positions 265 and 268, respectively, Phe^{115} and Ala^{117}. The only point to be recognized is that Ala and Phe would be on the opposite sides of any regular α-helix. All of this information is summarized diagrammatically in Figure 1.4. The conservation in the rhodopsins and G-protein-linked family as a whole is illustrated in Table 1.1.

Good indications of residues interacting with the chromophore of color pigments come from a comparison of the closely related red and green photoreceptors. It appears that the green cone pigment evolved before the division of vertebrates, since both man and the blind cave fish have homologous, repeated green pigment genes. Interestingly, they also have a single red pigment gene, that appears to have evolved at a more recent time, certainly after division of vertebrates into fish and mammals. The human red and green pigments are very similar; in total, there are only fifteen amino acid differences between them. By comparative modeling of the two pigments, it can be seen that several of these altered amino acids are

---

FIGURE 1.4. Helical wheel representation of transmembrane helices. Helical wheel arrangement of the intramembranous (TM 1–7) sections of bovine rhodopsin. The shaded residues represent amino acids that are either rigorously conserved throughout the rhodopsin family and, therefore, presumably of protein structural importance or have proposed functional roles that have been discussed in the text. Amino acids are shown as their single-letter code.

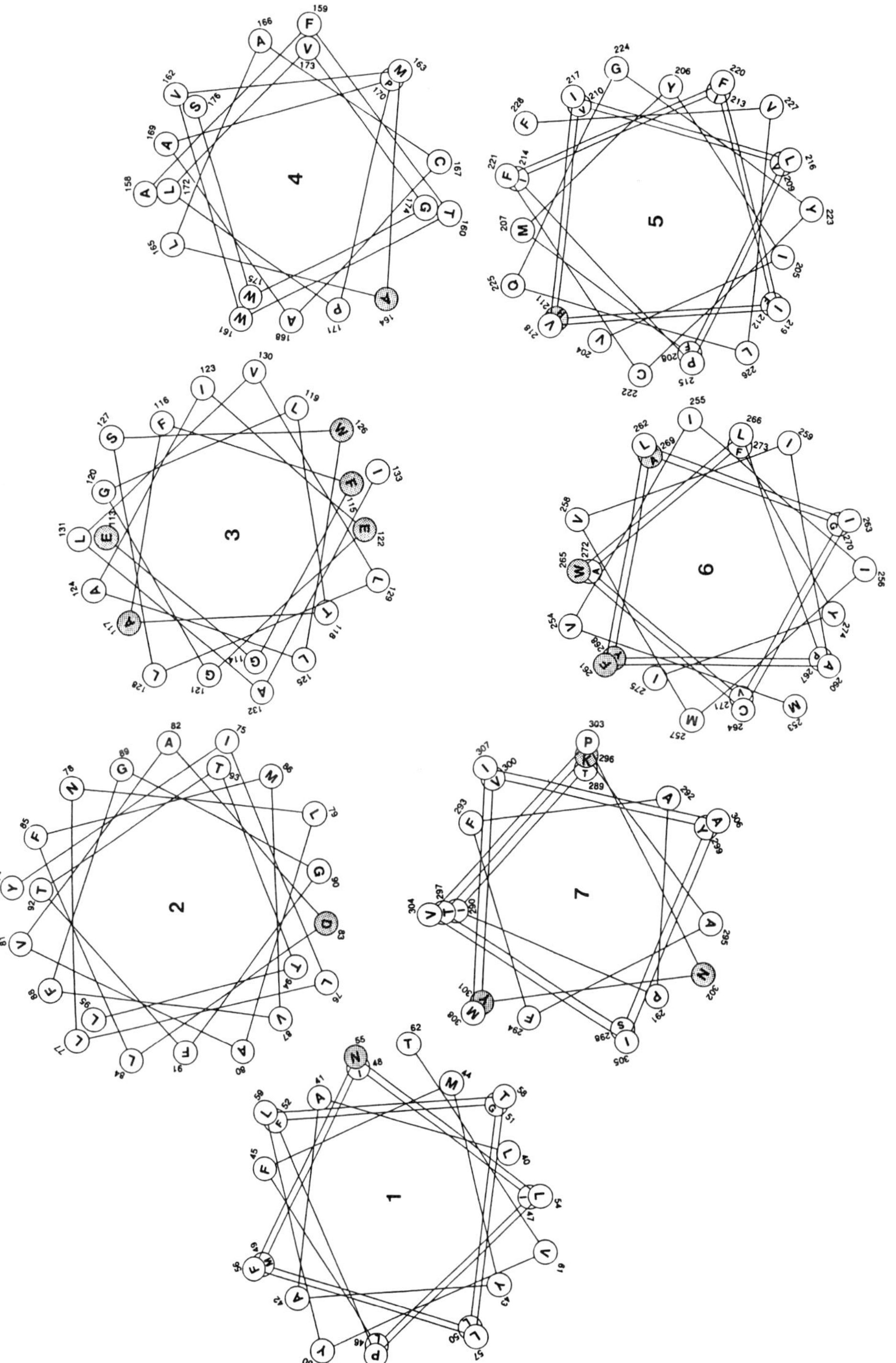

clustered in the putative chromophore binding site. Further evidence as to which residues were involved in determining the differences in absorption maximum of the pigments came from the partial gene sequence of the red–green hybrid of a color-blind individual (Neitz et al., 1989). This protein had the absorption maximum of the green pigment, despite containing the N-terminal half of the red pigment. Thus the differences between the two pigments can be attributed to residues in the C-terminal half of the protein. Comparison with the sequences of the red and green pigments of tamarins and squirrel monkeys (Neitz et al., 1991) suggests the wavelength switch probably just involves mutation at sites 180, 277, and 285 of the red pigment. The same substitutions appear to have occurred by convergent evolution at these positions in the blind cave fish and the human red pigment, namely, Ser^{180} for Ala, Tyr^{277} for Phe, and Thr^{285} for Ala. In the model, these residues appear in close spatial proximity to the retinal chromophore. Whether the conclusions drawn from these comparisons can be extended to the other visual pigments remains to be confirmed. Regeneration of red cone pigments or octopus rhodopsin with synthetic retinal isomers suggests similarity to the chromophore binding site of rhodopsin (Fukada et al., 1990; Koutalos et al., 1989). However, FT-IR spectroscopy of octopus and bovine rhodopsin has demonstrated that some marked differences also occur (Bagley et al., 1989).

3.2. Activation

After absorption of light, the conformation of rhodopsin is altered so that it is able to activate the G-protein transducin at the cytoplasmic membrane face of the disks. Various lines of evidence point to the meta-stable bleaching intermediate, with an absorption maximum of 380 nm (MII), as the photoactive derivative often known as R*. The identification of R* with MII was suggested at an early stage because, while the initial processes resulting from light absorption are restricted to the vicinity of the chromophore binding site, considerable conformational change is associated with the formation of MII. The alteration in protein structure associated with this intermediate has been demonstrated by the change in sensitivity of the protein to chemical agents (Findlay and Pappin, 1986), to in vivo modification (e.g., phosphorylation) (Kühn and Dreyer, 1972), to the binding character of the protein (e.g., of G-protein) (Emeis et al., 1982), and by the sensitivity of the meta-rhodopsin equilibrium to changes in pressure (Lamola et al., 1974). Long-range conformational changes of the whole protein structure are also supported by the sensitivity of the MI–MII equilibrium to the environment of rhodopsin (Applebury et al., 1974). Denaturing detergents tend to increase the rate of this transition and the ratio of MII present at any temperature, while reconstitution of the protein into lipid in which motional freedom is decreased appears to restrict the process (De Grip, 1982; Mitchell et al., 1991).

Although R* is characterized by the absorption spectrum of MII, the converse is not necessarily true. Notably, phosphorylation alters the ability of MII to activate transducin without affecting its absorption spectrum. Thus care should be taken in interpreting changes induced in the meta-rhodopsin equilibrium as reflecting changes in R*. It may be that, since the G-protein binding site is located some distance from the point of chromophore attachment, only a proportion of MII is normally in a configuration that binds G-protein.

The shorter wavelength of light absorption by MII suggests that the retinal Schiff base is deprotonated, and, recently, the finding that methylation of the Schiff base completely prevents formation of MII and phosphorylation of rhodopsin (Seckler and Rando, 1989) supports this. One interesting possibility is that the redistribution of charge resulting from changes in chromophore protonation triggers crucial conformational changes involved in activation. Because of its conservation throughout the superfamily, Asp^{83} appears a candidate for involvement in such a process. However, mutation of the Asp and Glu residues that are modeled as being located within the bilayer, and which would therefore be attractive candidates for transmitting such effects, does not seem to reduce the ability of photoproducts to activate G-protein (Sakmar et al., 1989). Moreover, although present in all other members of the superfamily, an equivalent to Asp^{83} is a Gly in the human blue and an Asn in the lamprey pigment. From the model, Asp^{83} appears to be in an unusually polar environment, sandwiched between two highly conserved asparagines from TM 1 and 7. The speculation is that Asp^{83} may stabilize this polar region, and that this region has functional significance in G-protein activation. Recent evidence that the function of this region may be still more complex comes not from rhodopsin but from studies of Na effects on α_2-adrenergic receptors, where mutation of the equivalent residue abolishes the influence of Na on agonist and antagonist binding (Horstman et al., 1990). In the native protein, the effects of Na are reciprocal to the effects of G-protein, and are allosteric to ligand. Since, in the rhodopsins, 11-*cis*-retinal is always bound (although as mentioned above, this ligand is best thought of as a covalently linked antagonist and, prior to illumination, rhodopsin is therefore locked in an inactive conformation), such allosteric effects may not be as important as in other G-protein-linked receptors. In turn, this may help to explain why the residue is not found in the lamprey and human blue pigment. Moreover, if Asp^{83} has lost its prime functional role in the rhodopsins, this may explain the replacement of one of the Asn's (proposed to interact with it) by an Asp in two *Drosophila* pigments, and the proximity of a His in TM 7 of the cephalopod rhodopsins.

In the vertebrate rhodopsins, chemical modification studies have demonstrated that the deprotonation of the Schiff base is a prerequisite for activation (Seckler and Rando, 1989). The formation of meta-II (MII) rhodopsin also involves the uptake of protons from solution (Matthews et al., 1963). Primary candidates for the residues that are protonated, at least transiently, are several of the transmembrane acidic residues (Asp^{83}, Glu^{122}, or Glu^{134}). In invertebrate rhodopsins, it is unclear as to whether Schiff base deprotonation is required for G-protein activation as only a small fraction of the photoproducts are deprotonated in vivo (Koutalos et al., 1990). However, in the absence of contradictory data, the assumption is that the mechanism will be largely the same, even though the photochemistry differs.

Substitution of retinal analogues has also been applied to examine the details of the activation process. A particularly significant finding was that removal of the 9-methyl (9-Me) group from retinal resulted in very poor activation of G-protein by light-activated rhodopsin analogues. FT-IR studies of the protein suggest that the crucial step that fails to occur in such rhodopsin derivatives is in the transition between the lumi- and MI spectral intermediates that precedes the final MI → MII activation stage. The indications from this particular study are that when the 9-Me group is missing, the isomerization of the chromophore is uncoupled from the changes in protein conformation. Presumably then, the 9-Me group may be involved in an important contact with the protein. One of the major changes that does not seem to occur in the absence of the 9-Me group is the twisting of an amide bond on the amino side of a proline (in the formation of the lumi equivalent) and its relaxation in MI (Ganter et al., 1989). This may indicate that there is a cis-trans isomerization of a proline in rhodopsin that is crucial for subsequent changes that lead to the deprotonation of the Schiff base and the formation of MII.

In the absence of definitive data involving protein structure and engineering, it is impossible to identify which residues or regions might be primarily involved in these conformational changes. One target for mutagenesis could be the short hydrophilic stretch which is C-terminal to the retinal attachment point (Lys^{296}). As discussed earlier, there are reasons for suggesting that TM 7 may not be completely regular. One possibility is that this hydrophilic region may act as a flexible loop responding to the conformational changes initiated by retinal isomerization (Findlay and Pappin, 1986; Renk and Crouch, 1989). This role may be common to most of the G-protein-linked receptors.

3.3. The G-Protein Binding Site

As discussed above, the interaction of rhodopsin with G-protein appears to be specific for the MII bleaching intermediate and, because of this, the position of the MI–MII equilibrium can be employed as a sensitive measure of the interaction. This measurement of the "extra"-MII induced by G-protein has been used to monitor the thermodynamics and kinetics of the activation of transducin (Emeis et al., 1982). A strong dependence of the amount of extra-MII with temperature indicated that the binding probably induced changes in conformation of one or both proteins involved in the interaction, most likely as a result of changes around the GTP binding site of transducin.

Competitive effects of peptides, both representing regions of rhodopsin sequence and parts of Gα, have been used to map the sites of interaction by studying their effect on this extra-MII (König et al., 1989). From these studies, the second and third cytoplasmic loops of rhodopsin and the C-terminal region between TM 7 and the double cysteine have been shown to be particularly important in this respect. Thus, more or less, the whole of the cytoplasmic surface appears to play a part in signal transduction. Recently, by combining peptides that represent different regions of the cytoplasmic surface, Hofmann and colleagues have proposed a three-site model for G-protein binding and activation. Binding to two sites seems to require dissociation of GDP and to be sufficient for generation of extra-MII, while binding to all three sites is required for activation of the G-protein (Kahlert et al., 1990).

The vertebrate rhodopsins interact with transducin, which is a member of the Gi/Go subgroup of Gα's, whereas the invertebrate rhodopsins appear to interact with the rather distantly related Gq subgroup (Strathmann and Simon, 1990). Thus, comparison of the sequences of invertebrate and vertebrate rhodopsins, and their comparison with those of other members of the superfamily, could be very instructive in identifying common and selective features involved in G-protein binding and activation. At the cytoplasmic end of TM 3, there is a tightly conserved motif E(D)-R-Y(W) that is found throughout the superfamily. Site-directed mutagenesis of this motif has revealed that it is indeed essential for G-protein binding (Franke et al., 1990). Thus, it may be that this region contributes a general G-protein interaction site. A second mutagenesis study also pinpointed this region as a determinant of rhodopsin folding (Janssen et al., 1991). It is worth considering the highly conserved Leu[68] (located in the first cytoplasmic loop) in both these contexts. However, the other cytoplasmic loops do not have such tight conservation; indeed, it is in these regions where the invertebrate rhodopsins are most similar to one another and most different from their vertebrate counterparts. Particular emphasis has fallen onto those regions of TM 5 and 6 which are closest to the bilayer surface. They are predicted to form helices continuous with the intramembranous domain and extending out from the bilayer (Findlay and Pappin, 1986). Mutation of Lys[248] to a Leu severely disrupts G-protein activation (Franke et al., 1988). The C-terminal region of this 5–6 loop has also been shown to specify the identity of G-protein activated by chimeric receptors (Liggett et al., 1991). When these regions were subjected to mutagenesis and the binding of G-protein assessed by measuring extra-MII, tight binding occurred but did not lead to G-protein activation (Franke et al., 1990). Thus the cytoplasmic surface is now beginning to be dissected into regions that bind G-proteins, regions that result in G-protein selectivity, and regions involved in activation.

3.4. Deactivation

Just as important as mechanisms that generate an active form of rhodopsin, the protein must also be rapidly deactivated in order to restore the retina to a resting state. The decay of MII is too slow to account for the required rate of deactivation (Baumann, 1972). Hence the observation that R* formation generated a substrate for phosphorylation by a kinase suggested that multiple phosphorylation may play an important role in inactivation (Wilden and Kühn, 1982). The serine and threonine residues that are phosphorylated by rhodopsin kinase in vertebrate rhodopsins occur after the double cysteines $Cys^{322, 323}$ (Thompson and Findlay, 1984). These cysteines have been proposed to be the site for palmitoylation (Ovchinnikov et al., 1988a), thereby effectively creating a fourth cytoplasmic loop. In the red and green color pigments, the double cysteine is absent. Thus, any fourth cytoplasmic loop may be stabilized by the interaction of the lysines that are present in this region of these proteins with negatively charged phospholipid headgroups or other regions of the protein. Whereas mutations of $Cys^{322, 323}$ had little effect on the ability of rhodopsin to activate transducin in detergent (Karnik et al., 1988), in the membrane, mutation of the equivalent palmitoylated residue of the β-adrenergic receptor markedly reduced its ability to activate Gs (O'Dowd et al., 1989). Palmitoylation of $Cys^{322, 323}$ therefore leaves an independent cytoplasmic tail that is subject to phosphorylation. Phosphorylated meta-rhodopsin was demonstrated to be a far weaker activator of G-protein than its unphosphorylated counterpart and, more recently, addition of the kinase and ATP proved sufficient to quench light-induced activity (Sitaramayya, 1986). Unusually among the protein kinases, that of rhodopsin always appears to be active. The sites on dark-adapted rhodopsin are cryptic, only exposed to modification by the kinase after the protein is activated. It has been argued that the dependence of kinase activity only on the state of rhodopsin should facilitate rapid deactivation since the active kinase concentration will not take time to build up after rhodopsin bleaching. However, rhodopsin-dependent activation of the kinase has also been reported (Aton, 1986).

The involvement of a soluble protein "arrestin" in inactivation was implied from its ability to interact with phosphorylated bleached rhodopsin. It has been proposed that this binding acts as a final stage of quenching light activity (Kühn et al., 1984; Bennett and Sitaramayya, 1988). Indeed, by monitoring the displacement of the MI–MII equilibrium, it has been shown that arrestin binds specifically to phosphorylated MII (Schleicher et al., 1989). The binding of arrestin, like that of transducin, has a very pronounced temperature dependence, suggesting that a fairly major protein rearrangement may be associated with it.

Among the G-protein receptors, the rhodopsins may form a special group with respect to deactivation. In other receptors, the activation state is primarily governed by the degree of receptor occupation by agonist. However, a kinase specific for β-adrenergic receptors plays a role in rapid homologous desensitization (Hausdorff et al., 1989), and binding of an arrestin homologue also appears essential (Lohse et al., 1990). Moreover, two arrestin-like proteins have been shown to be expressed in *Drosophila* eyes (Yamada et al., 1990; Hyde et al., 1990). These arrestins are themselves phosphorylated in a Ca^{2+}-dependent fashion.

There are still some unresolved problems in rhodopsin deactivation. At low bleaches, the amount of kinase is in vast excess of the active photoproduct, allowing rapid inactivation. However, although light adaptation increases the rate of inactivation as a function of background illumination, it would appear that such a background illumination progressively decreases the amount of kinase available to deactivate each individual bleached rhodopsin. The same problems appear to apply to arrestin to an even greater extent. The current idea is that this protein exerts its effects by complexing phosphorylated MII, thereby sterically preventing interaction of transducin. As the amount of arrestin is only about 2 to 5% that of rhodopsin (Hamm and Bownds, 1986), at higher-level bleaches no free arrestin should be present to fulfill this function.

Sequence comparison between phosphorylation sites reveals that there is little conservation either in the number or in the positions of the target serine and threonine residues. Moreover, this portion of the protein does not appear to have a major influence on the binding or activation of G-protein when not phosphorylated. Thus, the mechanism by which phosphorylation leads to deactivation of MII is also obscure. In the case of the squid and octopus rhodopsins, only two Ser/Thr residues are found in this region of the protein and only one of these is at an equivalent position. Instead, there are increased numbers of Asp and Glu residues at conserved positions. Therefore, it is tempting to speculate that these carboxylate residues substitute for phosphoserines and -threonines in an interaction with basic residues that contribute to the G-protein binding site.

The cephalopod rhodopsins are also distinguished by a long C-terminal proline-rich extension containing an elevenfold tandem repeat of a 5-residue consensus sequence (Glu-Gly-Tyr-Pro-Pro). Although not present in other members of the superfamily, this region bears similarity with regions of other proteins that are thought to interact with the cytoskeleton (Trimble and Scheller, 1988; Noegel et al., 1990). For squid protein, we also have evidence that the C-terminus may be involved in rhodopsin self-association. Again, it is tempting to speculate that the rapid and dramatic changes in cytoskeletal structure detected on light exposure of squid eyes may result from changes in interactions involving this region of rhodopsin.

4. Conclusions

Recent studies of rhodopsin have greatly increased our appreciation both of the 3-D structural interactions, and of how these relate to protein function. In general, the model built on the basis of the helix packing pattern of bacteriorhodopsin, which incorporated the orientation and topography arising from early labeling studies of the vertebrate proteins, has demonstrated applicability to all members of the superfamily. In particular, it is reassuring to see that the crude predictions made by the model about the chromophore and G-protein binding sites appear to hold. So far it has not been possible to incorporate much of the more physical data into the framework of the 3-D representation, because although it provides insights as to function, little structural precision (relative to the protein) is yet available. However, the model does help to target residues for further investigation, for example, the proline twisting of which is coupled to chromophore isomerization through the 9-Me group, or the asparagines that sandwich the highly conserved Asp^{83} or, again, the region adjacent to the retinal attachment point. Although sequence comparisons have, in general, been used for the most part in evolutionary studies, analysis of such data has been more useful in the case of the rhodopsins, in particular, and the G-protein superfamily, in general. It has, for example, established the principle of a conserved structural framework and highlighted residues of possible structural and functional significance. Thus, even in the absence of a high-resolution X-ray derived structure, considerable progress can be made toward the final goal of understanding the structure of rhodopsin and the molecular mechanisms through which its functional attributes are expressed. To be sure, of course, such a structure is essential.

References

Albert AD, Litman BJ (1978): Independent structural domains in the membrane protein bovine rhodopsin. *Biochemistry* 17:3893–3900

Al-Saleh S, Gore M, Akhtar M (1987): On the disulfide bonds of rhodopsins. *Biochem J* 246:131–137

Applebury ML, Zuckerman DM, Lamola AA, Jovin TM (1974): Rhodopsin. Purification and recombination with phospholipids assayed by the metarhodopsin I→ metarhodopsin II transition. *Biochemistry* 13:3448–3458

Aton BR (1986): Illumination of bovine photoreceptor membranes causes phosphorylation of both bleached and unbleached rhodopsin molecules. *Biochemistry* 25:677–680

Attwood TK, Eliopoulos EE, Findlay JBC (1991): Multiple sequence alignment of protein families showing low sequence homology—a methodological approach using database pattern-matching discriminators for G-protein-linked receptors. *Gene* 98:153–159

Baehr W, Falk JD, Bugra K, Triantafyllos JT, McGinnis JF (1988): Isolation and analysis of the mouse opsin gene. *FEBS Lett* 238:253–256

Bagley KA, Eisenstein L, Ebrey TG, Tsuda M (1989): A comparative-study of the infrared difference spectra for octopus and bovine rhodopsins and their batho-rhodopsin photointermediates. *Biochemistry* 28:3366–3373

Baumann C (1972): Kinetics of slow thermal reactions during the bleaching of rhodopsin in the perfused frog retina. *J Physiol (Lond)* 222:643–663

Bennett N, Sitaramayya A (1988): Inactivation of photoexcited rhodopsin in retinal rods—the roles of rhodopsin kinase and 48-kDa protein (arrestin). *Biochemistry* 27:1710–1715

Birge RR, Murray LP, Pierce BM, Akita H, Balogh-Nair V, Findsen LA, Nakanishi K (1985): Two-photon spectroscopy of locked 11-*cis*-rhodopsin: Evidence for a protonated Schiff base in a neutral protein binding site. *Proc Natl Acad Sci USA* 82:4117–4121

Brett M, Findlay JBC (1979): Investigation of the organization of rhodopsin in the sheep photoreceptor membrane by using cross-linking reagents. *Biochem J* 177:215–223

Brett M, Findlay JBC (1983): Isolation and characterisation of the CNBr peptides from the proteolytically derived N-terminal fragment of ovine opsin. *Biochem J* 211:661–670

Corless JM, McCaslin DR, Scott BL (1982): Two-dimensional rhodopsin crystals from disk membranes of frog retinal rod outer segments. *Proc Natl Acad Sci USA* 79:1116–1120

Cosson P, Lankford SP, Bonifacino JS, Klausner RD (1991) Membrane-protein association by potential intramembrane charge pairs. *Nature* 351:414–416

Cowman AF, Zuker CS, Rubin GM (1986): An opsin gene expressed in only one photoreceptor cell type of the Drosophila eye. *Cell* 44:705–710

Dalbey RE (1990): Positively charged residues are important determinants of membrane-protein topology. *Trends Biochem Sci* 15:253–257

Davison MD, Findlay JBC (1986): Identification of the sites in opsin modified by photoactivated azido [^{125}I]-iodobenzene. *Biochem J* 236:389–395

De Grip WJ (1982): Thermal stability of rhodopsin and opsin in some novel detergents. *Meth Enzymol* 81:256–265

De Groot HJM, Harbison GS, Herzfeld J, Griffin RG (1989): Nuclear magnetic-resonance study of the Schiff-base in bacteriorhodopsin—counterion effects on the ^{15}N shift anisotropy. *Biochemistry* 28:3346–3353

Dratz EA, Hargrave PA (1983): The structure of rhodopsin and the rod outer segment disk membrane. *Trends Biochem Sci* 8:128–131

Dratz EA, Van Breemen JFL, Kamps KMP, Keegstra W, Van Bruggen EFJ (1985): Two-dimensional crystallization of bovine rhodopsin. *Biochim Biophys Acta* 832:337–342

Dryja TP, McGee TL, Reichel E, Hahn LB, Cowley GS, Yandell DW, Sandberg MA, Berson EL (1990): A point mutation of the rhodopsin gene in one form of retinitis-pigmentosa. *Nature* 343:364–366

Emeis D, Kühn H, Reichert J, Hofmann KP (1982): Complex formation between metarhodopsin II and GTP-binding protein in bovine photoreceptor membranes leads to a shift of the photoproduct equilibrium. *FEBS Lett* 143:29–34

Engelman DM, Henderson R, McLachlan AD, Wallace BA (1980): Path of the polypeptide in bacteriorhodopsin. *Proc Natl Acad Sci USA* 77: 2023–2027

Finbow ME, Eliopoulos EE, Jackson PJ, Keen JN, Meagher L, Thompson P, Jones P, Findlay JBC (1991): Structure of a 16kD integral membrane protein that has identity to the putative proton channel of the vacuolar H^+-ATPase. *Protein Eng* 5:7–15

Findlay JBC, Brett M, Pappin DJC (1981): Primary structure of C-terminal functional sites in ovine rhodopsin *Nature* 293:314–316

Findlay JBC, Eliopoulos EE (1991): Three-dimensional modeling of G-protein-linked receptors. *Trends Pharmacol Sci* 11:492–499

Findlay JBC, Pappin DJC (1986): The opsin family of proteins. *Biochem J* 238:625–642

Findlay JBC, Pappin DJC, Eliopoulos EE (1988): The primary structure and molecular modelling of rhodopsin. *Progr Retinal Res* 7:63–68

Franke RR, König B, Sakmar TP, Khorana HG, Hofmann KP (1990): Rhodopsin mutants that bind but fail to activate transducin. *Science* 250:123–125

Franke RR, Sakmar TP, Oprian DD, Khorana HG (1988): A single amino-acid substitution in rhodopsin (lysine 248→leucine) prevents activation of transducin. *J Biol Chem* 263:2119–2122

Friedlander M, Blobel G (1985): Bovine opsin has more than one signal sequence *Nature* 318:338–343

Fryxell KJ, Meyerowitz EM (1987): An opsin gene that is expressed only in the R7 photoreceptor cell of *Drosophila. EMBO J* 6:443–451

Fukada Y, Okano T, Shichida Y, Yoshizawa T, Trehan A, Mead D, Denny M, Asato AE, Liu RSH (1990): Comparative-study on the chromophore binding-sites of rod and red-sensitive cone visual pigments by use of synthetic retinal isomers and analogues. *Biochemistry* 29:3133–3140

Ganter UM, Schmid ED, Perez-sala D, Rando RR, Siebert F (1989): Removal of the 9-methyl group of retinal inhibits signal transduction in the visual process—A Fourier-transform infrared and biochemical investigation. *Biochemistry* 28:5954–5962

Hall MD, Hoon MA, Ryba NJP, Pottinger JDD, Keen JN, Saibil HR, Findlay JBC (1991): Molecular-cloning and primary structure of squid *(Loligo forbesi)* rhodopsin, a phospholipase C-directed G-protein-linked receptor. *Biochem J* 274:35–40

Hamm HE, Bownds MD (1986): Protein complement of rod outer segments of frog retina. *Biochemistry* 25:4512–4523

Hargrave PA, Fong S-L, McDowell JH, Mas MT, Curtis DR, Wang JK, Juszczak E, Smith DP (1983): The structure of bovine rhodopsin. *Biophys Struct Mech* 9:235–244

Hausdorff WP, Bouvier M, O'Dowd BF, Irons GP, Caron MG, Lefkowitz RJ (1989): Phosphorylation sites on two domains of the β_2-adrenergic receptor are involved in distinct pathways of receptor desensitization. *J Biol Chem* 264:12657–12665

Henderson R, Baldwin JM, Ceska TA, Zemlin F, Beckmann E, Downing KH (1990): Model for the structure of bacteriorhodopsin based on high-resolution electron cryomicroscopy. *J Mol Biol* 213:899–929

Hisatomi O, Iwasa T, Tokunaga F, Yasui A (1991): Isolation and characterization of lamprey rhodopsin cDNA. *Biochem Biophys Res Commun* 174:1125–1132

Honig B, Dinur U, Nakanishi K, Balogh-Nair V, Grawinowicz M, Arnaboldi M, Motto MG (1979): An external point-charge model for wavelength regulation in visual pigments. *J Am Chem Soc* 101:7084–7086

Horstman DA, Brandon S, Wilson AL, Guyer CA, Cragoe EJ Jr, Limbird LE (1990): An aspartate conserved among G-protein receptors confers allosteric regulation of α_2-adrenergic receptors by sodium. *J Biol Chem* 265:21590–21595

Hubbard R, St. George RCC (1958): The rhodopsin system of the squid. *J Gen Physiol* 41:501–528

Huber A, Smith DP, Zuker CS, Paulsen R (1990): Opsin of *Calliphora* peripheral photoreceptors R1–6. Homology with *Drosophila* Rh1 and posttranslational processing. *J Biol Chem* 265:17906–17910

Hyde DR, Mecklenburg KL, Pollock JA, Vihtelic TS, Benzer S (1990): Twenty *Drosophila* visual-system cDNA clones—one is a homolog of human arrestin. *Proc Natl Acad Sci USA* 87:1008–1012

Janssen JJM, De Caluwé GLJ, De Grip WJ (1990): Asp_{83}, Glu_{113} and Glu_{134} are not specifically involved in Schiff-base protonation or wavelength regulation in bovine rhodopsin. *FEBS Lett* 260:113–118

Janssen JJM, Mulder WR, De Caluwé GLJ, Vlak JM, De Grip WJ (1991): In vitro expression of bovine opsin using recombinant baculovirus—the role of glutamic acid-(134) in opsin biosynthesis and glycosylation. *Biochim Biophys Acta* 1089:68–76

Kahlert M, König B, Hofmann KP (1990): Displacement of rhodopsin by GDP from three-loop interaction with transducin depends critically on the diphosphate β-position. *J Biol Chem* 265:18928–18932

Karnik SS, Khorana HG (1990): Assembly of functional rhodopsin requires a disulfide bond between cysteine residues 110 and 187. *J Biol Chem* 265:17520–17524

Karnik SS, Sakmar TP, Chen HB, Khorana HG (1988): Cysteine residues 110 and 187 are essential for the formation of correct structure in bovine rhodopsin. *Proc Natl Acad Sci USA* 85:8459–8463

Keen JN, Caceres I, Eliopoulos EE, Zagalsky PF, Findlay JBC (1991): Complete sequence and model for the A_2 subunit of the carotenoid pigment complex, crustacyanin. *Eur J Biochem* 197:407–417

König B, Arendt A, McDowell JH, Kahlert M, Hargrave PA, Hofmann KP (1989): Three cytoplasmic loops of rhodopsin interact with transducin. *Proc Natl Acad Sci USA* 86:6878–6882

Koutalos Y, Ebrey TG, Gilson HR, Honig B (1990): Octopus photoreceptor-membranes—surface-charge density and pK of the Schiff-base of the pigments. *Biophys J* 58:493–501

Koutalos Y, Ebrey TG, Tsuda M, Odashima K, Lien T, Park MH, Shimizu N, Derguini F, Nakanishi K, Gilson HR, Honig B (1989): Regeneration of bovine and octopus opsins in situ with natural and artificial retinals. *Biochemistry* 28:2732–2739

Kühn H, Dreyer WJ (1972): Light dependent phosphorylation of rhodopsin by ATP. *FEBS Lett* 20:1–6

Kühn H, Hall SW, Wilden U (1984): Light induced binding of 48-kDa protein to photoreceptor membranes is highly enhanced by phosphorylation of rhodopsin. *FEBS Lett* 176:473–478

Lamola AA, Yamane T, Zipp A (1974): Effects of detergents and high pressures upon the metarhodopsin I $\rightleftharpoons$ metarhodopsin II equilibrium. *Biochemistry* 13:738–745

Liggett SB, Caron MG, Lefkowitz RJ, Hnatowich M (1991): Coupling of a mutated form of the human β-2-adrenergic receptor to Gi and Gs—requirement for multiple cytoplasmic domains in the coupling process. *J Biol Chem* 266:4816–4821

Lohse MJ, Benovic JL, Codina J, Caron MG, Lefkowitz RJ (1990): β-Arrestin: A protein that regulates β-adrenergic receptor function. *Science* 248:1547–1550

Matthews RG, Hubbard R, Brown PK, Wald G (1963): Tautomeric forms of metarhodopsin. *J Gen Physiol* 47:215–240

McRee DE, Tainer JA, Meyer TE, Van Beeumen J, Cusanovich MA, Getzoff ED (1989): Crystallographic structure of a photoreceptor protein at 2.4Å resolution. *Proc Natl Acad Sci USA* 86:6533–6537

Miljanich GP, Brown MF, Mabrey-Gaud S, Dratz EA, Sturtevant JM (1985): Thermotropic behavior of retinal rod membranes and dispersions of extracted phospholipids. *J Membrane Biol* 85:79–86

Mitchell DC, Kibelbek J, Litman BJ (1991): Rhodopsin in dimyristoylphosphatidylcholine-reconstituted bilayers forms metarhodopsin II and activates G_t. *Biochemistry* 30:37–42

Mollevanger LCPJ, Kentgens APM, Pardoen JA, Courtin JML, Veeman WS, Lugtenburg J, De Grip WJ (1987): High-resolution solid state ^{13}C-NMR study of carbons C-5 and C-12 of the chromophore of bovine rhodopsin. Evidence for a 6-*S-cis* conformation with negative-charge perturbation near C-12. *Eur J Biochem* 163:9–14

Montell C, Jones K, Zuker CS, Rubin GM (1987): A second opsin gene expressed in the ultraviolet-sensitive R7 photoreceptor cells of *Drosophila melanogaster*. *J Neurosci* 7:1558–1566

Nakayama TA, Khorana HG (1990): Orientation of retinal in bovine rhodopsin determined by cross-linking using a photoactivatable analog of 11-*cis*-retinal. *J Biol Chem* 265:15762–15769

Nakayama TA, Khorana HG (1991): Mapping of the amino-acids in membrane-embedded helices that interact with the retinal chromophore in bovine rhodopsin. *J Biol Chem* 266:4269–4275

Nathans J (1990a): Determinants of visual pigment absorbance—role of charged amino-acids in the putative transmembrane segments. *Biochemistry* 29:937–942

Nathans J (1990b): Determinants of visual pigment absorbance. Identification of the retinylidene Schiff's base counterion in bovine rhodopsin. *Biochemistry* 29:9746–9752

Nathans J, Hogness DS (1984): Isolation and nucleotide sequence of the gene encoding human rhodopsin. *Proc Natl Acad Sci USA* 81:4851–4855

Nathans J, Thomas D, Hogness DS (1986): Molecular genetics of human color vision: The genes encoding blue, green, and red pigments. *Science* 232:193–202

Neitz J, Neitz M, Jacobs GH (1989): Analysis of fusion gene and encoded photopigment of colour-blind humans. *Nature* 342:679–682

Neitz M, Neitz J, Jacobs GH (1991): Spectral tuning of pigments underlying red-green color-vision. *Science* 252:971–974

Noegel AA, Gerisch G, Lottspeich F, Schleicher M (1990): A protein with homology to the C-terminal repeat sequence of *Octopus* rhodopsin and synaptophysin is a member of a multigene family in *Dictyostelium discoideum*. *FEBS Lett* 266:118–122

North ACT (1991): Structural homology in ligand-specific transport proteins. *Biochem Soc Symp* 90:35–48

O'Dowd BF, Hnatowich M, Caron MG, Lefkowitz RJ, Bouvier M (1989): Palmitoylation of the human β_2-adrenergic receptor. Mutation of Cys[341] in the carboxyl tail leads to an uncoupled nonpalymitoylated form of the receptor. *J Biol Chem* 264:7564–7569

O'Tousa JE, Baehr W, Martin RL, Hirsh J, Pak WL, Applebury ML (1985): The Drosophila *ninaE* gene encodes an opsin. *Cell* 40:839–850

Ovchinnikov Yu A (1982): Rhodopsin and bacterio-rhodopsin: Structure-function relationships. *FEBS Lett* 148:179–191

Ovchinnikov Yu A, Abdulaev NG, Bogachuk AS (1988a): Two adjacent cysteine residues in the C-terminal cytoplasmic fragment of bovine rhodopsin are palmitylated. *FEBS Lett* 230:1–5

Ovchinnikov Yu A, Abdulaev NG, Zolotarev AS, Artamonov ID, Bespalov IA, Dergachev AE, Tsuda M (1988b): Octopus rhodopsin—amino acid sequence deduced from cDNA. *FEBS Lett* 232:69–72

Pappin DJC, Eliopoulos E, Brett M, Findlay JBC (1984): A structural model for ovine rhodopsin. *Int J Biol Macromol* 6:73–76

Pappin DJC, Findlay JBC (1984): Sequence variability in the retinal-attachment domain of mammalian rhodopsins. *Biochem J* 217:605–613

Parry-Smith DJ, Attwood TK (1991): SOMAP: A novel interactive approach to multiple protein sequence alignment. *Comput Applic Biosci* 7:233–235

Renk G, Crouch RK (1989): Analog pigment studies of chromophore protein interactions in metarhodopsins. *Biochemistry* 28:907–912

Sakmar TP, Franke RR, Khorana HG (1989): Glutamic acid-113 serves as the retinylidene Schiff-base counterion in bovine rhodopsin. *Proc Natl Acad Sci USA* 86:8309–8313

Sakmar TP, Franke RR, Khorana HG (1991): The role of the retinylidene Schiff base counterion in rhodopsin in determining wavelength absorbence and Schiff base pK_a. *Proc Natl Acad Sci USA* 88:3079–3083

Schleicher A, Kühn H, Hofmann KP (1989): Kinetics, binding constant, and activation-energy of the 48-kDa protein rhodopsin complex by extra-metarhodopsin II. *Biochemistry* 28:1770–1775

Schneuwly S, Shortridge RD, Larrivee DC, Ono T, Ozaki M, Pak WL (1989): *Drosophila ninaA* gene encodes an eye-specific cyclophilin (cyclosporine a binding protein). *Proc Natl Acad Sci USA* 86:5390–5394

Seckler B, Rando RR (1989): Schiff-base deprotonation is mandatory for light-dependent rhodopsin phosphorylation. *Biochem J* 264:489–493

Sitaramayya A (1986): Rhodopsin kinase prepared from bovine rod disk membranes quenches light activation of cGMP phosphodiesterase in a reconstituted system. *Biochemistry* 25:5460–5468

Smith SO, Palings I, Copie V, Raleigh DP, Courtin J, Pardoen JA, Lugtenburg J, Mathies RA, Griffin RG (1987): Low-temperature solid-state ^{13}C NMR studies of the retinal chromophore in rhodopsin. *Biochemistry* 26:1606–1611

Smith SO, Palings I, Miley ME, Courtin J, De Groot H, Lugtenburg J, Mathies RA, Griffin RG (1990): Solid-state NMR-studies of the mechanism of the opsin shift in the visual pigment rhodopsin. *Biochemistry* 29:8158–8164

Stamnes MA, Shieh BH, Chuman L, Harris GL, Zuker CS (1991): The cyclophilin homolog ninaA is a tissue-specific integral membrane-protein required for the proper synthesis of a subset of Drosophila rhodopsins. *Cell* 65:219–227

Strader CD, Candelore MR, Hill WS, Dixon RAF, Sigal IS (1989): A single amino-acid substitution in the β-adrenergic-receptor promotes partial agonist activity from antagonists. *J Biol Chem* 264:16470–16477

Strathmann M, Simon MI (1990): G-protein diversity—a distinct class of α-subunits is present in vertebrates and invertebrates. *Proc Natl Acad Sci USA* 87:9113–9117

Stubbs GW, Smith HG, Litman BJ (1976): Alkyl glucosides as effective solubilizing agents for bovine rhodopsin. A comparison with several commonly used detergents. *Biochim Biophys Acta* 426:46–56

Takao M, Yasui A, Tokunaga F (1988): Isolation and sequence determination of the chicken rhodopsin gene. *Vision Res* 28:471–480

Thomas DD, Stryer L (1982): Transverse location of the retinal chromophore of rhodopsin in rod outer segment disc membranes. *J Mol Biol* 154:145–157

Thompson P, Findlay JBC (1984): Phosphorylation of ovine rhodopsin: Identification of the phosphorylated sites. *Biochem J* 220:773–780

Tokunaga F, Iwasa T, Miyagishi M, Kayada S (1990): Cloning of cDNA and amino-acid-sequence of one of chicken cone visual pigments. *Biochem Biophys Res Commun* 173:1212–1217

Trimble WS, Scheller RH (1988): Molecular-biology of synaptic vesicle-associated proteins. *Trends Neurosci* 11:241–242

Wald G (1968): Molecular basis of visual excitation. *Science* 162:230–239

Wald G, Brown PK (1950): The synthesis of rhodopsin from retinene$_1$. *Proc Natl Acad Sci USA* 36:84–92

Wilden U, Kühn H (1982): Light dependent phosphorylation of rhodopsin: Number of phosphorylation sites. *Biochemistry* 21:3014–3022

Yamada T, Takeuchi Y, Komori N, Kobayashi H, Sakai Y, Hotta Y, Matsumoto H (1990): A 49-kilodalton phosphoprotein in the *Drosophila* photoreceptor is an arrestin homolog. *Science* 248:483–486

Yokoyama R, Yokoyama S (1990a): Isolation, DNA-sequence and evolution of a color visual pigment gene of the blind cave fish *Astyanax fasciatus*. *Vision Res* 30:807–816

Yokoyama R, Yokoyama S (1990b): Convergent evolution of the red- and green-like visual pigment genes in fish, *Astyanax fasciatus,* and human. *Proc Natl Acad Sci USA* 87:9315–9318

Zagalsky PF, Eliopoulos EE, Findlay JBC (1990a): The architecture of invertebrate carotenoproteins. *Comp Biochem Physiol B* 97:1–18

Zagalsky PF, Eliopoulos EE, Findlay JBC (1991): The lobster carapace carotenoprotein, α-crustacyanin. A possible role for tryptophan in the bathochromic spectral shift of protein-bound astaxanthin. *Biochem J* 274:79–83

Zagalsky PF, Mummery RS, Eliopoulos EE, Findlay JBC (1990b): The quaternary structure of the lobster carapace carotenoprotein, crustacyanin. Studies using cross-linking agents. *Comp Biochem Physiol B* 97:837–848

Zhukovsky EA, Oprian DD (1989): Effect of carboxylic-acid side-chains on the absorption maximum of visual pigments. *Science* 246:928–930

Zhukovsky EA, Robinson PR, Oprian DD (1991): Transducin activation by rhodopsin without a covalent bond to the 11-*cis*-retinal chromophore. *Science* 251:558–560

Zuker CS, Cowman AF, Rubin GM (1985): Isolation and structure of a rhodopsin gene from *D. melanogaster*. *Cell* 40:851–858

Zuker CS, Montell C, Jones K, Laverty T, Rubin GM (1987): A rhodopsin gene expressed in photoreceptor cell R7 of the *Drosophila* eye—homologies with other signal-transducing molecules. *J Neurosci* 7:1550–1557

2

Structural, Functional, and Genetic Aspects of Receptors Coupled to G-Proteins

BRIAN O'DOWD, SHEILA COLLINS, MICHEL BOUVIER, MARC G. CARON, AND ROBERT J. LEFKOWITZ

1. Introduction

A remarkably diverse array of biologically active substances elicit their actions by interacting with cell surface receptors which are coupled via guanine nucleotide regulatory proteins (G-proteins) to specific biochemical effectors. Examples include neurotransmitters, hormones, many drugs, and even sensory stimuli such as light and odorants. Perhaps the most thoroughly studied example of such a receptor system is the β_2-adrenergic receptor for catecholamines which mediates stimulation of adenylyl cyclase via Gs. Within the past few years its primary structure has become known via molecular cloning, and substantial progress has also been made in learning how its unique structural features determine such specific functions as ligand binding and G-protein activation. Much has also been learned about the mechanisms by which its function is regulated. Moreover, a great deal of this information appears to be generally applicable to the other members of the broad family of G-protein-coupled receptors. This chapter reviews some of the most important information developed over the past several years.

2. Protein Sequences of G-Protein-Coupled Receptors

Primary structural similarities of the 19 G-protein-coupled receptors aligned in Figure 2.1 indicate a large and functionally diverse gene family. Each protein consists of a single polypeptide chain, and hydrophobicity plots suggest that these receptors span the lipid membrane seven times

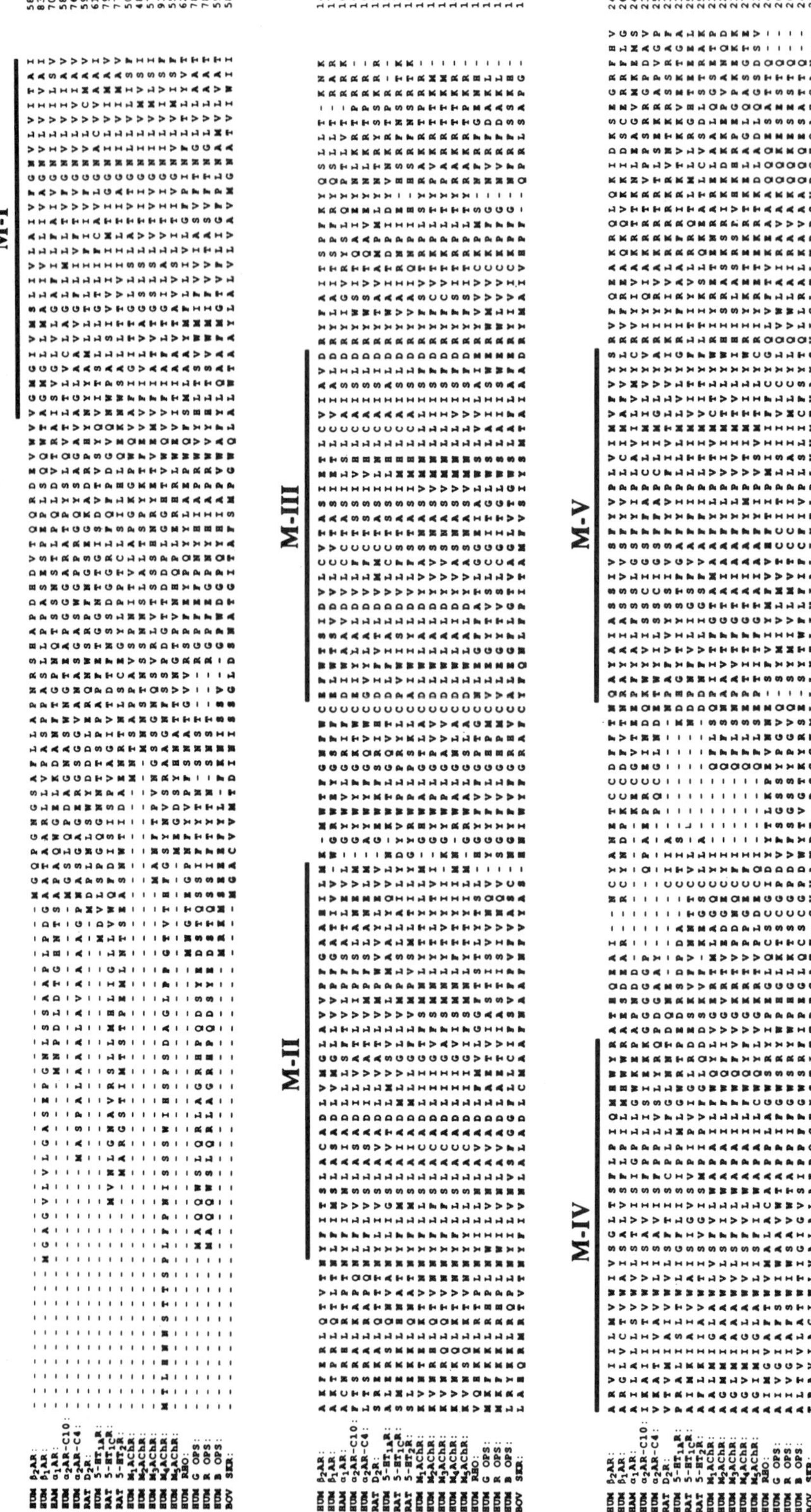

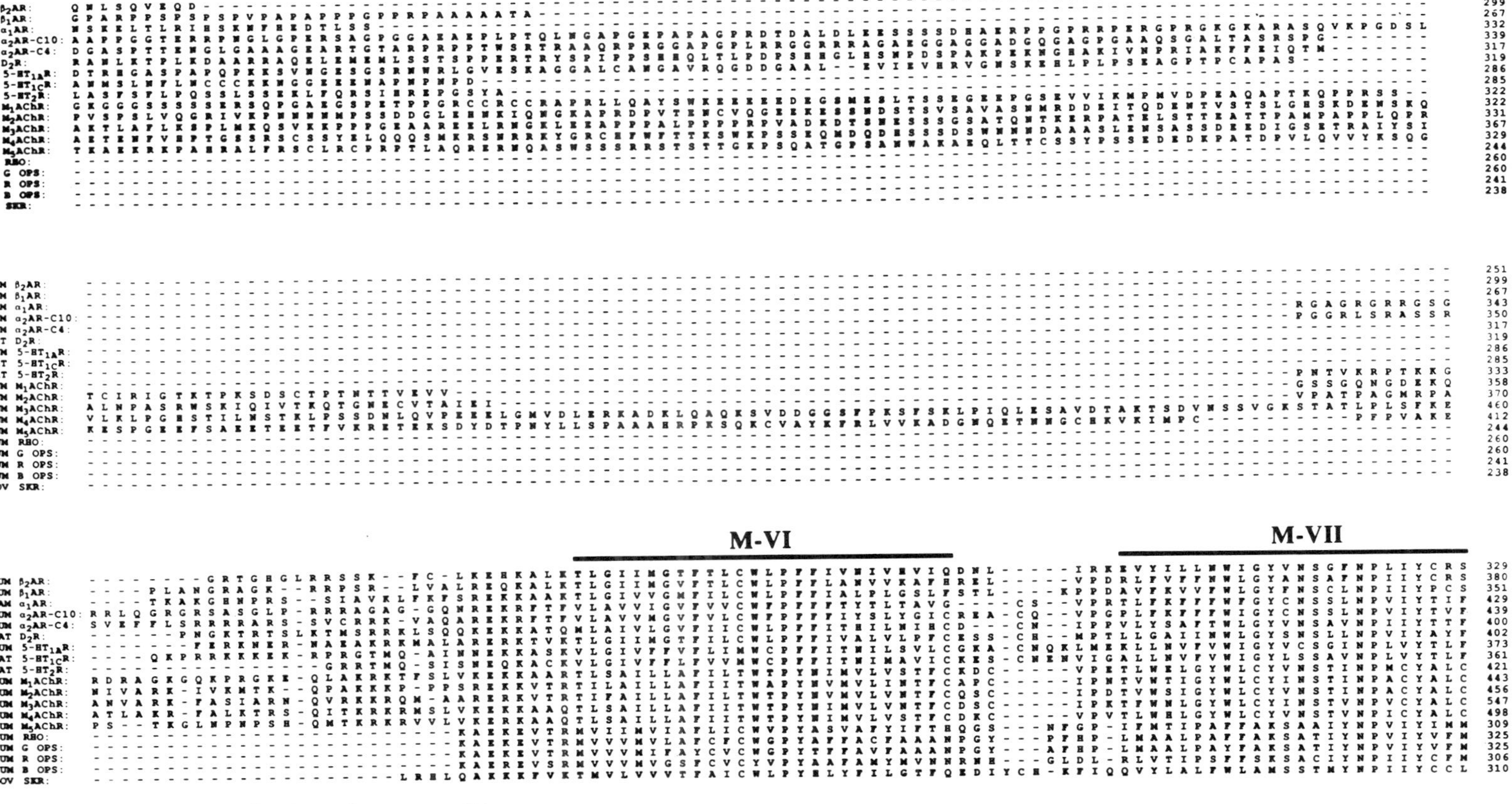

Figure 2.1. (pp. 32–34). For legend, see page 34.

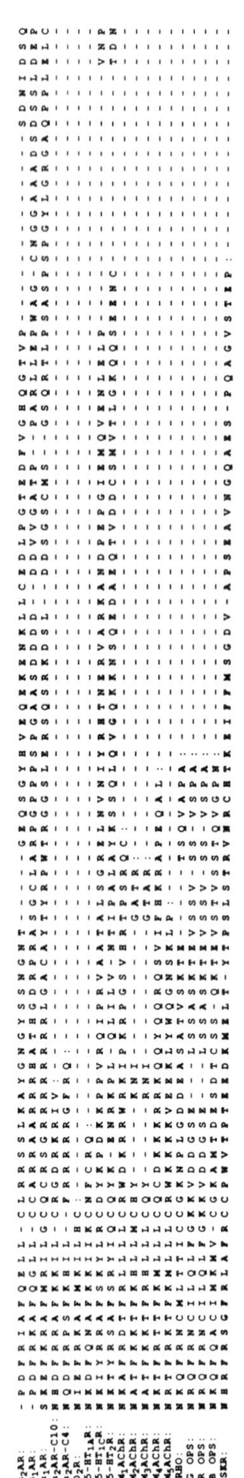
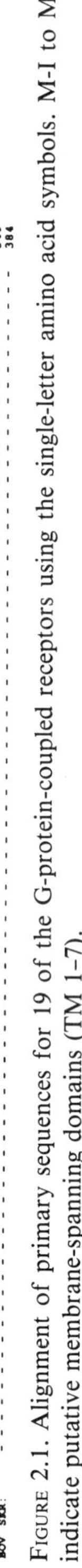

Figure 2.1. Alignment of primary sequences for 19 of the G-protein-coupled receptors using the single-letter amino acid symbols. M-I to M-VII indicate putative membrane-spanning domains (TM 1–7).

(Figure 2.2). This topographical organization is analogous to that demonstrated for bacteriorhodopsin (Dunn et al., 1981) and is not exclusive to the receptors aligned in Figure 2.1. It is also found in other, nonhomologous proteins including the mas oncogene product now identified as an angiotensin receptor (Jackson et al., 1988), 3-hydroxy-3-methyl-glutaryl coenzymeA reductase (HMG-CoA reductase) (Chin et al., 1984), mating factor receptors from *Saccharomyces* (Nakayama et al., 1985) and haloopsin (Schobert et al., 1988). The significance of the seven transmembrane motif is not fully clear, but it must be uniquely suited for transmitting a signal from the external surface, via a ligand-induced conformational change in the receptors to the internal surface of the plasma membrane.

The number of receptor proteins belonging to this family is increasing rapidly. Recently, the polymerase chain reaction (PCR), which is based on

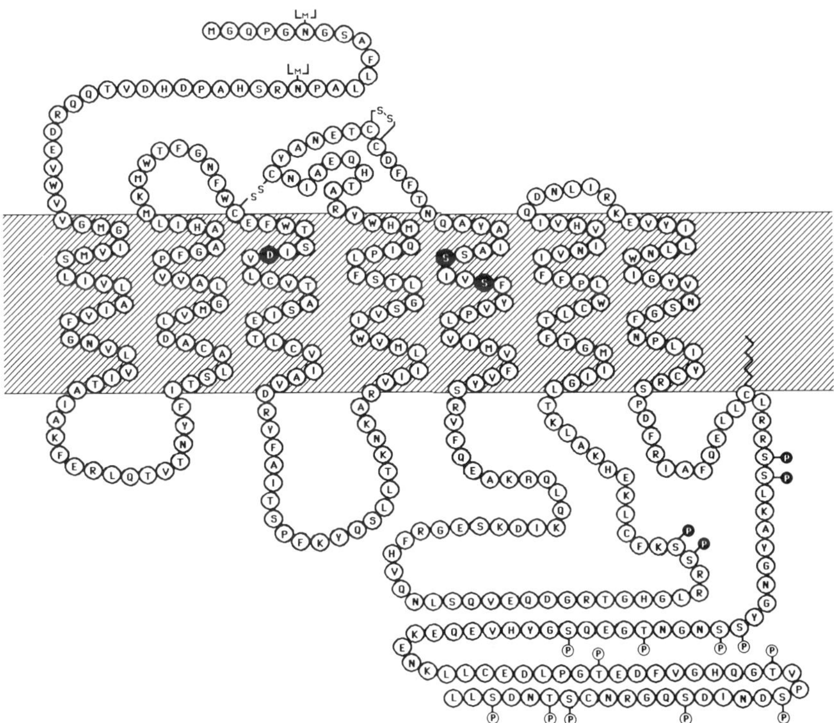

FIGURE 2.2. Representation of the topography of human β_2-AR in the plasma membrane. The aspartic acid highlighted in TM 3 has been proposed to act as the counterion for adrenergic ligands. Extracellular cysteine residues, which may be involved in disulfide bond formation, are indicated, as well as amino-terminal sites of glycosylation and putative sites of phosphorylation by PKA/protein kinase C and β-ARK. Fatty acylation of a cysteine residue in the C-terminal tail may anchor the amino-terminal segment of the tail to the plasma membrane.

the use of oligonucleotide primers homologous in sequence to known receptors, has been used to clone four new putative G-protein-coupled receptors from a dog thyroid cDNA library (Libert et al., 1989). Although the identities of these "orphan" proteins are presently unknown, functional studies and binding assays should eventually reveal the nature of these candidate receptors. Despite being cloned from a thyroid library, two of these putative receptors (RDC5 and RDC8) appear not to be efficiently expressed in this tissue (Libert et al., 1989). A possible explanation for this observation is provided in a report by Sarkar and Sommer (1989) who have recently demonstrated that there appears to be a basal rate of transcription of tissue-specific proteins in all tissues. Because of the sensitivity of the PCR technique, this level of transcription can be detected (Sarkar and Sommer, 1989).

Eleven amino acids are conserved in all of the receptors aligned in Figure 2.1 and all are found in or near the seven transmembrane (TM) segments (Figure 2.2). In fact, ten of these conserved residues are located on the side of the membrane closest to the cytoplasm, perhaps because these residues are for coupling to common cytoplasmic elements (e.g., G-proteins). When only those receptors capable of binding a protonated amine ligand are compared (i.e., adrenergic, muscarinic, serotonergic, and dopaminergic), the number of conserved residues increases considerably (Figure 2.3). Thirty-two residues are conserved in these types of receptors: 29 of these are found in or near the seven transmembrane segments, and 25 of these are again located on the side of the transmembrane segment nearest the cytoplasm. These conserved features are combined and highlighted in the model of the G-protein-coupled receptors shown in Figure 2.3. The same pattern of sequence conservation is apparent when conserved amino acid substitutions are also considered (Figure 2.3). Interestingly, only TM 3 has a majority of conserved substitutions (including one identical aspartic acid residue, Asp^{113} in β_2-AR) on the side of the receptor facing the outside of the cell (Figure 2.3). On the basis of site-specific mutagenesis experiments, Strader et al. (1987a) have proposed that Asp^{113} in β_2-AR acts as the counterion for the cationic amine group on the ligand. Thus, while the transmembrane segments of these receptors generally show the least degree of similarity on the extracellular side of the receptor in TM 3, only conserved substitutions were permitted in residues surrounding the aspartic acid counterion. Perhaps these conserved residues in TM 3 also constitute part of the ligand binding domain.

The only structural features that vary appreciably among the receptors aligned in Figure 2.1 are the sizes of their third cytoplasmic loops and carboxy-termini. Receptors with longer C-terminal tails tend to have shorter i3 loops and vice versa. A possible explanation for this pattern suggests that certain common structural or functional components of these receptors may be found either in the i3 loop or C-terminal tails. These regions character-

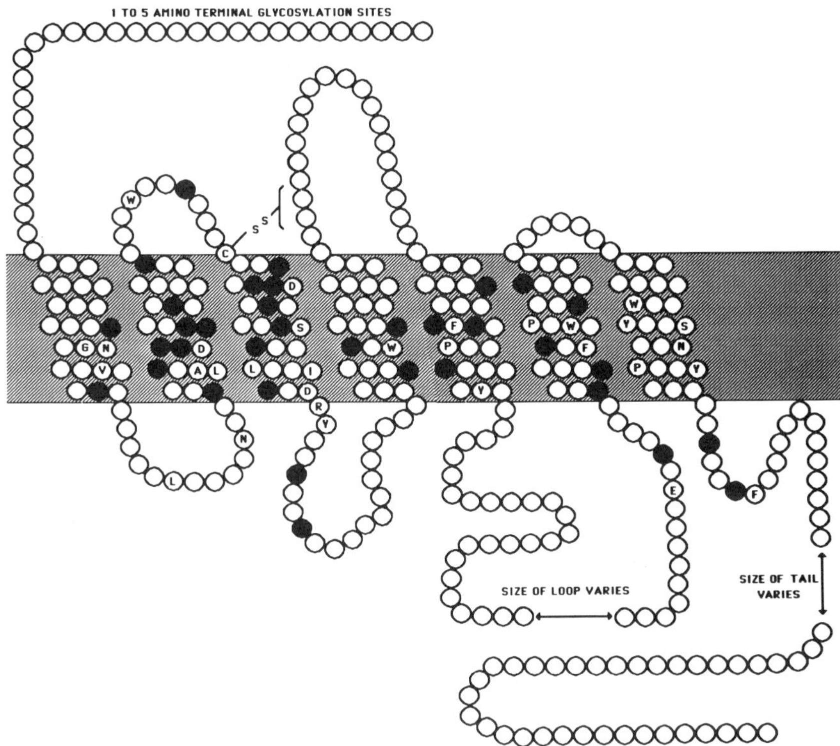

FIGURE 2.3. Distribution of conserved amino acids among the adrenergic, serotonergic, dopaminergic, and muscarinic receptors. Amino acids identical in all receptors are indicated by letters; conserved substitutions are indicated by black circles.

istically gain more acidic character with increased distance from the plasma membrane and generally share similar chemical composition. Recent evidence indicates that several of these receptors [α_2-adrenergic (Benovic et al., 1987b) and β_2-adrenergic (Benovic et al., 1986b), muscarinic (Kwatra et al., 1989), and rhodopsin (Benovic et al., 1986a)] can be phosphorylated by various kinases (see Sections 3.4 and 5.1 for more detailed discussion).

One other notable feature common to all of the receptors aligned in Figure 2.1 is the presence of at least one proline residue in TM 4. The adrenergic, dopaminergic, and serotonergic receptors also have a proline residue conserved in TM 2. The conservation of these prolines suggests that these residues contribute to some feature of the structure of these receptors. For example, each of these prolines could induce a kink in the transmembrane segment. From crystallographic studies of bacteriorhodopsin, Henderson and Unwin (1975) suggested that such a kink in the transmembrane segment could interlock amino acid side chains from adjacent segments.

For example, three of the transmembrane segments in bacteriorhodopsin are more strongly tilted than the others, promoting a left-handed supercoil conformation to that protein (Henderson and Unwin, 1975).

3. Receptor Modification by Several Posttranslational Events

3.1. Glycosylation

Each of the 19 receptor proteins aligned in Figure 2.1 (and also the angiotensin receptor) (Young et al., 1986) contains amino-terminal glycosylation sites. Each of the adrenergic receptors contains N-linked complex oligosaccharides (O'Dowd et al., 1989b). The hamster β_2-AR also contains high mannose-type oligosaccharides. For the canine and porcine brain D2-dopaminergic receptors, complex oligosaccharides have been found (Jarvie et al., 1988), while the D2-dopaminergic receptor isolated from the anterior pituitary contains hybrid oligosaccharides (Jarvie and Niznik, 1989). The homologues of the G-protein-coupled receptors recently cloned by PCR, RDC7 and RDC8 have very short NH_2-terminal domains which are devoid of the N-X Ser/Thr glycosylation motif (Libert et al., 1989), suggesting that these receptors are members of a new subclass of non-glycosylated G-protein-coupled receptors. The α_2-AR isolated from rat cerebral cortex may also represent a nonglycosylated G-protein-coupled receptor (Lanier et al., 1988). To date only the oligosaccharides attached to bovine rhodopsin have been studied by NMR techniques (Applebury and Hargrave, 1986).

The function of N-linked glycosylation in the β_2-adrenergic receptor has been studied by site-specific mutagenesis (Dixon et al., 1987), and it appears that glycosylation is not required for either ligand binding or functional coupling to G-proteins. However, glycosylation may play a role in both receptor trafficking through the cell and regulation (Terman and Insel, 1988).

3.2. Palmitoylation

An earlier report by O'Brien and colleagues (O'Brien and Zatz, 1984) indicated that rhodopsin was modified by the covalent attachment of the fatty acid, palmitic acid. More recently, Ovchinnikov and his group (Ovchinnikov et al., 1988) identified two adjacent cysteine residues (Cys^{322} and Cys^{323}) in the C-terminal tail of bovine rhodopsin which were, in fact, palmitoylated. In the alignment of amino acid sequences shown in Figure 2.1, 15 of the 19 receptors also have at least one cysteine in approximately the same position. However, of the "orphan" receptors isolated from the dog thyroid library only RDC7 has a cysteine residue in this region. Recently, we reported that the human β_2-AR is also palmitoylated, and that mutation of Cys^{341} to a Gly results in a nonpalmitoylated form of the receptor (O'Dowd et al., 1989a). In a model identical to that proposed for rhodopsin (Ovchinnikov et al., 1988), we suggested that this posttranslational modification by palmitoylation promoted the association of the amino-terminal portion of the C-terminal tail with the plasma membrane, thus forming a fourth inner (intracellular) loop (Figure 2.2). In this model, an agonist- or light-induced conformational change in the receptor might be mediated by release of the fatty acid group, thereby causing detachment of the amino-terminal region of the tail from the membrane. In each of the receptors shown in Figure 2.1, the conserved cysteine residue is close to a region of positively charged amino acids. Bach et al. (1988) have suggested that such a cluster of positively charged amino acids may define a type of consensus acylation site. Moreover, these positively charged residues would likely promote association of the C-terminal tail with the membrane via electrostatic interactions with the negatively charged phospholipid head-groups in the membrane.

We further established a role for palmitoylation of the human β_2-AR in receptor: Gs coupling (O'Dowd et al., 1989a). The mutant, nonpalmitoylated β_2-AR ($Cys^{341} \rightarrow$ Gly), was markedly reduced in its ability to form a guanyl nucleotide-sensitive, high-affinity, agonist-binding state characteristic of wild-type receptors, and also in its ability to activate adenylyl cyclase (O'Dowd et al., 1989a). Based on this data we suggested that palmitoylation may represent a general (but not necessarily universal) feature in members of the G-protein-coupled receptor family. It will be of interest to determine if other receptors lacking a cysteine residue at the equivalent position in the C-terminal tail undergo palmitoylation at cysteine residues located elsewhere.

3.3. Disulfide Bond Formation in β_2-AR

A structural feature shared by the receptors aligned in Figure 2.1 is the presence of conserved cysteine residues. Residue Cys^{106} in β_2-AR is conserved in every receptor. In addition, at least one cysteine residue is found in the extracellular loop between TM 4 and 5 (Cys^{184} in β_2-AR). Based on altered ligand binding characteristics of β_2-AR mutants lacking either Cys^{106} or Cys^{184}, Dixon et al. (1987) suggested that occurrence of a disulfide bond between the two residues in the native protein (Figure 2.2). More recently, it has been estimated (Dohlman et al., 1990) that altogether there are seven pairs of disulfide-bonded cysteines in the β_2-AR, with at least two of these pairs being essential for normal agonist binding. In a series of β_2-AR mutants in which valine residues were substituted for cysteines, agonist binding was most impaired in mutants which destroyed the vicinal Cys^{190} and Cys^{191} as well as Cys^{106} and Cys^{184}.

3.4. Phosphorylation

Since rhodopsin is phosphorylated on a series of serines and threonines in the cytoplasmic C-terminus of the protein by a specific retinal enzyme termed rhodopsin kinase, so by analogy is the β_2-AR phosphorylated. Serine or threonine residues in the vicinity of acidic amino acids are thought to be susceptible to phosphorylation by an enzyme called β-adrenergic receptor kinase (β-ARK) (Benovic et al., 1986a). In addition to the β_2-AR, the α_2-AR (Benovic et al., 1987b) and muscarinic (Kwatra et al., 1989) receptors, as well as rhodopsin itself (Benovic et al., 1986a), can be phosphorylated by β-ARK in a completely agonist- or light-dependent fashion. Such putative β-ARK sites, however, appear to be confined to either the middle portion of the i3 loop or the C-terminal tails in any given receptor, but not both (Figure 2.1).

Other kinases that are capable of phosphorylating the β_2-AR both in vitro and in vivo include the cAMP-dependent protein kinase (PKA) and protein kinase C (Benovic et al., 1988). Phosphorylation of the receptor by PKA and β-ARK appear to have important functional consequences (see Section 5.1 for a complete discussion).

4. Receptor Coupling to G-Proteins

A major goal of recent investigations into the structure and function of this receptor family has been the delineation of regions that determine G-protein recognition and coupling specificity. From a collection of chimeric α_2,β_2-adrenergic receptors, one construct, in which β_2-AR sequences encoding the entire third loop and contiguous sequence from TM 5 and 6 were placed in the context of the platelet α_2-AR (Kobilka et al., 1988), was capable of coupling to Gs (albeit at levels below the native β_2-AR). Similar functional analyses of chimeric muscarinic receptors have also indicated that regions of the i3 loop are involved in the coupling of muscarinic acetylcholine receptors 1 and 2 (M1 and M2) with different G-proteins (Kubo et al., 1988). Substitution of β_2-AR amino acid residues in the carboxyl-most portion of the i3 loop (263–274, 267–274, and 271–274) with sequence from platelet α_2-AR produced three mutant β-adrenergic receptors with profound impairment of their ability to couple to Gs (O'Dowd et al., 1988; Figure 2.4). Thus, it appears that some of the amino acids that confer G-protein specificity reside in this region. Paradoxically, a deletion mutant of the β_2-AR, which removed the 11 amino acid deletion of residues 263–274, produced a mutant β_2-AR with only a modest effect on the receptor's ability to couple with Gs, while a mutant β_2-AR with a *smaller* deletion in this region (267–273) was associated with a *greater* impairment in coupling ability. To understand these data, we hypothesized that a realignment of Arg^{259}–Arg^{260} in the 11 amino acid-deletion mutant recon-

β_2AR SPECIES	COUPLING IMPAIRMENT	C-TERMINUS OF INTRACELULAR LOOP III		TMS-VI
		252	267	
WILD TYPE	0%	- - G R T G H G L R R S S K F C L K E	[H K] A L K	T - -
D263-273 + S259-260	~100%	- - - - - - - - - - - - - - - G R T G H G L	G A S S	T - -
D263-273	~15%	- - - - - - - - - - - - - - - G R T G H G L	[R R] S S	T - -
D267-273	~50%	- - - - - - - - - - G R T G H G L	[R R] S S K F C L	T - -
S267-274	~75%	- - G R T G H G L R R S S K F C L R E	[K R] F T F	V - -
S271-274	~65%	- - G R T G H G L R R S S K F C L K E	[H K] F T F	V - -

FIGURE 2.4. Amino acid residues at the carboxy-terminus of the i3 loop of β_2-AR following various deletion (D) and substitution (S) mutations. Percentage coupling impairment refers to the decrease in the maximal level of agonist-induced adenylyl cyclase activity.

structed an important charge distribution (Figure 2.3), which approximated that of the wild type, thus allowing Gs coupling to occur. An additional series of mutant β_2-ARs bearing substitutions of platelet α_2-AR sequence, but now in the amino-terminal region of the i3 loop were still capable of efficient coupling to Gs (O'Dowd et al., 1988). However, substitution of two basic residues in β_2-AR (conserved in other receptors aligned in Figure 2.1) 226–228 (A-K-R) with unrelated and uncharged sequence (G-A-G) resulted in virtually complete abolition of coupling. Thus, it appears that the role of this region of the receptor is to bind common elements in the different G-proteins, but not to determine specificity.

5. Regulation of β_2-AR-Mediated Transmembrane Signaling

As discussed in the previous sections, hormonal signaling across biological membranes can involve a complex set of interactions between receptors and distinct cellular components. As we will now discuss, the responsiveness of such transmembrane signaling pathways to hormonal stimulation is subject to dynamic regulatory processes. The adenylyl cyclase-coupled β_2-AR system has been one of the most extensively studied in this regard. For the purpose of our discussion we will focus on the mechanisms involved in agonist-induced desensitization of the β_2-AR itself. This phenomenon, also known as refractoriness or tolerance, is characterized by the fact that the intensity of the adenylyl cyclase stimulation wanes over time despite the presence of a constant concentration of the agonist.

Desensitization of adenylyl cyclase following catecholamine stimulation was, at first, thought to be related solely to the decrease in β-AR number ("downregulation") which followed prolonged stimulation (Lefkowitz, 1979). However, it soon became evident that agonist-induced desensitization is much more complex. At least three distinct processes may be involved in this phenomenon: (1) a rapid (minutes) functional uncoupling of the receptor from adenylyl cyclase activation, (2) a rapid (minutes) sequestration or internalization of the receptor away from the cell surface, and (3) a slower (hours) "downregulation" or loss of the receptor.

5.1. Functional Uncoupling

In the last few years, the concept has emerged that rapid uncoupling of the β_2-AR from adenylyl cyclase activation precedes the physical sequestration of the β_2-AR, and that this process itself contributes to the loss of hormonal responsiveness. Harden et al. (1980) demonstrated in human astrocytoma cells that a significant decrease in agonist-induced adenylyl cyclase activity preceded the appearance of receptor sequestration by ~ 1 min. Moreover, blockade of receptor sequestration by various manipulations (Hertel et al., 1985, 1986; Homburger et al., 1980) failed to prevent desensitization.

Other evidence for a functional uncoupling of the β_2-AR from Gs includes the observation that agonist-induced desensitization in S49 lymphoma cells is associated with a decreased ratio of the high affinity vs. low affinity form of the receptor for agonist (Strasser et al., 1986). Indeed, the activity of an agonist to stabilize the high-affinity form of the β_2-AR is believed to represent the formation of a ternary agonist–receptor–Gs complex (DeLean et al., 1980). Consistent with the notion that uncoupling is distinct from sequestration are the observations (Kassis and Fishman, 1984; Kassis and Sullivan, 1986) that β_2-ARs from desensitized cells such as HeLa, cyc$^-$ S49 lymphoma, and C6 rat glioma have a decreased ability to stimulate adenylyl cyclase, as assessed by membrane fusion with cells devoid of β_2-AR. Similarly, when partially purified receptors from control or agonist-desensitized frog erythrocytes were reconstituted in phospholipid vesicles, their functionality was diminished — again assessed by fusion with cells deficient in receptor (Strulovici et al., 1984). These observations suggested that an alteration of the receptor itself could lead to uncoupling and, hence, desensitization. Moreover, a structural modification of the receptor was suggested to accompany desensitization, since photoaffinity-labeled β-ARs from desensitized cells showed an altered electrophoretic mobility when compared to control cells (Stadel et al., 1982).

By metabolically labeling avian erythrocytes with [^{32}P]inorganic phosphate prior to desensitization, it was shown that stoichiometric (~ 2 mol of phosphate/mole of receptor) phosphorylation of the β-AR accompanied desensitization (Stadel et al., 1983a; Sibley et al., 1984). The occurrence of β-AR phosphorylation during agonist-induced desensitization was then confirmed using frog erythrocytes (Sibley et al., 1985) and S49 mouse lymphoma cells (Strasser et al., 1986). The stoichiometry of phosphorylation was strongly correlated with the extent of desensitization (Stadel et al., 1983a; Sibley et al., 1984) and the time courses of these two events were very similar (Sibley et al., 1985). Taken together these observations suggested a causal relationship between phosphorylation of the β-AR and its desensitization.

Moreover, membrane-permeable analogues of cAMP were found to cause submaximal phosphorylation of the β-AR as well as a partial desensitization of the β-adrenergic stimulated adenylyl cyclase activity in avian (Sibley et al., 1985) and amphibian (Sibley et al., 1985) erythrocytes, suggesting a role for PKA in this event. Consistent with this notion, isoproterenol-induced desensitization and phosphorylation of the β-AR in turkey erythrocyte lysates were each inhibited up to 50% by a specific inhibitor of PKA (Nambi et al., 1985). More recently, Clark and colleagues (1988) have reported that in S49 wild-type lymphoma cells, desensitization of epinephrine-stimulated adenylyl cyclase was induced by pretreatment with a low concentration of epinephrine (5–50 nM). However, in the kin⁻ variant of the S49 cells, which lacks PKA activity, desensitization was absent. These observations further support a role for PKA in the development of agonist-induced desensitization.

Other evidence for PKA-mediated phosphorylation of the β-AR in the desensitization process has come from in vitro experiments. The purified catalytic subunit of the PKA can promote the addition of up to 2 mol of phosphate per mole of purified hamster lung β_2-AR (Benovic et al., 1985, Bouvier et al., 1987), while the addition of β-adrenergic agonist to the reaction mixture was found to enhance the rate of this phosphorylation. These results suggested that agonist occupancy of the receptor makes it a better substrate for PKA. Moreover, it has been demonstrated that receptor which is phosphorylated by PKA has a reduced ability to stimulate the guanine triphosphate hydrolase (GTPase) activity of Gs in a reconstituted system, confirming that phosphorylation of receptor by this kinase may impair its functionality (Benovic et al., 1985).

Although phosphorylation of the β_2-AR by PKA appears to be involved in the desensitization process, it cannot fully account for the entire desensitization pattern. As mentioned previously, cAMP analogues lead to only partial desensitization and phosphorylation, and PKA inhibitors only blocked 50% of the desensitization and receptor phosphorylation. In other experiments, desensitization to *low* doses of epinephrine was absent in kin⁻ S49 cells (see above) while desensitization to *higher* doses was observed in these cells (Clark et al., 1988). Similarly, other studies using the S49 cell variants kin⁻ or cyc⁻ (the latter lacking Gsα) concluded that neither cAMP production nor PKA activation is required for β_2-AR phosphorylation and desensitization to take place. It was thus speculated that a PKA may be involved in the phosphorylation of the β_2-AR during desensitization.

Benovic et al. (1987a,b) first identified and purified such a kinase which phosphorylates purified β_2-AR in a totally agonist-dependent manner. This enzyme (β-ARK), which is independent of cAMP, cGMP, Ca^{2+}/ calmodulin, or phospholipids, has now been cloned (Benovic et al., 1989b). Two highly homologous but distinct clones were obtained, suggesting the presence of isoenzymes and possibly a family of receptor kinase genes.

As discussed previously, the β-adrenergic-stimulated adenylyl cyclase system shares several features with the rhodopsin-stimulated cGMP phosphodiesterase transduction pathway. Namely, in both cases specific receptor kinases, the rhodopsin kinase and β-ARK, have been shown to phosphorylate their respective substrates in a stimulus-dependent fashion (Kuhn and Dreyer, 1972; Benovic et al., 1986a,b). Phosphopeptide analysis and sequencing of phosphorhodopsin (Thompson and Findley, 1984) have demonstrated that multiple serine and threonine residues located near the carboxyl-terminus of the visual pigment constitute the major site of phosphorylation for the rhodopsin kinase. For the β_2-AR, Dohlman et al. (1987) have shown that following carboxypeptidase treatment most of the sites phosphorylated by β-ARK were lost. This enzymatic cleavage was shown to remove the cytoplasmic tail of the receptor which contains a serine- and threonine-rich segment near the carboxyl-terminus. It therefore appears that the serine- and threonine-rich segment of the carboxyl-terminal portion of the β_2-AR harbors the major phosphorylation sites for β-ARK.

In an attempt to address the functional significance of these putative phosphorylation sites, Benovic et al. (1987a) also showed that phosphorylation of purified hamster β_2-AR by pure β-ARK in vitro led to a modest ($\sim 16\%$) reduction of the ability of the receptor to stimulate the GTPase activity of Gs in a reconstituted system. The addition of pure retinal arrestin to the phosphorylated β_2-AR further reduced (by 41%) the ability of the receptor to stimulate the GTPase activity of reconstituted Gs. Since arrestin contributes to the inactivating effect of rhodopsin phosphorylation by rhodopsin kinase in the light transduction system (Wilden et al., 1986), a protein analogous to arrestin has been postulated to mediate the functional consequences of receptor phosphorylation by β-ARK. This hypothesis is supported by the recent cloning of a distinct β-arrestin gene (Lohse et al., 1990).

Using site-directed mutagenesis and an intact cell system, the role of these β-ARK phosphorylation sites in agonist-induced desensitization was explored. Bouvier et al. (1988) generated cell lines expressing mutant β_2-AR that lack the serine- and threonine-rich carboxyl-terminus. To achieve this goal, two different mutants were produced. In one mutant the terminal 48 most carboxyl residues were truncated, leading to a β_2-AR with a shorter cytoplasmic tail lacking most of the serines and threonines. In a second mutant, the 11 serines and threonines of the C-terminal tail of the receptor were selectively mutated to alanine or glycine. In cells expressing either of these mutated receptors, agonist-induced phosphorylation was found to be dramatically reduced as compared with cells expressing wild-type receptors (Bouvier et al., 1988; Hausdorff et al., 1989). These data therefore suggest that these serines and threonines represent major sites for agonist-induced phosphorylation and strongly implicate β-ARK as the enzyme responsible for this phosphorylation in whole cells. Moreover, cells expressing these mutated receptors display a much delayed onset of agonist-induced desensitization as compared with cells expressing the wild-type receptor (Bouvier et al., 1988). This suggests that phosphorylation of the carboxyl-terminus of the receptor by β-ARK is a crucial event in the early stages of agonist-induced desensitization. However, a normal desensitization pattern was observed following longer agonist exposure (Strader et al., 1987b; Kobilka et al., 1987a; Bouvier et al., 1988). Since sequestration of these mutated receptors was reported to be normal (Strader et al., 1987b; Hausdorff et al., 1989) or even increased (Bouvier et al., 1988), the agonist-induced phosphorylation may be a triggering event for the early uncoupling that precedes sequestration of the receptor.

Lohse et al. (1989) recently confirmed the importance of β-ARK-mediated phosphorylation of the β_2-AR in rapid agonist-induced desensitization. By using permeabilized human epidermoid carcinoma A431 cells they were able to show that heparin, a known inhibitor of β-ARK in vitro (Benovic et al., 1989a), drastically inhibited both the agonist-induced phosphorylation and desensitization of the β_2-AR.

The data presented thus far indicate that phosphorylation of the β_2-AR by two distinct protein kinases, PKA and β-ARK, may contribute to the development of rapid agonist-induced desensitization. How these two kinases contribute to the functional uncoupling of the receptor from the stimulation of adenylyl cyclase, and how they together produce an integrated set of cellular signals remains unclear. Hausdorff et al. (1989) have utilized three mutant β_2-AR genes encoding the receptor lacking putative phosphorylation sites for PKA and/or β-ARK to explore the role of receptor phosphorylation by specific kinases in agonist-induced desensitization. Agonist stimulation of cell lines expressing these mutants showed that low levels (10 nM) were found to preferentially induce phosphorylation of the receptor at PKA sites, whereas higher levels (2 μM) promoted phosphorylation on both the PKA and the β-ARK sites. These observations are consistent with the known characteristics of these kinases. Indeed, at 10 nM isoproterenol the adenylyl cyclase would be maximally stimulated in whole cells, thereby fully activating PKA. On the other hand, at such agonist (isoproterenol) concentrations only $\sim 10\%$ of the receptors are occupied and therefore able to serve as a substrate for β-ARK. At higher stimulation levels (2 μM isoproterenol) more than 90% of the β_2-AR should be bound by agonist and readily available for phosphorylation by β-ARK.

It was shown in the same study that the rapid desensitization induced by a low level of stimulation was blocked by mutating the PKA phosphorylation sites, but was not affected when the β-ARK sites were missing. At higher agonist concentration, such desensitization was blocked by mutating either the PKA or β-ARK phosphorylation sites.

These observations are consistent with the finding of Clark et al. (1988) in S49 lymphoma cells. In their studies, desensitization induced in wild-type cells with low doses of epinephrine (5–50 nM) was absent in the kin$^-$ variant of S49. At higher agonist concentrations, desensitization could be observed in both cell types.

Using a different approach, Lohse et al. (1990) confirmed the existence of a distinctive role for PKA and β-ARK in the desensitization induced by different levels of stimulation. In permeabilized A431 cells, heparin (a β-ARK inhibitor) could prevent the desensitization induced by high levels of stimulation, whereas only the heat-stable PKA inhibitor could block the desensitization induced by low agonist levels. It therefore appears that phosphorylation of the β_2-AR by PKA and β-ARK on distinct domains may contribute to the desensitization brought about by different levels of stimulation.

5.2. Sequestration

Evidence that agonists could induce a physical sequestration of the β_2-AR came from the observation that preincubation of cells with an agonist converts a high proportion of the receptors to a form that displays a lower apparent affinity for the agonist as assessed by radioligand binding studies on intact cells (Putman and Molinoff 1980; Toews et al., 1983; Insel et al., 1983). This shift to lower affinity was interpreted as a slow equilibrium of the hydrophilic agonist with a population of receptors sequestered in some lipophilic environment. More recently, development of the hydrophilic antagonist [^{3}H]CGP 12177 has confirmed that exposure to β-adrenergic agonists leads to a rapid decrease in a number of β-AR at the cell surface without changing the total number of receptors recognized by a hydrophobic ligand. These data supported the notion that the receptor was sequestered into an intracellular compartment. Chuang and Costa (1979) provided the first evidence that such sequestration may correspond to the internalization of β-AR. In these studies, exposure of frog erythrocytes to agonist caused a decrease in the number of β-AR found in the plasma membrane fraction, which was paralleled by an increase in the receptor number found in a cytosolic fraction. This agonist-induced cellular redistribution of the β-AR was found to be blocked by concanavalin A (Chuang et al., 1980). In mammalian astrocytoma and C6 glioma cells, Harden et al. (1980) demonstrated that isoproterenol incubation promoted the translocation of a significant portion of the β-AR into a membrane fraction sedimenting at a lighter sucrose density than the plasma membrane. These "light" membrane vesicles were devoid of adenylyl cyclase activity. Similar observations were made by Stadel et al. (1983b) in frog erythrocytes.

While a significant body of evidence suggests that cellular redistribution of the β-AR occurs following agonist stimulation, and with a time course similar to the development of desensitization, other studies show that loss of hormonal responsiveness is not uniquely the consequence of receptor sequestration. First, desensitization was found to precede the appearance of receptor in the light vesicle fraction upon agonist stimulation. This is consistent with the observation, previously discussed, that several treatments which block receptor sequestration failed to prevent desensitization. Moreover, Clark et al. observed that EC_{50} for agonist-induced desensitization was ~tenfold lower than the EC_{50} for agonist-induced sequestration. The fact that agonist-induced sequestration does not seem to be required for the appearance of desensitization does not exclude the possibility that such receptor redistribution might contribute to the regulation of hormonal responsiveness under a specific set of circumstances. While the role played by phosphorylation of β_2-AR in desensitization has been clearly established, this posttranslational modification does not appear to be involved in the onset of agonist-induced sequestration. This conclusion derives from the fact that mutated β_2-AR lacking the carboxyl-terminal β-ARK phosphorylation sites (Bouvier et al., 1988, Strader et al., 1987b), the consensus sequences for PKA phosphorylation (Hausdorff et al., 1989), or both (Hausdorff et al., 1989) did not impair sequestration.

Strader and colleagues (Cheung et al., 1988) recently proposed that structural features of β_2-AR involved in its interaction with Gs are also required for receptor sequestration. They reported that deletions which inhibit the ability of the receptor to couple to Gs prevented agonist-mediated sequestration. This same study reported that increasingly large truncations of the carboxyl-terminus of the receptor ultimately lead to a reduction in the rate of receptor sequestration. Unfortunately, the authors assessed β_2-AR sequestration by measuring only the cell surface receptor (using [^{3}H]CGP 12177 as the radioligand). The decrease in cell surface receptor without information on the total number of cellular receptors may, in fact, reflect downregulation of the receptor rather than sequestration. This would be more in line with previous studies since normal sequestration has been reported in cyc$^-$ (Mahan et al., 1985; Clark et al., 1985), unc and H21a (Mahan et al., 1985), and S49 mutant cells. All of these cells show impaired β_2-AR–Gs coupling. These results clearly indicate that Gs is not required for β_2-AR sequestration. However, a blunted downregulation was observed in cyc$^-$ and unc variants, suggesting that receptor coupling to Gs may play a role in the downregulation process.

5.3. Downregulation or Receptor Loss

Downregulation is defined by a loss of total receptor binding sites in the cell with an accompanying loss in effector stimulation. In this case, loss of binding sites cannot be explained by sequestration of the β_2-AR into the intracellular compartment, as binding to hydrophobic as well as hydrophilic ligands is reduced. Therefore, downregulation is clearly distinct from sequestration (as defined earlier). Moreover, whereas uncoupling and sequestration are very rapid events which take place within minutes after agonist stimulation, downregulation occurs following prolonged stimulation (hours). Proteolytic degradation of the receptor has been proposed to contribute to the process of downregulation. Indeed, in some systems, recovery to a normal level of cell β_2-AR following prolonged agonist exposure is slow and requires new protein synthesis (Marshema et al., 1980; Doss et al., 1981; Frederich et al., 1983). In other cases, however, seemingly contradictory results have been obtained, in that recovery of β_2-AR binding sites following downregulation is more rapid and appears independent of protein synthesis (Frederich et al., 1983; Su et al., 1980; Homburger et al., 1984). In the latter case, such downregulation would result not from degradation of β_2-AR, but rather from a translocation or conformational modification of the receptor which precludes its recognition by the classic hydrophobic antagonists. This hypothesis is supported by the observation of Strader et al. (1984) that, following downregulation in frog erythrocytes, β-AR could be recovered by solubilization of the cells with digitonin.

In general, molecular mechanisms underlying downregulation are poorly understood. Studies using the variants of the S49 lymphoma cell series (Mahan et al., 1985; Su et al., 1980; Shear et al., 1976) suggest that receptor Gs coupling, rather than the generation of cAMP and activation of PKA, is a key determinant in downregulation. Mutants lacking PKA activity (kin$^-$) or with perturbed Gs-effector coupling (H21a) displayed normal β_2-AR downregulation, but mutants with perturbed receptor–Gs (cyc$^-$, unc) coupling showed blunted downregulation of β_2-AR. More recently, the role of cAMP in downregulation has been reevaluated. It was found that cAMP analogues or agents which elevate intracellular cAMP levels, such as IBMX and forskolin, could induce downregulation of β_2-AR in Chinese hamster fibroblasts expressing the human β_2-AR (Bouvier et al., 1989), and also in DDT$_1$MF^{-2} hamster smooth muscle cells (Collins et al., 1989; Hadcock et al., 1989a,b). Phosphorylation of β_2-AR by PKA, however, does not seem to be the key event responsible for this cAMP-mediated downregulation. Indeed, in cells expressing a mutated β_2-AR and in which cAMP-induced phosphorylation of β_2-AR was abolished, the downregulation induced by cAMP was only moderately slowed (Bouvier et al., 1989). As we will discuss in the next section, regulation at a different level appears responsible for this downregulation.

The simplest way to visualize the phenomena of sequestration and downregulation is to imagine that these processes are sequentially ordered in the following manner. First, the receptors are sequestered into vesicles which are still accessible to hydrophobic but not to hydrophilic ligands. Second, the receptors are translocated into a different cellular compartment where they are unavailable for binding, even with hydrophobic ligands (but from which they can still be recycled to the plasma membrane in a fully functional conformation). Finally, the receptors are degraded in lysosomes. Two pieces of evidence, however, argue against this simple sequential model. First, it was shown that atypical partial agonists can induce β-AR downregulation but not sequestration (Neve et al., 1985). Second, cAMP-induced downregulation is not accompanied by an increased sequestration of β_2-AR (Bouvier et al., 1989). These observations therefore suggest that sequestration and downregulation might represent distinct pathways in the regulation of receptor number.

6. Role of Receptor Gene Expression in Desensitization

From the results discussed above, it appears that cAMP-mediated receptor phosphorylation contributes to the process of downregulation promoted by either agonists or cAMP, but is not the sole factor involved in its appearance. In addition to the regulation of β-adrenergic receptor responsiveness resulting from phosphorylation by PKA, additional cAMP-mediated mechanisms, presumably also mediated by PKA, may be operating. Since cAMP has been shown to affect mRNA levels for a number of genes (Roesler et al., 1988), attempts have been made to further dissect the role of cAMP in regulating β_2-AR number and responsiveness in these cell systems by examining steady-state levels of β_2-AR mRNA during the course of downregulation. An initial correlation between downregulation of β_2-AR and a change in β_2-AR mRNA levels was made in the DDT$_1$MF-2 smooth muscle cell line (Hadcock and Malbon, 1988; Collins et al., 1989). Hadcock and Malbon (1988) showed that 1 μM isoproterenol promoted a rapid downregulation of β_2-AR from the cell (reaching maximum by 1 hr), which was followed by a more gradual decline in β_2-AR mRNA levels to less than 50% of their initial values. Collins et al. (1989), using a lower concentration (100 nM) of epinephrine or isoproterenol, observed an equivalent reduction of β_2-AR mRNA. This decrease was in parallel with the loss in receptors. Both groups examined the role of cAMP in mediating the decrease in β_2-AR mRNA. Hadcock and Malbon (1988) showed that forskolin and cholera toxin, agents which are capable of elevating intracellular levels of cAMP, also promoted downregulation of mRNA for the receptor, but to a lesser extent than isoproterenol. Using both forskolin and the membrane-permeable analogue of cAMP, dibutyryl-cAMP, Collins et al. (1989) found decreases in β_2-AR mRNA that were essentially equivalent to those obtained with epinephrine.

A recent study (Bouvier et al., 1989) attempted to dissect the role of cAMP in promoting β_2-AR downregulation, and to identify discrete steps involved in this process. As mentioned above, these studies were conducted by treating hamster fibroblast cells expressing a transfected human β_2-AR gene with the cAMP analogue, dibutyryl-cAMP. As described earlier, phosphorylation of β_2-AR appeared to be one step involved in the cAMP-dependent mechanism of downregulation. However, when considering the time period over which a significant reduction in receptor number becomes apparent during downregulation, perhaps the more significant finding was, once again, a dramatic decrease in mRNA levels which preceded the downregulation of the receptor (Figure 2.5). Identical results were obtained with forskolin. Since the half-life of β_2-AR in the plasma membrane has been variably estimated to be between 12 to 28 hr (Neve and Molinoff, 1986; Mahan and Insel, 1986), these findings suggest that cAMP-promoted downregulation of receptor mRNA levels may be responsible for the reduced density of receptors at later times. Utilizing the S49 mouse lymphoma series of mutants (Hadcock et al., 1989a), a good correlation was observed between the ability to activate the PKA pathway and decreases in β_2-AR mRNA levels. Wild-type S49 cells displayed a 50% reduction in β_2-AR mRNA in response to agonist or forskolin. In contrast, mutants such as cyc$^-$ and unc, which are unable to activate adenylyl cyclase due to lack of coupling between receptor and Gs, or kin$^-$, which lacks PKA activity, showed no decrement in β_2-AR mRNA. Accordingly, binding of the β-AR antagonist [^{125}I]cyanopindolol was unchanged in kin$^-$ cells after 48 hr of forskolin, while wild-type cells showed a 40% reduction in receptor levels. Unexpectedly, the novel mutant H21a, whose inability to elevate intracellular cAMP levels derives from defective coupling between Gs and cyclase, did show substantial decreases in levels of β_2-AR mRNA, albeit to a more modest ($\sim$25%) extent than wild-type cells. This finding lead the authors to propose that receptor-Gs activation of a PKA-independent pathway, such as calcium and potassium channels (Yatani et al., 1987, 1988; Mattera et al., 1989), may also serve to regulate β_2-AR mRNA levels. Unfortunately, no other binding data were presented in this study for either H21a or other mutants in order to correlate receptor density with mRNA levels during downregulation.

With the mechanisms responsible for the cAMP-induced downregulation of β_2-AR message levels unknown, indirect evidence allowed us to make some predictions. An effect on gene transcription is unlikely in the transfected cells used in these studies since they express the β_2-AR under the control of the SV40 promoter and not the original β_2-AR promoter. While the SV40 early promoter has been shown in some cells to be modestly stimulated by cAMP (Imagawa et al., 1987), we observed a rapid decrease in β_2-AR mRNA levels. Thus, an effect of cAMP on posttranscriptional events, such as β_2-AR mRNA stabilization, must be considered. This idea is supported by the finding that, for a number of cellular genes, cAMP affects mRNA stability (Jungmann et al., 1983; Hod and Hanson, 1988; Smith and Liu, 1988). Moreover, for another G-protein-coupled receptor, the thyrotropin-releasing hormone receptor, receptor downregulation was accompanied by a rapid decrease in the level of its mRNA (Oron et al., 1987), and was suggested to result from posttranscriptional control. Recently, evidence

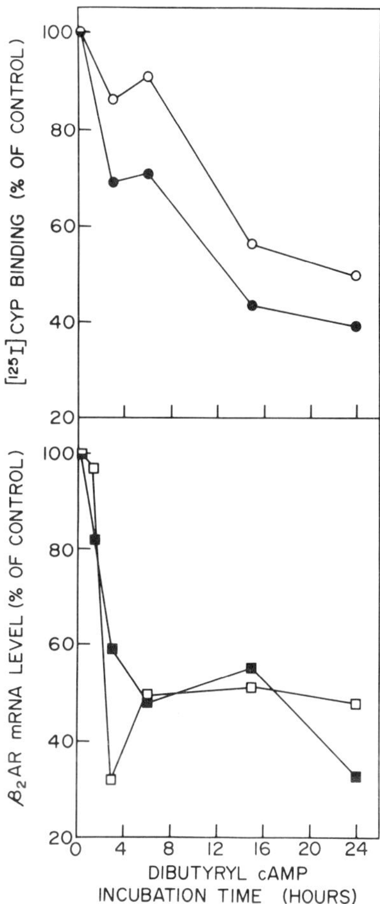

FIGURE 2.5. Effect of dibutyryl-cAMP treatment on β_2-AR number and β_2-AR mRNA levels in cells expressing wild-type (●, ■) β_2-AR or a mutant (○, □) β_2-AR lacking sites for phosphorylation by PKA. Cells were incubated with dibutyryl-cAMP (1 mM) for various periods of time at 37°C. β_2-AR density was determined by measuring specific [^{125}I]cyanopindolol binding in plasma membranes. The mRNA levels were measured in the same cells in parallel by dot-blot hybridization to ^{32}P-labeled probes specific for β_2-AR. Levels of cytoplasmic β-actin mRNA (which did not change throughout treatment) were also monitored. The data are expressed as percentage of values assessed in untreated cells. Reprinted with permission of the American Society for Biochemistry and Molecular Biology, from Bouvier et al. (1989).

has been obtained for an agonist-dependent destabilization of the β_2-AR mRNA (Hadcock et al., 1989b).

In an effort to elaborate upon the recent findings implicating cAMP as playing multiple regulatory roles in the desensitization process, more recent studies examining β_2-AR gene expression during the early phase of agonist-promoted desensitization have led to some unexpected observations (Collins et al., 1989). In DDT_1MF-2 cells (which express the β_2-AR under the control of its own promoter), short-term (minutes) exposure to elevated levels of intracellular cAMP, produced either by β-adrenergic agonists or the direct addition of cAMP analogues, promoted a rapid increase in the steady-state level of β_2-AR mRNA. This three- to fourfold increase in β_2-AR mRNA was evident 30–90 min after the addition of the hormone. No change in actin mRNA was apparent. This enhancement rapidly returned to control levels, whereupon downregulation of both the receptor and its mRNA was observed over the 24-hr period, as described earlier.

To explore the mechanisms involved in this complex cAMP-dependent regulation of β_2-AR mRNA, recall that changes in the steady-state level of expression of a gene result from alterations either in the rate of transcription, the rate of degradation of the message, or a combination of the two processes. Cyclic AMP has been shown to regulate a number of cellular genes by enhancing the rate of their transcription. For several cAMP-regulated genes, there is evidence for changes in both of these pathways (Jungmann et al., 1983; Hod and Hanson, 1988; Smith and Liu, 1988). Nuclear run-off transcription studies in DDT_1MF-2 cells following 30-min exposure to agonist show a specific increase in the rate of β_2-AR gene transcription that is in good agreement with the increased steady-state mRNA levels measured in these same experiments. Therefore, the rapid increase in β_2-AR mRNA in response to cAMP is due to enhanced transcription of the β_2-AR gene. This conclusion was further supported by detailed experiments which examined β_2-AR mRNA stability in cells pretreated for 30 min with agonist or media alone, indicating that, at least at these early times, the rate of turnover of β_2-AR mRNA was not altered (Collins et al., 1989).

For many genes whose transcription is modulated by cAMP, the 5'-flanking promoter regions contain specific sequences which have been identified as critical for this regulation (Roesler et al., 1988). These sequences, termed cAMP response elements (CREs), contain a variation of the palindromic sequence motif TGACGTCA (Table 2.1). This sequence is recognized by phosphoprotein called CREB (CRE binding protein). This protein has been purified (Yamamoto et al., 1988) and cloned (Hoeffler et al., 1988; Gonzalez et al., 1989); its ability to stimulate transcription of target genes is modified by phosphorylation (Gonzalez and Montminy, 1989). It is interesting to note that many of the genes regulated by cAMP encode protein products which are themselves regulated or whose secretion is evoked by hormones that elevate cAMP levels (Roesler et al., 1988). As shown in Figure 2.5, sequences in the β_2-AR promoter region which correspond to CREs are present between -50 to -60 bp from the start site of transcription. Promoter–reporter gene fusions were constructed between the 5'-flanking region of the human β_2-AR and the coding sequences of the bacterial chloramphenicol acetyltransferase (CAT) gene. When introduced

TABLE 2.1. 5′-Flanking region elements responsible for cAMP regulation consensus sequence: TGACGTCA

PEPCK	−90	TTACGTCA	−83
CG α subunit	−143	TGACGTCA	−136
	−123	TGACGTCA	−116
Somatostatin	−48	TGACGTCA	−41
VIP	−76	TGACGTCT	−69
		or	
	−86	CGTCAtactgTGACGtc	−70
TH	−45	TGACGTCA	−38
c-fos			
Human	−63	GTGACGTTC	−56
Mouse	−68	GTGACGTA	−61
proENK	−93	CTGCGTCA	−86
		or	
	−104	TGGCGTAgggccTGCGTCA	−86
β_2-AR	−57	GTACGTCA	−50

Key to Abbreviations: PEPCK = phosphoenolpyruvate carboxykinase; CG = chorionic gonadotropin; VIP = vasoactive intestinal peptide; TH = tyrosine hydroxylase; proENK = proenkephalin.

into a suitable host cell, activation of the β_2-AR promoter is reflected in increased CAT activity. When test plasmids containing varying lengths of β_2-AR promoter were introduced by DNA-mediated gene transfer (Gorman, 1985) into rat C6 glioma cells, CAT activity from both constructs is enhanced three- to fourfold over basal expression by forskolin plus IBMX or IBMX alone, but not by TPA (tumor-promoting agent) (Figure 2.6). These results established that the promoter region of the β_2-AR contains elements responsible for the transcriptional enhancing activity of cAMP as observed by nuclear run-off transcription assays. To define this element more precisely, we have now shown (Collins et al., 1990) by site-directed mutagenesis, gel-shift analysis, and DNA footprinting, that the sequence GTACGTCA in the β_2-AR promoter is a CRE which is recognized and stimulated by the CREB transcription factor. This sequence thus serves as an enhancer of β_2-AR gene transcription. Further, the generality of this mode of regulation is suggested by the presence of comparable CRE sequences in the 5′-flanking regions of the three mammalian β_2-AR genes cloned to date (Kobilka et al., 1987b; Allen et al., 1988; Nakada et al., 1989). The role of this sequence in basal and tissue-specific expression of the β_2-AR will be important to establish, as the CREs from several genes serve a crucial role in their expression (Roesler et al., 1988).

Thus, two distinct cAMP-dependent mechanisms, in addition to an apparent cAMP-independent mechanism, contribute to the downregulation of β_2-AR. First, cAMP-dependent phosphorylation of the β_2-AR by PKA may enhance the rate of downregulation, especially in the first few hours, presumably by shortening the half-life of the receptors in the cell membrane. A second regulatory action of cAMP, the mechanism of which is not

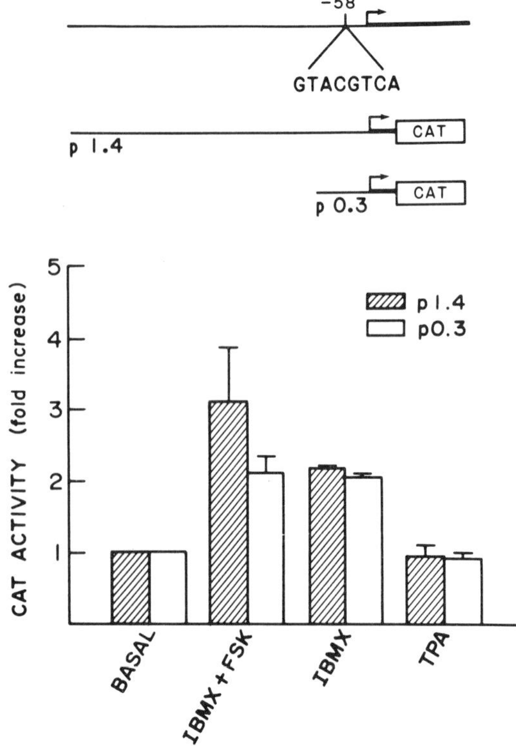

FIGURE 2.6. Cyclic AMP stimulation of β_2-AR gene expression in C6 glioma cells transfected with β_2-AR promoter-CAT gene fusions. Schematic diagram of the β_2-AR 5′-flanking region indicating the location and sequence of the CRE at -58 to -50 bp relative to the start site of transcription (arrow). Rat C6 glioma cells were transfected with CAT plasmid DNAs containing 1335 bp (p1.4) or 235 bp (p0.3) of human β_2-AR 5′-flanking DNA (10 μg DNA per 100-mm dish) by the calcium phosphate precipitation method. CAT enzyme activity (which is a measure of β_2-AR promoter function) was determined in cell extracts 18 to 20 hr after the addition of forskolin (25 mM), IBMX (0.25 mM), or TPA (100 nM). The data shown are the average of four separate experiments each performed in duplicate. Slight differences in CAT activity observed between p1.4 and p0.3 in response to IBMX plus forskolin were not statistically significant. Reprinted with permission of the National Academy of Science, from Collins et al. (1989).

currently understood, leads to a reduction in the stability and steady-state level of β_2-AR mRNA, consequently lowering receptor density.

In summary, the molecular cloning of G-protein-coupled receptors has not only highlighted the many common structural features among this large (and growing) family of transmembrane proteins, but has provided the opportunity to distinguish the properties of the receptors which specify their preferential coupling to the various G-proteins, and the pathways which regulate individual receptor subtypes. While receptor phosphorylation has been most thoroughly explored for the β_2-AR, it appears to be a general strategy for the regulation of transmembrane signaling. The ability to provide complex and finely tuned regulation of the β_2-AR through the action of multiple kinases, such as PKA, protein kinase C and receptor-specific kinases, such as β-ARK, has only recently become apparent. It will be interesting to compare these results with studies of other G-protein-coupled receptors. While much of the recent progress in understanding the regulation of these receptors has involved manipulating their structure for the purpose of directly testing mechanisms of desensitization and G-protein coupling, recent studies of transcriptional regulation and expression of receptor subtypes is revealing the complex and integral nature of the mechanisms controlling signal transduction by this family of receptors.

References

Allen JM, Baetge EE, Abrass IB, Palmiter RD (1988): Isoproterenol response following transfection of the mouse β_2-adrenergic receptor gene into Y1 cells. *EMBO J* 7:133–138

Applebury ML, Hargrave PA (1986): Molecular biology of the visual pigments. *Vision Res* 26:1881–1895

Bach R, Konigsberg WH, Nemerson Y (1988): Human tissue factor contains thioester-linked palmitate and stearate on the cytoplasmic half-cystine. *Biochemistry* 27:4227–4231

Benovic JL, Pike LJ, Cerione RA, Staniszewski C, Yoshimasa T, Codina J, Birnbaumer L, Caron MG, Lefkowitz RJ (1985): Phosphorylation of the mammalian β-adrenergic receptor by cyclic AMP-dependent protein kinase: Regulation of the rate of receptor phosphorylation and dephosphorylation by agonist occupancy and effects on coupling of the receptor to the stimulatory guanine nucleotide regulatory protein. *J Biol Chem* 260:7094–7101

Benovic JL, Mayor F Jr, Somers RL, Caron MG, Lefkowitz RJ (1986a): Light-dependent phosphorylation of rhodopsin by β-adrenergic receptor kinase. *Nature* 321:869–872

Benovic JL, Strasser RH, Caron MG, Lefkowitz RJ (1986b): β-Adrenergic receptor kinase: Identification of a novel protein kinase which phosphorylates the agonist-occupied form of the receptor. *Proc Natl Acad Sci USA* 83:2797–2801

Benovic JL, Kuhn H, Weyland I, Codina J, Caron MG, Lefkowitz RJ (1987a): Functional desensitization of the isolated β-adrenergic receptor by the β-adrenergic receptor kinase: Potential role of an analog of the retinal protein arrestin (48 kDa protein). *Proc Natl Acad Sci USA* 84:8879–8882

Benovic JL, Regan JW, Matsui H, Mayor FM Jr, Cotecchia S, Leeb-Lundberg LMF, Caron MG, Lefkowitz RJ (1987b): Agonist-dependent phosphorylation of the α_2-adrenergic receptor by the β-adrenergic receptor kinase. *J Biol Chem* 262:17251–17253

Benovic JL, Bouvier M, Caron MG, Lefkowitz RJ (1988): Regulation of adenylyl cyclase-coupled β-adrenergic receptors. *Annu Rev Cell Biol* 4:405–428

Benovic JL, Stone WC, Caron MG, Lefkowitz RJ (1989a): Inhibition of the β-adrenergic receptor kinase by polyanions. *J Biol Chem* 264:6707–6710

Benovic JL, DeBlasi A, Stone WC, Caron MG, Lefkowitz RJ (1989b): β-Adrenergic receptor kinase: Primary structure delineates a multigene family. *Science* 246:235–240

Bouvier M, Leeb-Lundberg LM, Benovic JL, Caron MG, Lefkowitz RJ (1987): Regulation of adrenergic receptor function by phosphorylation. II. Effects of agonist occupancy on phosphorylation of α_1 and β_2-adrenergic receptors by protein kinase C and the cyclic AMP-dependent protein kinase. *J Biol Chem* 262:3106–3113

Bouvier M, Hausdorff WP, DeBlasi A, O'Dowd BF, Kobilka BK, Caron MB, Lefkowitz RJ (1988): Mutations of the β_2-adrenergic receptor which remove phosphorylation sites delay the onset of agonist promoted desensitization. *Nature* 333:370–373

Bouvier M, Collins S, O'Dowd BF, Campbell PT, DeBlasi A, Kobilka BK, MacGregor C, Irons GP, Caron MG, Lefkowitz RJ (1989): Two distinct pathways for cAMP mediated down regulation of the β_2-adrenergic receptor: Phosphorylation of the receptor and regulation of its mRNA level. *J Biol Chem* 264:16786–16792

Cheung AH, Sigal IS, Dixon RAF, Strader CD (1988): Agonist-promoted sequestration of the β_2-adrenergic receptor requires regions involved in functional coupling with G_s. *Mol Pharmacol* 34:132–128

Chin DJ, Gil G, Russell DW, Liscum L, Luskey KL, Basu SK, Okayama H, Berg P, Goldstein JL, Brown MS (1984): Nucleotide sequence of 3-hydroxy-3-methyl-glutaryl coenzyme A reductase, a glycoprotein of endoplasmic reticulum. *Nature* 30:613–617

Chuang DM, Costa E (1979): Evidence for internalization of the recognition site of β-adrenergic receptors during receptor subsensitivity induced by (−)isoproterenol. *Proc Natl Acad Sci USA* 76:3024–3028

Chuang DM, Kinnier WJ, Farber L, Costa E (1980): A biochemical study of receptor internalization during β-adrenergic receptor desensitization in frog erythrocytes. *Mol Pharmacol* 18:348–355

Clark RB, Friedman S, Prashad N, Kuoho AE (1985): Epinephrine induced sequestration of the βAR in cultured S49-WT cyc⁻ lymphoma. *J Cyclic Nucleotide Protein Phosphor Res* 10:97–119

Clark RB, Kunkel MW, Friedman J, Coka TS, Johnson JA (1988): Activation of cAMP-dependent protein kinase is required for heterologous desensitization of adenylyl cyclase in S49 wild-type lymphoma cells. *Proc Natl Acad Sci USA* 85:1442–1446

Collins S, Bouvier M, Bolanowski MA, Caron MG, Lefkowitz RJ (1989): Cyclic AMP stimulates transcription of the β_2-adrenergic receptor gene in response to short term agonist exposure. *Proc Natl Acad Sci USA* 86:4853–4857

Collins S, Altschmied J, Herbsmen O, Caron MG, Mellon PL, Lefkowitz RJ (1990): Cyclic AMP response element in the β_2-adrenergic receptor gene confers transcriptional autoregulation by cAMP. *J Biol Chem* 265:19330–19335

DeLean A, Stadel JM, Lefkowitz RJ (1980): A ternary complex model explains the agonist-specific binding properties of the adenylate-cyclase coupled β-adrenergic receptor. *J Biol Chem* 255:7108–7117

Dixon RAF, Sigal IS, Candelore MR, Register RB, Scattergood W, Rands E, Strader CD (1987): Structural features required for ligand binding to the β-adrenergic receptor. *EMBO J* 6:3269–3275

Dohlman HG, Bouvier M, Benovic JL, Caron MG, Lefkowitz RJ (1987): The multiple membrane spanning topography of the β-adrenergic receptor. Localization of the sites of binding, glycosylation and regulatory phosphorylation by limited proteolysis. *J Biol Chem* 262:14282–14288

Dohlman HG, Caron MG, DeBlasi A, Frielle T, Lefkowitz RJ (1990): Role of extracellular disulfide-bonded cysteines in the ligand binding function of the β_2-adrenergic receptor. *Biochemistry* 29:2335–2342

Doss RC, Perkins JP, Harden TK (1981): Recovery of β-adrenergic receptor following long term exposure of astrocytoma cells to catecholamine: Role of protein synthesis. *J Biol Chem* 256:12281–12286

Dunn R, McCoy J, Simsek M, Majumdar A, Chang SH, Rajbhandary UL, Khorana HG (1981): The bacteriorhodopsin gene. *Proc Natl Acad Sci USA* 78:6744–6748

Frederich RC Jr, Waldo GL, Harden TK, Perkins JP (1983): Characterization of agonist-induced β-adrenergic receptor specific desensitization in C62B glioma cells. *J Cyclic Nucleotide Protein Phosphor Res* 9:103–118

Gonzalez GA, Montminy MR (1989): Cyclic AMP stimulates somatostatin gene transcription by phosphorylational CREB at serine 133. *Cell* 59:675–580

Gonzalez GA, Yamamoto KK, Fischer WH, Karr D, Menzel P, Biggs WH III, Vale WW, Montminy MR (1989): A cluster of phosphorylation sites on the cAMP-regulated factor CREB predicted by its sequence. *Nature* 337:749–752

Gorman C (1985): High efficiency gene transfer into mammalian cells. In: *DNA Cloning,* Volume II, Glower DM, ed. IRL Press Ltd.

Hadcock JR, Malbon CC (1988): Down-regulation of β-adrenergic receptors: Agonist-induced reduction in receptor mRNA levels. *Proc Natl Acad Sci USA* 85:5021–5025

Hadcock JR, Ros M, Malbon CC (1989a): Agonist regulation of β-adrenergic receptor mRNA. Analysis in S49 mouse lymphoma mutants. *J Biol Chem* 264:13956–13961

Hadcock JR, Wang H, Malbon CC (1989b): Agonist-induced destabilization of β-adrenergic receptor mRNA. Attenuation of glucocorticoid induced u-regulation of β-adrenergic receptors. *J Biol Chem* 264:19928–19933

Harden TK, Cotton CV, Waldo GL, Lutton JK, Perkins JP (1980): Catecholamine-induced alteration in sedimentation behavior of membrane bound β-adrenergic receptors. *Science* 210:441–443

Hausdorff WP, Bouvier M, O'Dowd BF, Irons GP, Caron MG, Lefkowitz RJ (1989): Phosphorylation sites on two domains of the β_2-adrenergic receptor are involved in distinct pathways of receptor desensitization. *J Biol Chem* 264:12657–12665

Henderson R, Unwin PN (1975): Three dimensional model of purple membranes obtained by electron microscopy. *Nature* 257:28–32

Hertel C, Muller P, Portenier M, Staehelin M (1983): Determination of the desensitization of β-adrenergic receptors by [^{3}H]CGP-12177. *Biochem J* 216:669–674

Hertel C, Coulter SJ, Perkins JP (1985): Comparison of catecholamine-induced internalization of adrenergic receptors and receptor-mediated endocytosis of epidermal growth factor in human astrocytoma cells: Inhibition by phenylarsine oxide. *J Biol Chem* 260:12547–12553

Hertel C, Coulter SJ, Perkins JP (1986): The involvement of cellular ATP in receptor-mediated internalization of epidermal growth factor and hormone-induced internalization of β-adrenergic receptors. *J Biol Chem* 261:5974–5980

Hod Y, Hanson RW (1988): Cyclic AMP stabilizes the mRNA for phosphoenol pyruvate carboxykinase (GTP) agonist degradation. *J Biol Chem* 263:7747–7752

Hoeffler JP, Meyer TE, Yun Y, Jameson JL, Habener JF (1988): Cyclic AMP-responsive DNA-binding protein-structure based on a cloned placental cDNA. *Science* 242:1430–1433

Homburger V, Lucase M, Cantau B, Perit J, Bockaert J (1980): Further evidence that desensitization of β-adrenergic sensitive adenylyl cyclase proceeds in two steps: Modification of the coupling and loss of β-adrenergic receptor. *J Biol Chem* 255:10436–10444

Homburger V, Pantaloni C, Lucas M, Coglan H, Bockaert J (1984): β-Adrenergic receptor repopulation of C6 glioma cells after irreversible blockade and down regulation. *J Cell Physiol* 121:589–597

Imagawa M, Chiu R, Karin M (1987): Transcription factor AP-2 mediates induction by two different signal transduction pathways: protein kinase C and cAMP. *Cell* 51:251–260

Insel PA, Mahan LL, Motulsky HJ, Stoolman LM, Koachman AM (1983): Time dependent decrease in binding affinity of agonist for β-adrenergic receptors of intact S49 lymphoma cells: A mechanism of desensitization. *J Biol Chem* 258:13597–13605

Jackson TR, Blair LAC, Marshall J, Goedert M, Hanley MR (1988): The *mas* oncogene encodes an angiotensin receptor. *Nature* 335:437–440

Jarvie KR, Niznik HB, Seeman P (1988): Dopamine D_2 receptor binding subunits of $M_r \approx 140,000$ and 94,000 in brain: Deglycosylation yields a common unit of $M_r \approx 44,000$. *Mol Pharmacol* 34:91–97

Jarvie KR, Niznik HB (1989): Deglycosylation and proteolysis of photolabeled D_2 dopamine receptors of the porcine anterior pituitary. *J Biochem* 106:17–22

Jungmann RA, Kelley DC, Miles MF, Milkowski DM (1983): Cyclic AMP regulation of lactate dehydrogenease. *J Biol Chem* 258:5312–5318

Kassis S, Fishman PH (1984): Functional alteration of the β-adrenergic receptor during desensitization of mammalian adenylyl cyclase by β-agonist. *Proc Natl Acad Sci USA* 81:6686–6690

Kassis S, Sullivan M (1986): Desensitization of the mammalian β-adrenergic receptor: Analysis of receptor redistribution on nonlinear sucrose gradients. *J Cyclic Nucleotide Protein Phosphor Res* 11:35–46

Kobilka BK, MacGregor C, Daniel K, Kobilka TS, Caron MG, Lefkowitz RJ (1987a): Functional activity and regulations of human β_2-adrenergic receptors expressed in Xenopus oocytes. *J Biol Chem* 262:15796–15802

Kobilka BK, Frielle T, Dohlman HG, Bolanowski MA, Dixon RAF, Keller P, Caron MG, Lefkowitz RJ (1987b): Delineation of the intronless nature of the genes for the human and hamster β_2-adrenergic receptor and their putative promoter regions. *J Biol Chem* 262:7321–7327

Kobilka BK, Kobilka TS, Daniel K, Regan JW, Caron MG, Lefkowitz RJ (1988): Chimeric α_2-, β_2-adrenergic receptors: Delineation of domains involved in effector coupling and ligand binding specificity. *Science* 240:1310–1316

Kubo T, Bujo H, Akiba I, Nakai J, Mishina M, Numa S (1988): Location of a region of the muscarinic acetylcholine receptor involved in selective effector coupling. *FEBS Lett* 241:119–125

Kuhn H, Dreyer WJ (1972): Light-dependent phosphorylation of rhodopsin by ATP. *FEBS Lett* 20:1–6

Kwatra MM, Benovic JL, Caron MG, Lefkowitz RJ, Hosey MM (1989): Phosphorylation of chick heart muscarinic cholinergic receptors by the β-adrenergic receptor kinase. *Biochemistry* 28:4543–4547

Lanier SM, Homcy CJ, Patenaude C, Graham RM (1988): Identification of structurally distinct α_2-adrenergic receptors. *J Biol Chem* 263:14491–14496

Lefkowitz RJ (1979): Direct radioligand binding studies of adrenergic receptors: Biochemical, physiological and clinical implications. *Ann Intern Med* 91:450–558

Libert F, Parmentier M, Lefort A, Dinsart C, Van Sande JV, Maenhaut C, Simons M-J, Dumont JE, Vassart G (1989): Selective amplification and cloning of four new members of the G protein-coupled receptor family. *Science* 244:569–572

Lohse MJ, Lefkowitz RJ, Caron MG, Benovic JL (1989): Inhibition of β-adrenergic receptor kinase prevents homologous desensitization of β_2-adrenergic receptors. *Proc Natl Acad Sci USA* 86:3011–3015

Lohse MJ, Benovic JL, Codina J, Caron MG, Lefkowitz RJ (1990): β-Arrestin: A protein that regulates β-adrenergic receptor function. *Science* 248:1547–1550

Mahan LC, Koachman AM, Insel PA (1985): Genetic analysis of β-adrenergic receptor internalization and downregulation. *Proc Natl Acad Sci USA* 82:129–133

Mahan LC, Insel PA (1986): Expression of β-adrenergic receptors in synchronous and asynchronous S49 lymphoma cells. I. Receptor metabolism after irreversible blockade of receptors and in cells traversing the cell cyclase. *Mol Pharmacol* 29:7–15

Marshema I, Thompson WS, Robinson GA, Strada JJ (1980): Loss and restoration of sensitivity to epinephrine in cultured BHK cells: Effect of inhibition of RNA and protein synthesis. *Mol Pharmacol* 18:370–378

Mattera R, Graziano MP, Yattani A, Zhov Z, Graf R, Codina J, Birnbaumer L, Gilman A, Brown AM (1989): Splice variants of the α subunit of the G protein G_s activate both adenylyl cyclase and calcium channels. *Science* 243:804–807

Nakada MT, Haskell KM, Ecker DJ, Stadel JM, Crooke ST (1989): Genetic regulation of β_2-adrenergic receptors in 3T3-L1 fibroblasts. *Biochem J* 260:53–59

Nakayama N, Miyajima A, Arai K (1985): Nucleotide sequences of STE2 and STE3, cell type-specific sterile genes from *Saccharomyces cerevisiae*. *EMBO J* 4:2643–2648

Nambi P, Peters JR, Sibley DR, Lefkowitz RJ (1985): Desensitization of the turkey erythrocyte β-adrenergic receptor in a cell-free system: Evidence that multiple protein kinases can phosphorylate and desensitize the receptor. *J Biol Chem* 260:2165–2171

Neve KA, Barret DA, Molinoff PB (1985): Selective regulation of β-1 and β-2 adrenergic receptors by atypical agonists. *J Pharmacol Exptl Ther* 235:657–664

Neve KA, Molinoff PB (1986): Turnover of β_1- and β_2-adrenergic receptors after down regulation or irreversible blockade. *Mol Pharmacol* 30:104–111

O'Brien PJ, Zatz M (1984): Acylation of bovine rhodopsin by [^{3}H]palmitic acid. *J Biol Chem* 259:5054–5057

O'Dowd BF, Hnatowich M, Regan JW, Leader WM, Caron MG, Lefkowitz RJ (1988): Site-directed mutagenesis of the cytoplasmic domains of the human β_2-adrenergic receptor: Localization of regions involved in G protein-receptor coupling. *J Biol Chem* 263:15985–15992

O'Dowd BF, Hnatowich M, Caron MG, Lefkowitz RJ, Bouvier M (1989a): Palmitoylation of the human β_2-adrenergic receptor. *J Biol Chem* 264:7564–7569

O'Dowd BF, Lefkowitz RJ, Caron MG (1989b): Structure of the adrenergic and related receptors. *Annu Rev Neurosci* 12:67–83

Oron Y, Straub RE, Traktman P, Gershengorn MD (1987): Decreased TRH receptor mRNA activity precedes homologous down regulation: Assay in oocytes. *Science* 238:1406–1408

Ovchinnikov YA, Abdulaev NG, Bogachuk AS (1988): Two adjacent cysteine residues in the C-terminal cytoplasmic fragment of bovine rhodopsin are palmitylated. *FEBS Lett* 230:1–5

Putman RN, Molinoff PB (1980): Interactions of agonists and antagonists with β-adrenergic receptors on intact L6 muscle cells. *J Cyclic Nucleotide Protein Phosphor Res* 6:421–435

Roesler WJ, Vandenbark GR, Hanson RW (1988): Cyclic AMP and the induction of eukaryotic gene transcription. *J Biol Chem* 263:9063–9066

Sarkar G, Sommer SS (1989): Access to a messenger RNA sequence or its protein product is not limited by tissue or species specificity. *Science* 244:331–334

Schobert B, Lanyi JK, Oesterheld D (1988): Structure and orientation of halorhodopsin in the membrane: A proteolytic fragmentation study. *EMBO J* 4:905–911

Shear M, Insel PA, Melmon KL, Coffino P (1976): Agonist-specific refractoriness induced by isoproterenol: Studies with mutant cells. *J Biol Chem* 251:7572–7576

Sibley DR, Peters JR, Nambi P, Caron MG, Lefkowitz RJ (1984): Desensitization of turkey erythrocyte adenylate cyclase: β-adrenergic receptor phosphorylation is correlated with attenuation of adenylate cyclase activity. *J Biol Chem* 259:9742–9749

Sibley DR, Strasser RH, Caron MG, Lefkowitz RJ (1985): Homologous desensitization of adenylyl cyclase is associated with phosphorylation of the β-adrenergic receptor. *J Biol Chem* 260:3883–3886

Smith JD, Liu A Y-C (1988): Increased turnover of the messenger RNA encoding tyrosine aminotransferase can account for the desensitization and de-induction of tyrosine aminotransferase by 8-bromo-cyclic AMP treatment and removal. *EMBO J* 7:3711–3716

Stadel JM, Nambi P, Lavin TN, Heald SL, Caron MG, Lefkowitz RJ (1982): Catecholamine-induced desensitization of turkey erythrocyte adenylyl cyclase: Structural alterations in the β-adrenergic receptor revealed by photoaffinity labeling. *J Biol Chem* 257:9242–9245

Stadel JM, Nambi P, Shorr RGL, Sawyer DF, Caron MG, Lefkowitz RJ (1983a): Catecholamine-induced desensitization of turkey erythrocyte adenylyl cyclase is associated with phosphorylation of the β-adrenergic receptor. *Proc Natl Acad Sci USA* 80:3173–3177

Stadel JM, Strulovici B, Nambi P, Lavin TN, Briggs MM, Caron MG, Lefkowitz RJ (1983b): Desensitization of the β-adrenergic receptor of frog erythrocytes: Recovery and characterization of the down regulated receptors in sequestered vesicles. *J Biol Chem* 258:3032–3038

Strader CD, Sibley DR, Lefkowitz RJ (1984): Association of sequestered β_2-adrenergic receptors with the plasma membrane: A novel mechanism for receptor down regulation. *Life Sci* 35:1601–1610

Strader CD, Sigal IS, Register R, Candelore MR, Rands E, Dixon RAF (1987a):

Identification of residues required for ligand binding to the β-adrenergic receptor. *Proc Natl Acad Sci USA* 94:4384–4388

Strader CD, Sigal IS, Blake AD, Cheung AH, Register RB, Rands E, Zemcik BA, Candelore MR, Dixon RAF (1987b): The carboxyl terminus of the hamster β-adrenergic receptor expressed in mouse L cells is not required for receptor sequestration. *Cell* 49:855–863

Strasser RH, Sibley DR, Lefkowitz RJ (1986): A novel catecholamine activated adenosine cyclic 3′,5′-phosphate independent pathway for β-adrenergic receptor phosphorylation in wild type and mutant S49 lymphoma cells: Mechanism of homologous desensitization of adenylyl cyclase. *Biochemistry* 25:1371–1377

Strulovici B, Cerione RA, Kilpatrick BF, Caron MG, Lefkowitz RJ (1984): Direct demonstration of impaired functionality of a purified desensitized β-adrenergic receptor in a reconstituted system. *Science* 225:837–840

Su YF, Harden TK, Perkins JP (1980): Catecholamine specific desensitization of adenylyl cyclase: Evidence for a multiple step process. *J Biol Chem* 255:7410–7419

Terman BI, Insel PA (1988): Use of 1-deoxymannogirimycin to show that complex oligosaccharides regulate cellular distribution of the α_1-adrenergic receptor glycoprotein in BC_3H_1 muscle cells. *Mol Pharmacol* 34:8–14

Thompson P, Findley JBC (1984): Phosphorylation of bovine rhodopsin. Identification of the phosphorylated sites. *Biochem J* 220:773–780

Toews ML, Harden TK, Perkins JP (1983): High-affinity binding of agonists to β-adrenergic receptors on intact cells. *Proc Natl Acad Sci USA* 80:3553–3557

Wilden U, Hall SW, Kühn H (1986): Phosphodiesterase activation of photoexcited rhodopsin is quenched when rhodopsin is phosphorylated and binds the intrinsic 48kDa protein of rod outer segments. *Proc Natl Acad Sci USA* 83:1174–1178

Yamamoto KK, Gonzalez GA, Biggs WH III, Montminy MR (1988): Phosphorylation-induced binding and transcriptional efficacy of nuclear factor CREB. *Nature* 334:494–498

Yatani A, Codina J, Brown AM, Birnbaumer L (1987): Direct activation of mammalian atrial muscarinic K channels by a human erythrocyte pertussis toxin-sensitive G protein, G_k. *Science* 235:207–211

Yatani A, Imoto Y, Codina J, Hamilton SL, Brown AM, Birnbaumer L (1988): The stimulatory G-protein of adenylyl cyclase, G_s, also stimulate dihydropyridine-sensitive Ca^{2+} channels. *J Biol Chem* 263:9887–9895

Young D, Waitches G, Birchmeier C, Fasano O, Wigler M (1986): Isolation and characterization of a new cellular oncogene encoding a protein with multiple potential transmembrane domains. *Cell* 45:711–719

3

Genetic Analysis of the β-Adrenergic Receptor

CATHERINE D. STRADER AND RICHARD A. F. DIXON

Many hormone and neurotransmitter receptors mediate their signals through interaction with guanine nucleotide-binding regulatory proteins (G-proteins) (see Gilman, 1987, for review). As described elsewhere in this volume, the cloning of the genes for these receptor proteins has revealed a large degree of structural homology within this family of receptors. The homology among the receptors is concentrated within the hydrophobic cores of the proteins, formed by the seven putative transmembrane helices, whereas the extracellular and intracellular hydrophilic loop regions are more divergent. Although the hormones and neurotransmitters which couple various receptors to G-protein-mediated signaling pathways represent a diverse group of small molecule and peptide ligands, there are striking parallels in their mechanisms of signal transduction. Thus, the structural similarities among G-protein-coupled receptors probably reflect similar molecular interactions which are responsible for their common mechanism of action. For this reason, the structural and functional information obtained for one of these receptor proteins should be applicable to the entire family of G-protein-coupled receptors.

The β-adrenergic receptor (β-AR), which binds catecholamine agonists and activates the G-protein Gs to stimulate adenylyl cyclase, has served as a useful model system for defining the structure–function relationships of G-protein-coupled receptors (Strader et al., 1989a). Genetic analysis of the β-AR has led to the identification of specific structural elements involved in the processes of ligand binding and G-protein coupling. Analogies with other receptors of this class are now proving useful in understanding the molecular basis for receptor activation. The approach taken in our laboratory to define the structure–function relationships of the β-AR has involved mutagenesis of the β-AR gene and expression of the mutant receptor in mammalian cells, followed by biochemical and pharmacological analysis of the resulting mutant protein. The use of a combination of three different mutagenesis strategies, each with its own constraints and benefits, has revealed molecular interactions involved in ligand binding and Gs activation by the β-AR and by other G-protein-linked receptors. As described below, these different mutagenesis strategies have proved valuable in addressing different types of questions regarding the structure–function relationships of these receptors.

1. Genetic Analysis of Receptor Function

Hydropathicity analysis of G-protein-coupled receptors reveals the presence of discrete, alternating hydrophilic and hydrophobic domains. Based on the structural model which has been developed for rhodopsin, these domains are postulated to form extramembranous and transmembrane regions, respectively (Findlay and Pappin, 1986). The presence of glycosylation sites in the N-terminal hydrophilic segment of the β-AR (Rands et al., 1990) and rhodopsin (Hargrave et al., 1984) suggests that the N-terminus of the protein is exposed extracellularly. Thus, if each of the seven hydrophobic regions represents a transmembrane helix, the receptor would span the membrane seven times, leaving the C-terminus exposed intracellularly. The initial genetic approach to defining the function of these discrete structural domains of the β-AR was to systematically delete the DNA encoding each of these hydrophilic and hydrophobic regions from the β-AR gene and assess the effect of each deletion on the structure and function of the resulting mutant receptor, expressed in mammalian cells (Dixon et al., 1987a,b). This approach would be expected to identify regions of the receptor which are *not* required for structural or functional interactions, based on the rationale that any region of the protein which can be deleted without affecting a particular functional process cannot be essential for that function. Thus, deletion mutagenesis serves as a rapid initial screen to rule out regions of the protein that are not required for the structural integrity or functional activity of the protein. Regions that cannot be dispensed with by deletion mutagenesis are therefore candidates for further investigation. Another genetic approach which has been used for defining the structure–function relationships of receptors is the construction of chimeric receptor proteins, in which various regions from two related receptors are combined and the pharmacological phenotype of the resulting hybrid protein determined (Kubo et al., 1988; Frielle et al., 1988). As with deletion mutagenesis, the type of information obtained with this technique allows the assignment of functional properties to relatively large domains of the protein. This approach has proved most useful in the determination of the structural basis for G-protein coupling, although some information about the ligand binding domain has also been obtained. More specific structural information can be obtained through site-directed mutagenesis, in which one or a few of the amino acid residues thought to be involved in a particular molecular interaction are substituted with other residues, thus specifically altering the chemical nature of amino acid side chains. This approach has been particularly useful in identifying residues which interact with distinct functional moieties on β-adrenergic agonists and antagonists, as will be described below.

In analyzing the results of mutagenesis experiments, it is critical to be able to distinguish functionally important mutations from those that merely affect the tertiary structure of the protein. The failure to analyze the structural integrity of mutant receptors could lead to incorrect interpretations of the effects of a mutation. For the β-AR, the ability of mutant receptors to fold correctly and be inserted into the plasma membrane of the cell has been assessed by SDS-PAGE followed by protein immunoblotting or photoaffinity labeling (Dixon et al., 1987b). This approach is based on the observation that the wild-type β-AR expressed in mammalian cells or found endogenously in tissues is a glycoprotein which appears on SDS-PAGE as a broad band centered at an apparent molecular weight of 67 kDa. Enzymatic or genetic removal of the carbohydrate moieties from the receptor results in a decrease in the apparent molecular weight of the protein and a concomitant sharpening of the band on the gel. Thus, the failure of a mutant receptor to be fully glycosylated or to be correctly inserted into the plasma membrane can be detected by protein immunoblotting experiments as a decrease in the apparent molecular weight of the protein or as an absence of immunoreactive protein, respectively. Such an alteration of the appearance of the immunoreactive receptor on SDS-PAGE is diagnostic of an adverse effect of the mutation on the overall folding of the receptor, and the functional consequences of such a mutation cannot be accurately determined (Dixon et al., 1987b). When large quantities of mutant receptors become available, calorimetric or other biophysical analytical techniques will be substituted for this relatively crude approach, to refine the determination of the tertiary structure of the mutant proteins.

The initial deletion analysis of the β-AR gene revealed that most of the hydrophilic regions of the protein could be removed without affecting the tertiary structure or the ligand binding characteristics of the receptor (Dixon et al., 1987a,b). In contrast, the hydrophobic domains and the hydrophilic regions immediately adjacent to the predicted transmembrane helices of the β-AR were essential for the normal folding and processing of the protein, reflected in a reduction in the levels of glycosylated receptor. The observation that mutant receptors having deletions of the extracellular and intracellular hydrophilic domains had normal agonist and antagonist binding properties demonstrated that these regions are not required for ligand binding to occur and indicates that the ligand binding domain of the receptor must involve residues within the hydrophobic core of the protein

(Dixon et al., 1987a). The disruption of the tertiary structure of the receptor upon deletion of the hydrophobic domains prevented further dissection of the contributions of individual hydrophobic regions to the binding pocket by deletion mutagenesis. Thus, we used site-directed mutagenesis of specific residues within these hydrophobic regions of the receptor to further probe the molecular interactions involved in ligand binding to the β-AR.

2. Ligand Binding Interactions

The discovery that the ligand binding site of the β-AR involves residues within the hydrophobic domain of the receptor suggests the existence of a binding pocket buried within the membrane bilayer, formed by contacts with amino acid side chains in several of the transmembrane helices. This hypothesis is consistent with the known orientation of the ligand retinal in the binding pocket of rhodopsin (Findlay and Pappin, 1986). Retinal binds to rhodopsin through a Schiff base linkage of the aldehyde moiety on retinal to the amine side chain of Lys^{296} in TM 7 of the protein. Biophysical analysis has suggested that retinal is buried in a hydrophobic pocket of the protein, within the membrane bilayer. The results of the deletion mutagenesis analysis of the β-AR suggest that the binding pocket of this protein may be similarly buried within the transmembrane region, as illustrated by the model for the binding of the agonist isoproterenol shown in Figure 3.1. Specific points of contact between the ligand and the receptor may be inferred from the results of structure–activity studies of adrenergic ligands (Main and Tucker, 1985). Such studies have shown the protonated amine group and the hydroxyl substituent on the chiral β-carbon to be critical for both agonist and antagonist binding affinity. Thus, the receptor would be expected to contribute an acidic side chain to form an ion pair with the amine of the ligand and a hydrogen bond donor or acceptor to interact with the β-hydroxyl substituent on the ligand. These binding interactions would be important for both agonist and antagonist binding to the receptor. Agonist ligands are further characterized by a catechol ring, requiring two hydrogen bonding interactions, as well as aromatic interactions, with the receptor (Figure 3.1). The substituent on the amine group is important in defining the β_1 or β_2 subtype of the receptor with respect to the endogenous agonists epinephrine and norepinephrine. For epinephrine, the amino substituent is a methyl group, whereas norepinephrine contains an unsubstituted primary amine. The sensitivity of the receptor to the size of the amino substituent suggests an interaction at this position, as well. Antagonists are characterized by an increase in the distance between the amine and the aromatic ring, usually achieved by the substitution of a phenoxymethylene moiety for the phenyl ring of the agonists. The aromatic ring systems of antagonists are also more hydrophobic than those found on

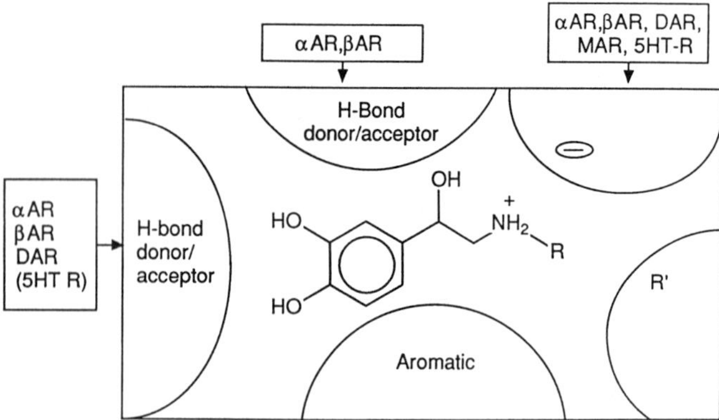

FIGURE 3.1. Map of receptor–ligand interactions within the binding pocket of the β-AR. A catecholamine ligand is shown in a hypothetical binding site within the β-AR, intercalated among the transmembrane helices of the receptor. Each of the large circles represents a transmembrane helix, labeled with the type of binding interaction it is postulated to contribute. Other G-protein-coupled receptors which would be expected to have similar interactions with their ligands, based on the structures of ligands which they bind, are designated in boxes next to the helix. α-AR, α-adrenergic receptor; β-AR, β-adrenergic receptor; DAR, dopamine receptor; MAR, muscarinic acetylcholine receptor; 5HT R, serotonin receptor. Reprinted with permission of Elsevier Science Publishing Co., New York, from Strader et al. (1989d).

agonists, with fused ring systems being common at this position, suggesting the presence of strong antagonist-specific hydrophobic interactions with amino acid residues within the binding site of the receptor.

Further insights into the structure of the ligand binding site of the β-AR arise from an examination of the structures of ligands which bind to other G-protein-coupled receptors (cf. Strader et al., 1989a). Although endogenous hormones and neurotransmitters show specificity for a particular receptor type, many of the ligands share some common structural features, perhaps indicating the presence of similar molecular interactions with the receptors. Therefore, one might expect that the binding sites of receptors which recognize similar ligands would contain similar amino acid residues at positions that represent points of contact with the common functional moiety on the ligand. Hence, as outlined in Figure 3.1, the amino acid residue(s) which interact with the protonated amine group of β-adrenergic ligands might be conserved with other receptors which bind protonated amines, such as the α-adrenergic receptor (α-AR), muscarinic acetylcholine receptors, dopamine receptor, and serotonin receptor (5-HTR), but not with the receptors for peptides or rhodopsin. By the same reasoning, the residue(s) forming hydrogen bonds with the β-hydroxyl group on the ligand should be common to the α- and β-AR, whereas those residues hydrogen bonding to the catechol hydroxyl groups should be present in the α-AR, β-AR, and dopamine receptor. As more G-protein-coupled receptors are cloned and their sequences determined, the bank of available information upon which to base such analogies is increasing, with the ultimate goal of predicting the types of ligands that will bind to a particular receptor directly from the primary sequence of the receptor protein.

Based on the pharmacological "map" shown in Figure 3.1, specific mutations were introduced into the β-AR gene to identify amino acid residues important for ligand binding. To localize the site of interaction of the protonated amine group of the ligand, neutral residues were substituted for conserved acidic amino acids located within the transmembrane core of the receptor (Strader et al., 1987a, 1988). Asp[113], in the third hydrophobic domain of the receptor, was found to be required for both agonist and antagonist binding to occur. Substitution of a Glu residue at this position resulted in a 100- to 300-fold reduction in the affinity of the receptor for agonists, as determined by the ability of the mutant receptor to stimulate adenylyl cyclase activity. Substitution of an Asn residue for Asp[113] resulted in a 10,000-fold decrease in agonist affinity, suggesting that the acidity of the Asp residue was important in mediating ligand interactions. Substitution of Asn or Glu for Asp[113] also decreased the affinity of the receptor for antagonists by >10,000-fold, as assessed by radioligand binding and antagonist-mediated inhibition of agonist-stimulated adenylyl cyclase activity, indicating that the Asp residue at this position is required for both agonist and antagonist binding to the receptor. In contrast, replacement of Asp[113] did not affect the V_{max} of agonist activation of the receptor, indicating that Gs coupling was not affected, nor was the β_2 subtype of the receptor altered by this mutation. Substitution of Ala, His, Lys, Arg, Ser, Phe, or Cys at position 113 resulted in receptors that had no detectable binding activity at the experimentally accessible concentrations of radiolabeled antagonist. Thus, it appears that agonists and antagonists are anchored to the β-AR through an interaction with Asp[113] in the TM 3 of the receptor, probably through the formation of an ion pair between the protonated amine group of the ligand and the carboxylate side chain of Asp[113].

Interestingly, the substitution of a Glu residue for Asp[113] resulted in partial agonist activity of several adrenergic antagonists, such as alprenolol and pindolol (Strader et al., 1989b). These ligands, which were full antagonists at the wild-type β-AR, stimulated adenylyl cyclase activation by [Glu[113]] β-AR to levels of 10 to 35% of the maximal level seen with NaF. Thus, the substitution of a Glu for Asp[113], altering the distance between the negative charge and the protein backbone, allowed the β-AR to assume its active conformation upon binding antagonist ligands, which normally do not activate the wild-type receptor. These results argue for the existence of overlapping binding sites for agonists and antagonists, and suggest that the positioning of the negative charge at position 113 is important in determining the orientation of the ligand in the binding pocket of the receptor. Examination of the sequences of other G-protein-coupled receptors reveals that an Asp residue in a position analogous to Asp[113] of the β-AR is present in all receptors of this class which bind protonated amine ligands, including all known subtypes of the α-adrenergic, β-adrenergic, dopamine, serotonin, and muscarinic acetylcholine receptors, but is absent from other G-protein-linked receptors whose endogenous ligands are not amines, suggesting that the presence of an Asp residue at this position is diagnostic of a receptor that binds amine ligands (Strader et al., 1988).

A similar strategy of combined genetic and biochemical analysis was employed to define the roles of amino acid residues whose side chains are potential hydrogen bond donors or acceptors. Two Ser residues, at positions 204 and 207 in the TM 5 of the β-AR, were found to be critical for the binding and activation of the receptor by catecholamine agonists (Strader et al., 1989c). Replacement of either Ser[204] or Ser[207] by an Ala residue resulted in a reduction in the affinity and efficacy of catecholamine agonists at the receptor, with no alteration in antagonist binding. The agonist specificity of these effects would be consistent with an interaction of these Ser residues with the catechol hydroxyl groups of adrenergic agonists, rather than an interaction with the β-hydroxyl group, which is common to both agonist and antagonist ligands. Furthermore, the effects of substitution of Ser[204] or Ser[207] on receptor–agonist interactions could be selectively mimicked by substitution of the meta or para hydroxyl substituents, respectively, on the catechol ring of the agonist. These data suggest the presence of a specific interaction, possibly a hydrogen bond, between the hydroxyl side chain of Ser[204] and the *m*-catechol hydroxyl group of the agonist, with a second hydrogen bond being formed between Ser[207] and the *p*-catechol hydroxyl substituent. If, as has been postulated, TM 5 of the β-AR forms a transmembrane α-helix, then Ser[204] and Ser[207] would be located approximately one helical turn apart, almost directly above one another in the membrane. Such a configuration would allow both of the catechol hydroxyl groups of the ligand to bind to the receptor simultaneously, consistent with the measured binding energies of the agonists. Further support for the assignment of the catechol hydroxyl binding interactions to the side chains of Ser[204] and Ser[207] arises from the observation that Ser residues are present at the analogous positions in all known catecholamine-binding receptors (α-AR, β-AR, and dopamine receptor), but not in those receptors which do not bind catechol ligands.

A model for the ligand binding site of the β-AR which is consistent with the genetic data presented above is shown in Figure 3.2 (Dixon et al., 1989a; Strader et al., 1989a). The ligand binds within the hydrophobic transmembrane core of the receptor, with the binding energy contributed by multiple interactions between the functional groups on the ligand and specific amino acid side chains on the receptor. Both agonists and antagonists are anchored to the receptor through an ionic interaction between the carboxylate side chain of Asp[113] in TM 3 and the protonated amine group of the ligand. The catechol substituents on the aromatic ring of the agonist appear to form

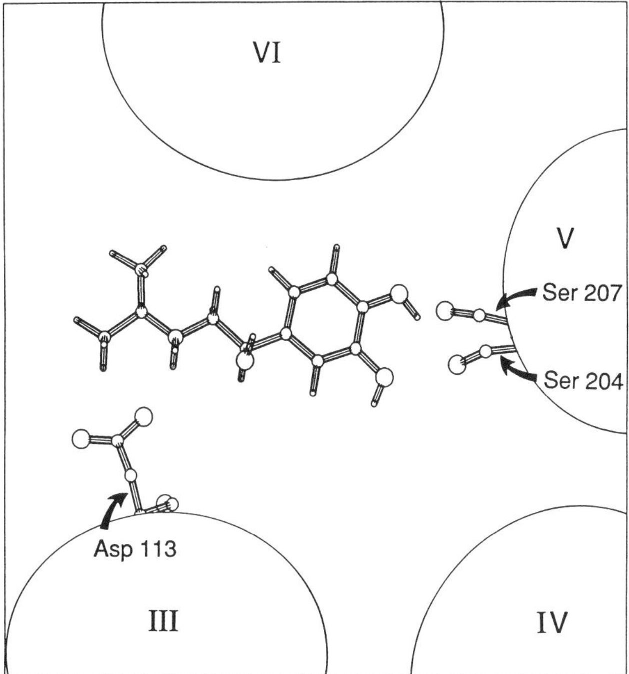

FIGURE 3.2. Model of the ligand binding site of the β-AR. Isoproterenol is shown in the binding pocket, interacting with several of the transmembrane helices, which are labeled with Roman numerals. The disposition of the helices is according to the structure of bacteriorhodopsin. The side chains of Asp^{113} in TM 3 (III in figure) and Ser^{204} and Ser^{207} in TM 5 (V in figure) are shown in their postulated interactions with the functional moieties of the ligand. Reprinted with permission of the American Society for Biochemistry and Molecular Biology, from Strader et al. (1989c).

hydrogen bonds with the hydroxyl side chains of Ser^{204} and Ser^{207} in TM 5. Additional hydrophobic interactions link the aromatic moieties of agonists and antagonists to the receptor. These interactions, which have yet to be identified, are important in distinguishing agonists from antagonists. Preliminary results have implicated Phe^{289} and Phe^{290} in TM 6 and Tyr^{326} in TM 7 in the binding of agonists, perhaps through interactions with the aromatic ring (Dixon et al., 1989a). Amino acid residues which contribute to the binding of the large aromatic rings of adrenergic antagonists have not yet been identified, although they appear to be different from those residues mentioned above which interact with the catechol agonist ring. The hydrogen bond which has been postulated between the receptor and the β-hydroxyl group, which is critical for the binding of both agonists and antagonists, has also not yet been identified. Current efforts in our laboratory are directed at the elucidation of the molecular basis for these critical binding interactions.

The application of the strategy for site-directed mutagenesis described above to defining the binding interactions involved in the determination of receptor subtype specificity has been less successful, primarily because the biochemical basis of the subtype-specific interaction is less well understood. As previously mentioned, the subtype selectivity of the β-AR for its endogenous agonists is based on the relative affinity of the receptor in recognizing the presence (epinephrine) or absence (norepinephrine) of a methyl substituent on the amine nitrogen of the ligand. Systematic replacement of each of the residues of the β_2-AR with the analogous amino acid from the β_1-AR failed to identify any single substitution capable of altering the subtype specificity of the receptor (Dixon et al., 1989b). Further analysis of this interaction was performed using chimeric β_1/β_2-adrenergic receptors. Construction of hybrid receptors in which increasingly long regions of the N-terminal domain of the β_2-AR were attached to the remaining C-terminal region of the β_1-AR revealed that the relative affinities of the receptor for epinephrine and norepinephrine changed from a β_1 to a β_2 phenotype when TM 1–4 were derived from the β_2-AR (Frielle et al., 1988). Subsequent refinement of these results demonstrated that the region important for recognition of the subtype-specific amine substituent is distributed between TM 4 and 5, with minor contributions from other regions of the receptor (Dixon et al., 1989b). Therefore, the combined use of chimeric receptors and molecular replacements has demonstrated that β-receptor subtype selectivity for its endogenous agonists probably arises from a subtype-specific conformation of the receptor, rather than from any single molecular interaction between the ligand and a subtype-specific amino acid side chain on the protein.

Analysis of the structure–activity relationships of synthetic subtype-selective adrenergic antagonists has shown that the structural features contributing to the subtype specificity of these antagonists differ from those which convey subtype specificity to the endogenous catecholamine agonists. Thus, a not unexpected finding of the mutagenesis studies was that the mutant receptors which display an altered subtype phenotype with respect to agonists do not necessarily show an alteration in the subtype specificity with respect to antagonists (Frielle et al., 1988; Dixon et al., 1989b). These results demonstrate that the synthetic antagonists are interacting with other regions of the receptor which differ between the β_1 and β_2 receptors but which do not directly contribute to the recognition of the amine substituent of the ligand.

3. G-Protein Interactions

The chimeric receptor approach has proved most useful in elucidating the sites of G-protein interaction on the receptor. Initial deletion mutagenesis of the β-AR revealed that, in contrast to the apparent intramembrane localization of the binding site of the receptor, interaction with Gs appears to require amino acid residues within the putative i3 hydrophilic loop (Dixon et al., 1987a; Strader et al., 1987b). The analogous region of rhodopsin had previously been implicated, by proteolytic analysis, in its interaction with transducin (Findlay and Pappin, 1986). The cytoplasmic exposure of this region of the receptor would be consistent with the localization of the G-protein on the inner face of the plasma membrane, and the intracellular exposure of this hydrophilic domain of the β-AR has been demonstrated immunocytochemically using monoclonal antipeptide antibodies specific for this region (Aoki et al., 1989). Whereas a large deletion of 34 amino acids from this loop of the β-AR completely abolished the ability of the receptor to couple to the Gs–adenylyl cyclase effector system, the ligand binding and protein folding characteristics of the receptor were not affected by the deletion (Dixon et al., 1987a). Further refinement utilizing smaller deletions within this hydrophilic domain revealed that regions in the central portion of the loop were not required for Gs coupling to occur (Strader et al., 1987b). However, deletion of a stretch of 12 residues at the C-terminal portion of this loop severely attenuated the ability of the receptor to couple to Gs, and no G-protein activation was observed with a mutant receptor from which 8 amino acid residues near the N-terminus of this region were deleted (Strader et al., 1987b; Cheung et al., 1989). Examination of the primary sequences of various G-protein-coupled receptors reveals no striking sequence identity within these regions, even among receptors known to couple to the same G-protein, suggesting that the determinants for G-protein specificity may reside elsewhere on the protein. Deletions within the i2 loop, or deletion of the majority of the C-terminal tail of the protein had no effect on the ability of the receptor to stimulate Gs, demonstrating that these regions are not required for G-protein coupling (Dixon et al., 1989a). However, the i1 loop and the region of the C-terminal tail immediately adjacent to the plasma membrane could not be deleted without adversely affecting the folding of the protein (Dixon et al., 1989a). Therefore, the significance of these domains could not be investigated by deletion mutagenesis and these regions remain candidates for involvement in G-protein coupling.

Further insights into the structural basis for receptor–G-protein interactions have arisen from the use of hybrid receptors. In studies from several laboratories, chimeric receptors have been constructed in which domains from two receptors which couple to different G-proteins are combined, in order to determine the source of the G-protein specificity. Hybrids containing TM 5 and 6 and the connecting i3 loop of the β-AR (couples to Gs to stimulate adenylyl cyclase) and the remaining domains of the α_2-AR (couples to Gi to inhibit adenylyl cyclase) showed a small amount of adenylyl cyclase stimulation, suggesting that the determinant of G-protein specificity lies, at least in part, within the region of the receptor spanned by

TM 5 and 6 (Kobilka et al., 1988). However, a smaller scale hybrid, in which a short region at the N-terminus of the i3 loop of the β-AR was substituted with the α_2-AR coupled to Gs (O'Dowd et al., 1988), and a similar substitution of m1 (couples to Gp to activate phospholipase C) (Dixon et al., 1989a) failed to show detectable G-protein coupling. A similarly small substitution of the N-terminal region of the C-terminal tail of the β-AR with sequence from m1 also resulted in a hybrid receptor which coupled to Gs (Dixon et al., 1989a). Thus, small molecular replacements in these intracellular regions have not defined the basis for G-protein specificity. A more extensive analysis using chimeric m1/m2 suggested that the presence of the i3 loop of each receptor subtype is sufficient to determine the G-protein specificity of a hybrid receptor (Kubo et al., 1988). Further analysis using small molecular replacements and single-residue substitutions should more completely define the molecular determinants of G-protein specificity.

The regions at the N- and C-terminal portions of the i3 loop of the β-AR, which were shown to be essential for Gs coupling, would be predicted by secondary structure prediction algorithms to form amphipathic α-helices (Dixon et al., 1989a; Cheung et al., 1989). Thus, these regions might form amphipathic cytoplasmic extensions of TM 5 and 6. In combination with the information obtained from the genetic analysis of the ligand binding site described above, these observations suggest a possible mechanism for agonist activation of the β-AR, as outlined in Figure 3.3. According to this model, agonists bind to the receptor through an ionic interaction of the amine group with the acidic side chain of Asp^{113} in TM 3 (Strader et al., 1987b, 1988) and through hydrogen bonding interactions between the catechol hydroxyl substituents and the hydroxyl side chains of Ser^{204} and Ser^{207} in TM 5 (Strader et al., 1989c). The interactions with the Ser residues in TM 5, which are agonist-specific, would be expected to cause conformational changes in TM 5. These conformational changes could then cause a reorientation of the amphipathic cytoplasmic domain at the bottom of the helix, altering the conformation of this region and leading to its interaction with Gs. Since antagonists do not interact with the Ser residues in TM 5, they would not promote a conformational change in this region, and Gs activation would not be catalyzed. Further definition of the specific molecular interactions involved in ligand binding and receptor-mediated signal transduction will require a combination of genetic analysis and biophysical characterization of the precise conformational alterations which occur during these processes.

4. Conclusions

The cloning of the genes for G-protein-linked receptors and the subsequent genetic analysis of the proteins has led to the tentative identification of

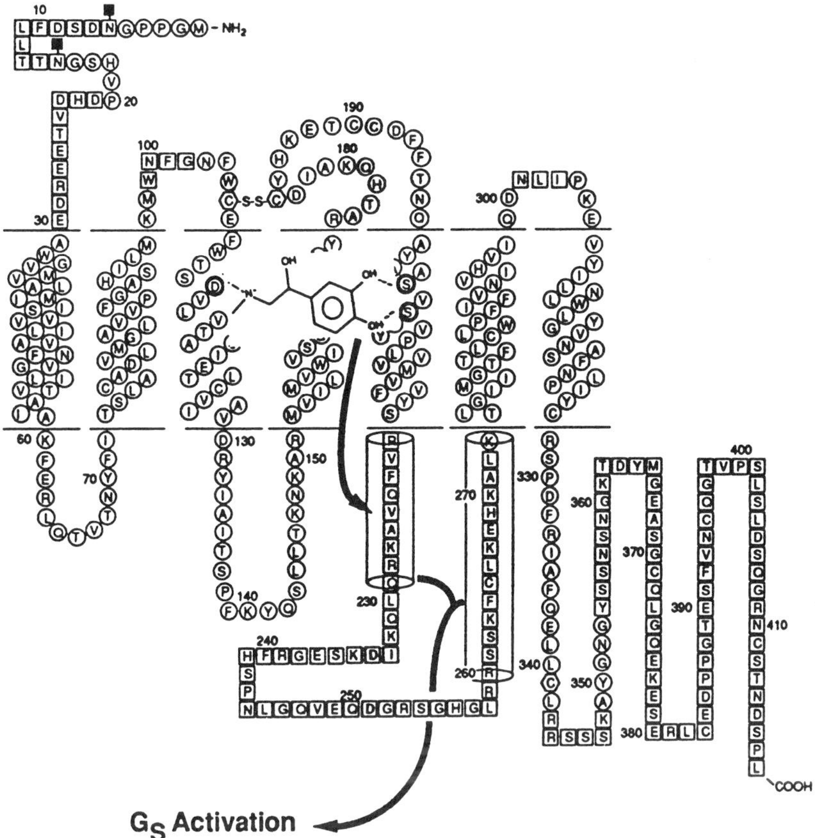

FIGURE 3.3. Model for agonist activation of the β-AR. The extracellular space is at the top of the page, and the cytoplasmic region at the bottom. Amino acid residues are identified by the single-letter code, and are numbered starting at the N-terminus. The agonist epinephrine is shown in the binding site, with the amine group linked to Asp[113] in TM 3, and the catechol hydroxyls linked to Ser[204] and Ser[207] in TM 5. The large cylinders delineate the putative amphiphilic helices in the i3 loop which could not be deleted without affecting G-protein coupling. The black arrows indicate that conformational changes in TM 5 could be transmitted to these amphiphilic regions, leading to G-protein coupling.

some of the molecular interactions involved in ligand binding and G-protein coupling to these receptors. A model has been developed for the ligand binding site of the β-AR which is similar to that postulated for rhodopsin. The similarities in the primary sequences of other receptors of this class, combined with the structural similarities in the ligands which interact with these various receptors, suggest that similar interactions may be involved in the binding of other G-protein-coupled receptors to their ligands. A complete knowledge of the structure of the ligand binding and G-protein-coupling domains of the β-AR and other G-protein-linked receptors should lead to the rational design of new classes of therapeutic agents which act at these receptors.

Acknowledgments

We are grateful to the late Dr. I. S. Sigal for his collaboration and many insights during the work reviewed in this chapter.

References

Aoki C, Zemcik BA, Strader CD, Pickel VM (1989): Cytoplasmic loop of β-adrenergic receptors: Synaptic and intracellular localization and relation to catecholaminergic neurons in the nuclei of the solitary tracts. *Brain Res* 493:331–347

Cheung AH, Sigal IS, Dixon RAF, Strader CD (1989): Agonist-promoted sequestration of the β_2-adrenergic receptor requires regions involved in functional coupling with Gs. *Mol Pharmacol* 34:132–138

Dixon RAF, Sigal IS, Rands E, Register RB, Candelore MR, Blake AD, Strader CD (1987a): Ligand binding to the β-adrenergic receptor involves its rhodopsin-like core. *Nature* 326:73–77

Dixon RAF, Sigal IS, Candelore MR, Register RB, Scattergood W, Rands E, Strader CD (1987b): Structural features required for ligand binding to the β-adrenergic receptor. *EMBO J* 6:3269–3275

Dixon RAF, Sigal IS, Strader CD (1989a): Structure-function analysis of the β-adrenergic receptor. *Cold Spring Harbor Symp Quant Biol* 53:487–497

Dixon RAF, Hill WS, Candelore MR, Rands E, Diehl RE, Marshall MS, Sigal IS, Strader CD (1989b): Genetic analysis of the molecular basis for β-adrenergic receptor subtype specificity. *Proteins* 6:267–274

Findlay JBC, Pappin DJC (1986): The opsin family of proteins. *Biochem J* 238:625–642

Frielle, T, Daniel, KW, Caron, MG, Lefkowitz, RJ (1988): Structural basis of β-adrenergic receptor subtype specificity studied with chimeric β_1/β_2-adrenergic receptors. *Proc Natl Acad Sci USA* 85:9494–9498

Gilman AG (1987): G-Proteins: Transducers of receptor-generated signals. *Annu Rev Biochem* 56:615–624

Hargrave P, McDowell JH, Feldman RJ, Atkinson PH, Rao JKM, Argos P (1984): Rhodopsin's protein and carbohydrate structure, selected aspects. *Vision Res* 24:1487–1493

Kobilka BK, Kobilka TS, Daniel K, Regan JW, Caron MG, Lefkowitz RJ (1988): Chimeric α_2-,β_2-adrenergic receptors: Delineation of domains involved in effector coupling and ligand binding specificity. *Science* 240:1310–1316

Kubo T, Bujo H, Akiba I, Nakai J, Mishina M, Numa S (1988): Location of a region of the muscarinic acetylcholine receptor involved in selective effector coupling. *FEBS Lett* 241:119–125

Main BG, Tucker H (1985): Recent advances in β-adrenergic blocking agents. *Prog Med Chem* 22:122–143

O'Dowd BF, Hnatowich M, Regan JW, Leader WM, Caron MG, Lefkowitz RJ (1988): Site-directed mutagenesis of the cytoplasmic domains of the human β_2-adrenergic receptor. *J Biol Chem* 263:15985–15992

Rands E, Candelore MR, Cheung AH, Strader CD, and Dixon RAF (1990): Mutational analysis of β-adrenergic receptor glycosylation: N-Linked sugar addition is required for receptor transport, but not function. *J Biol Chem* 265:10759–10764

Strader CD, Sigal IS, Register RB, Candelore MR, Rands E, and Dixon RAF (1987a): Identification of residues required for ligand binding to the β-adrenergic receptor. *Proc Natl Acad Sci* USA 84:4384–4388

Strader CD, Dixon RAF, Cheung AH, Candelore MR, Blake AD, Sigal IS (1987b): Mutations that uncouple the β-adrenergic receptor from Gs and increase agonist affinity. *J Biol Chem* 262:16439–16443

Strader CD, Sigal IS, Candelore MR, Rands E, Hill WS, Dixon RAF (1988): Conserved aspartic acid residues 79 and 113 of the β-adrenergic receptor have different roles in receptor function. *J Biol Chem* 263:10267–10271

Strader CD, Sigal IS, Dixon RAF (1989a): Structural basis of β-adrenergic receptor function. *FASEB J* 3:1825–1832

Strader CD, Candelore MR, Hill WS, Dixon RAF, Sigal IS (1989b): A single amino acid substitution in the β-adrenergic receptor promotes partial agonist activity from antagonists. *J Biol Chem* 264:16470–16477

Strader CD, Candelore MR, Hill WS, Sigal IS, Dixon RAF (1989c): Identification of two serine residues involved in agonist activation of the β-adrenergic receptor. *J Biol Chem* 264:13572–13578

Strader CD, Sigal IS, Dixon RAF (1989d): Genetic approaches to the determination of structure-function relationships of G-protein-coupled receptors. *Trends Pharmacol Sci* Suppl, 26–30

4

The α-Adrenergic Receptors: New Subtypes, Pharmacology, and Coupling Mechanisms

John W. Regan and Susanna Cotecchia

If evolution truly moves in leaps and bounds rather than by gradual change, then the evolution of our understanding of the alpha (α) receptors is in one of these periods of dynamic transition. In 1982 most pharmacologists were content with two α-adrenergic receptor subtypes, although some fringe elements were clamoring for more (cf. McGrath, 1982). At present, biochemical techniques have led to the positive identification of six subtypes with the prospect of more being discovered in the near future. At this time, the pharmacological and physiological significance of these new receptor subtypes is largely unknown, and discovering their raison d'être promises to be exciting. This chapter examines the initial α-receptor classification and work that led to the first purification and cloning of the α-adrenergic receptors. It will also review recent studies on the identification of new α-receptor subtypes and what is presently known about their biochemical and pharmacological characteristics.

1. Original Classification and Pharmacology

The α-adrenergic receptors were first differentiated from the β-adrenergic receptors by pharmacological criteria which showed that the order of potency for a series of agonists was different for these two adrenergic receptor subtypes (Ahlquist, 1948). Twenty-six years later it was appreciated that heterogeneity was present within the α-receptor subclass itself (Dubocovich and Langer, 1974). This was shown by the different concentrations of phenoxybenzamine that were needed to block the presynaptic release of neurotransmitters as compared with the postsynaptic response of increased perfusion pressure in cat spleen. For a time it appeared that the α-receptors could be distinguished by anatomical location, that is, pre- versus postsynaptic. However, further pharmacological considerations showed that receptors with presynaptic characteristics were present on platelets, in frog skin, and in other places that were decidedly unsynaptic (Berthelsen and Pettinger, 1977). The development of radioligand binding techniques, coupled with the introduction of [^{3}H]yohimbine and of [^{3}H]prazosin, corroborated this notion and by 1978 the division of α-adrenergic receptors was recognized in terms of the α_1- and the α_2-receptor subtypes (Starke, 1981).

As alluded to above, the α-adrenergic antagonists, yohimbine and prazosin, proved useful for the α-adrenergic receptor classification scheme. Yohimbine, an alkaloid present in *Rauwolfia serpentina* and other plant species, is a potent antagonist that is selective for the α_2-receptor subtype. Another diastereomer, rauwolscine (also known as α-yohimbine), is the same in this regard. Interestingly, however, a third diastereomer named corynanthine (there are 32 possible stereoisomers of yohimbine; see Lambert et al., 1978) is selective for the α_1-adrenergic receptor subtype. This combination of isomers, differing only in configuration, aided the classification of the α-receptor subtypes (Weitzell et al., 1979). Besides corynanthine, antagonists that are selective for α_1-receptors include prazosin, BE 2254 (also known as HEAT) and WB 4101.[1] The latter compound, however, has had a checkered history beginning as a ligand selective for the antagonist state of the α-receptor (U'Prichard et al., 1977), then as an α_1-selective antagonist (Haga and Haga, 1980), followed by evidence that it was nonselective for the α-adrenergic receptors (Hoffman and Lefkowitz, 1980). Presently, WB 4101 is useful for the classification of α_1-receptor subtypes (Morrow and Cresse, 1986). There is an organic basis for WB 4101's past that will be partially clarified later in this chapter.

Agonists that have been valuable for the α_1/α_2 classification include clonidine, phenylephrine, and methoxamine. Clonidine, an α_2-selective imidazoline, has therapeutic use as a centrally acting antihypertensive and has an extensive literature (Timmermans et al., 1980). Other α_2-selective imidazolines that have been used to classify α-receptors include oxymetazoline and UK 14304. The agonists phenylephrine and methoxamine are relatively α_1-selective, although, not to the same extent that clonidine is for the α_2-receptors.

---

[1] Key to abbreviations: WB 4101, 2-[(2', 6'-dimethoxyphenoxyethyl)aminomethyl]-1, 4-benzodioxan; BE 2254, 2-[β-(4-hydroxyphenyl)ethylaminomethyl]tetralone; UK 14304, 5-bromo-6-N-(2,4,5-dihydroimidazolyl)quinoxaline; HPLC, high-pressure liquid chromatography; SDS-PAGE, sodium dodecyl sulfate polyacrylamide gel electrophoresis; M_r, apparent molecular weight; G-protein, guanine nucleotide-binding regulatory protein; SKF 104078, 6-chloro-9-[(3-methyl-2-butenyl)oxy]3-methyl-1H-2,3,4,5-tetrahydro-3-benzazepine; [^{125}I]APDQ, 4-amino-6,7-dimethoxy-2-[4-[5-(4-azido-3-[^{125}I]iodophenyl)pentanoyl]-1-piperazinyl]quinazoline.

2. Initial Molecular Identification

2.1. Affinity Labeling

The molecular nature of both the α_1- and α_2-adrenergic receptors was first characterized by SDS-PAGE following radiolabeling of the receptor proteins with [^{3}H]phenoxybenzamine (Kunos et al., 1983; Regan et al., 1984). Interestingly, phenoxybenzamine, a chemically reactive β-chloroalkylamine, was also the first ligand used for the subclassification of the α-adrenergic receptors (Dubocovich and Langer, 1974). Although selective for the α_1-adrenergic receptors, it also has moderately high affinity for the α_2-adrenergic receptors. Using rat hepatic membranes, [^{3}H]phenoxybenzamine covalently labeled peptides with molecular weights of 59,000 and 80,000 (Kunos et al., 1983). Since the liver contains predominantly α_1-adrenergic receptors, it was assumed that these labeled peptides represented the receptor, but the relationship of these two peptides to each other and/or to other receptors was unclear. With the development of [^{125}I]APDQ, a higher affinity and more selective photoaffinity ligand (Leeb-Lundberg et al., 1984), the M_r 59,000 species was clearly shown to be a degradation product of the 80,000 peptide. [^{125}I]APDQ was also used to label α_1-receptors in rat cerebral cortex (Leeb-Lundberg et al., 1983) and in rat, guinea pig, and rabbit spleen, and in rabbit lung and aortic smooth muscle cells (Leeb-Lundberg et al., 1984). In all cases, the predominant species that was labeled had an M_r of approximately 80,000 and an α_1-adrenergic pharmacology. These results suggested that α_1-receptors were relatively homogeneous and that they might be encoded by a similar gene regardless of the tissue or species.

[^{3}H]Phenoxybenzamine was also used to label human platelet α_2-adrenergic receptors but not until they had been partially purified by affinity chromatography (Regan et al., 1984). A peptide with an M_r of approximately 61,000 was labeled, which had the expected pharmacology of an α_2-adrenergic receptor. This was confirmed with a specific photoaffinity ligand that labeled α_2-adrenergic receptors in human platelet membranes and in rabbit basolateral membranes (Regan et al., 1986a). Besides showing the constancy of the receptor between tissues and species, these results showed that solubilization per se did not cause major changes in the apparent structure of the receptor—at least with respect to the ligand binding site.

2.2. Purification

2.2.1. α_1-Adrenergic Receptors

α_1-Adrenergic receptors were first clearly purified from a hamster smooth muscle cell line (DDT$_1$MF-2), using a combination of affinity chromatography, lectin chromatography, and size-exclusion HPLC (Lomasney et al., 1986). The nonspecific labeling of protein in the purified material with the [^{125}I]Bolton Hunter reagent identified a protein with an M_r of 80,000. This same protein was also labeled with the α_1-photoaffinity ligand, [^{125}I]APDQ. Competition binding studies showed that an α_1-adrenergic receptor had been purified and that its apparent molecular weight, as judged by SDS-PAGE, was consistent with the specific activity data (binding activity/milligram of protein).

Purification of hepatic α_1-adrenergic receptors had also been reported using a similar combination of affinity chromatography and HPLC (Graham et al., 1982). These studies reported an M_r of 59,000, which was consistent with the proteolytic α_1-adrenergic receptor fragment identified subsequently by photoaffinity labeling (Leeb-Lundberg et al., 1984). Coomassie staining of SDS-PAGE gels, however, suggested that the purified M_r 59,000 peptide might not represent the receptor. In particular, the Coomassie-stained gel showed a sharp band, whereas the M_r 59,000 fragment identified by photoaffinity labeling was quite diffuse. Competition binding data were consistent with an α_1-adrenergic receptor, but identification of the M_r 59,000 protein by affinity labeling was not reported.

2.2.2. α_2-Adrenergic Receptors

The purification of α_2-adrenergic receptors from human platelets required a five-step chromatographic procedure involving affinity, heparin-agarose, and lectin-agarose chromatography (Regan et al., 1986b; earlier studies on the solubilization and purification of α_2-receptors have been reviewed, Regan, 1988). Following purification, SDS-PAGE of [^{125}I]Bolton Hunter labeled protein showed one band with an M_r of approximately 64,000. This same protein was also specifically labeled with [^{3}H]phenoxybenzamine. Competition binding studies with the purified protein were consistent with an α_2-adrenergic receptor, and reconstitution studies showed that this M_r 64,000 protein contained the molecular features needed for signal transduction as well as binding (Cerione et al., 1986). Together, these studies were consistent with the idea that the receptor protein identified by affinity labeling in membranes was the same as that which could be solubilized and then purified.

Although both the α_1- and α_2-adrenergic receptors had been purified, the quantities that were actually isolated were less than minuscule. For example, from 18 g of membrane protein only about 2 μg of purified α_2-adrenergic receptor were obtained. For the α_1-receptor, 300 mg of membrane protein yielded approximately 1 μg of receptor protein. It was clear that detailed structural studies of these receptors would not occur without significant improvements: the most feasible being toward cloning and the use of recombinant DNA techniques to express larger quantities of these proteins. The prospect of cloning, however, was daunting, especially since initial studies of the α_1 and of other G-protein-coupled receptors (the β_2, avian β_1, and M1) indicated that their amino-termini were blocked. The consequence was that enough receptor would need to be purified to survive an initial round of chemical or enzymatic cleavage followed by the isolation of peptide fragments. Experience with the β (Dixon et al., 1986; Yarden et al., 1986) and muscarinic receptors (Kubo et al., 1986; Peralta et al., 1987) suggested that approximately 50 μg (1 nmol) would be required.

2.3. Cloning

2.3.1. α_2-ADRENERGIC RECEPTORS

Initial attempts to obtain 50 μg of purified protein were frustrated by difficulties with the scale-up and with the consistency of the affinity chromatography step (reviewed, Regan and Matsui, 1990). A new affinity resin based on yohimbinic acid (Repaske et al., 1987) greatly improved this situation. Over a 2-month period, 800–1000 pmol of α_2-receptor activity were purified from approximately 100 L of platelet concentrate. The receptors were cleaved with cyanogen bromide and, using a combination of size exclusion and reverse-phase HPLC, 145 pmol of one peptide were obtained. A sequence of 27 amino acids from this peptide was identified and used to design synthetic oligonucleotide probes for screening a human genomic library (Kobilka et al., 1987b). Three identical clones were obtained by plaque hybridization analysis and a 5.5-kb BamHI fragment was present which contained a 1350-base open reading frame that encoded the human platelet α_2-adrenergic receptor.[2] The predicted molecular weight of this protein, 49,400, was significantly less than that observed for the purified protein (approximately 64,000). The most likely explanation for this is that the mature protein is glycosylated (Lanier et al., 1988; Regan, 1988) and may be subject to other posttranslational modifications such as fatty acylation or phosphorylation. In addition to increasing the size of the receptor, these modifications can also cause the receptor to move anomalously during SDS-PAGE. Figure 4.1 shows SDS-polyacrylamide gels of the human platelet α_2-adrenergic receptor following in vitro translation of RNA encoding the receptor and following photoaffinity labeling of the purified receptor. This translation system does not add the N-linked complex carbohydrates to the newly synthesized receptor and thus its size (approx-

---

[2]A sequencing error is present in the published sequence of the cloned human platelet α_2-adrenergic receptor (Kobilka et al., 1987b). This error affects the deduced sequence of 31 amino acids in the third inner (intracellular) loop. The correct amino acid sequence starting with residue 333 and ending with 363 is: PRRGPGATGIGTPAAGPGEERVGAAKASRWR. The corresponding nucleotide sequence (997–1089) is: ccg cgg cgc ggg ccg ggg gcg acg ggg atc ggg acg ccg gct gca ggg ccg ggg gag gag cgc gtc ggg gct gcc aag gcg tcg cgc tgg cgc (see Guyer et al., 1990).

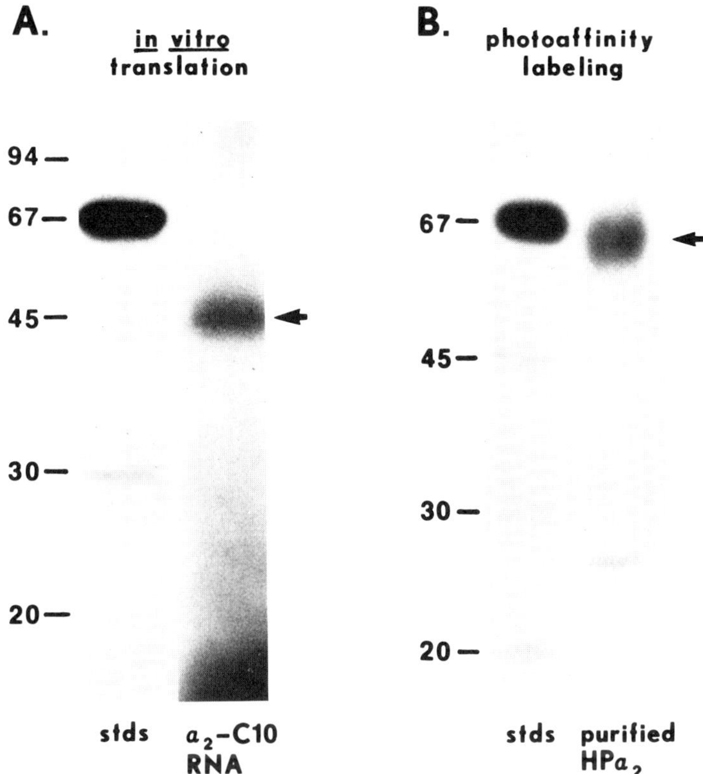

FIGURE 4.1. (A) In vitro translation and (B) photoaffinity labeling of the α_2-C10 (human platelet) adrenergic receptor. RNA encoding the α_2-C10 adrenergic receptor was synthesized from its corresponding cDNA in the vector pSP65 using SP6 RNA polymerase. The RNA was translated in the presence of L-[^{35}S]methionine using rat reticulocyte lysate and dog pancreatic microsomes as supplied (NEN-Dupont). The translation reaction was then electrophoresed on a 10% polyacrylamide-SDS gel and was fixed and prepared for fluorography. Photoaffinity labeling was done as described (Matsui et al., 1989) using the α_2-selective photoaffinity ligand [^{3}H]SKF 102229 and purified human platelet α_2-adrenergic receptors.

imately 47,000) is close to that predicted by the deduced amino acid sequence. In contrast, the size of the purified and photoaffinity-labeled receptor is significantly larger (approximately 66,000).

RNA prepared from the cloned human platelet α_2-receptor DNA was injected into *Xenopus laevis* oocytes to see if it encoded a functional α_2-adrenergic receptor (Kobilka et al., 1987b). Two days after the injection, membranes were prepared and the binding of [^{3}H]yohimbine was examined. Whereas uninjected oocytes and oocytes injected with β_2-receptor RNA showed no specific [^{3}H]yohimbine binding, those injected with α_2

RNA yielded saturable binding that was competed for by appropriate α_2-adrenergic ligands. The actual expression of the receptor was relatively low and averaged approximately 400 fmol/mg membrane protein, or 1–2 fmol per oocyte. Although β_2-adrenergic receptor stimulation in transfected oocytes stimulated adenylyl cyclase (Kobilka et al., 1987a), repeated attempts to inhibit adenylyl cyclase proved unsuccessful. Perhaps the low expression or the absence of an appropriate G-protein was responsible. Inability to inhibit adenylyl cyclase was not intrinsic to the expressed receptor since this was subsequently demonstrated in transfected mammalian cells (Fraser et al., 1989; Cotecchia et al., 1990a).

The cloning of the α_2-adrenergic receptor DNA, therefore, showed that the M_r 64,000 protein, as purified and expressed naturally in human platelets, was encoded by a gene consisting of a single exon and that the essential features with respect to its functional properties were contained in this single subunit structure. The cloning also showed a deduced amino acid sequence that contained conserved residues and general structural motifs that were similar to the opsins and to the β-adrenergic and muscarinic receptors that had been cloned previously (see Dohlman et al., 1987). These data supported the hypothesis that most, if not all, of the receptors that were coupled to G-proteins were members of a family of structurally and functionally related proteins that were derived from some common ancestor.

2.3.2. α_1-ADRENERGIC RECEPTORS

The cloning of the α_1-adrenergic receptor followed a period of several months in which receptors were purified from a total of 1600 L of suspension culture so that enough material for one chemical cleavage and peptide isolation could be done (Cotecchia et al., 1988). From approximately 1 nmol of purified receptor, three peptides were isolated and sequenced. Figure 4.2 shows an SDS-polyacrylamide gel of the receptor and some of the peptides generated by cyanogen bromide (CNBr) cleavage as well as the overall cloning strategy. A 43-base oligonucleotide probe was designed and used to screen a hamster genomic DNA library. A genomic library was used because the genes for the β_2- and α_2-adrenergic receptors were previously found to lack introns in their coding sequences. The gene for the α_1-adrenergic receptor, however, is different and contains at least one intron. When this became evident, a hamster cDNA library was screened and a complete 2-kb cDNA was cloned (Cotecchia et al., 1988).

This 2-kb clone contained an open reading frame that coded for a protein consisting of 515 amino acids. The deduced molecular weight of 56,000 was in good agreement with that observed by SDS-PAGE after deglycosylation of the native α_1-adrenergic receptor (Sawutz et al., 1987). To determine the binding and coupling characteristics encoded by this α_1-receptor cDNA, the coding region was ligated into an expression vector which was then used to

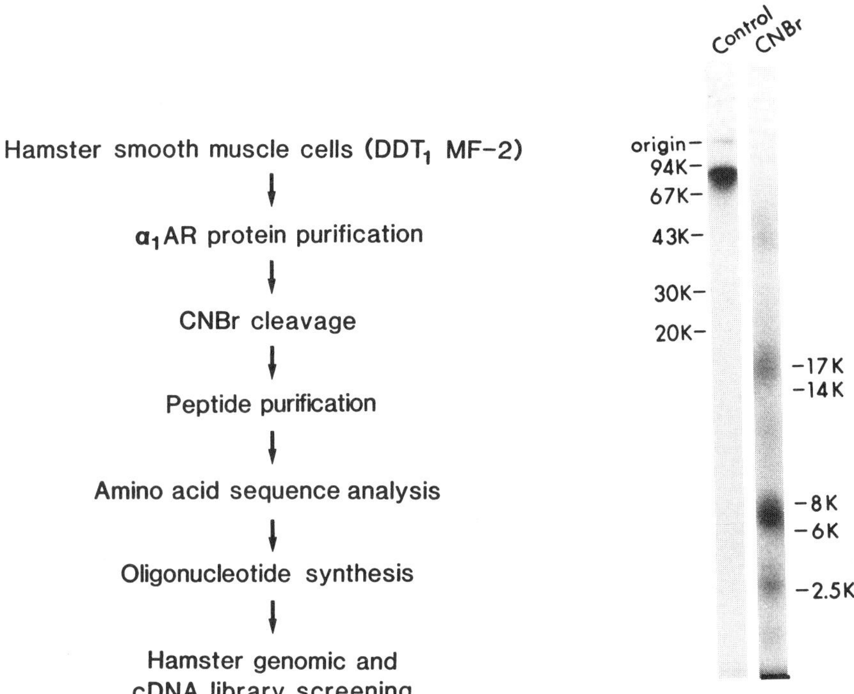

FIGURE 4.2. Cloning strategy for the α_1-adrenergic receptor from DDT$_1$MF-2 cells and the ^{125}I-labeled receptor after the initial purification (control) and after cleavage with cyanogen bromide (CNBr). Approximately 10 pmol of the purified receptor were radioiodinated with [^{125}I]Bolton Hunter reagent and samples were electrophoresed before and after treatment with CNBr on a 8 M urea/10% polyacrylamide-SDS gel. The sample was fixed and autoradiographs were obtained. Reprinted with permission of the National Academy of Science, from Cotecchia et al. (1988).

transfect mammalian cells. Following expression in COS-7 cells, the α_1-selective antagonist [^{125}I]HEAT bound with high affinity. This binding was competed by appropriate α_1-agonists and antagonists. By prelabeling transfected cells with ^{32}P$_i$, it was shown that norepinephrine stimulated the turnover of phosphatidylinositol. This effect, which could be blocked with antagonists, was consistent with the expected coupling of α_1-adrenergic receptors to the stimulation of the phosphatidylinositol signal transduction pathway. Thus, as for the α_2-, and β-adrenergic receptors, all the known functional properties of the receptor with respect to binding and signal transduction were present in DNA coding for a single polypeptide chain.

2.3.3. GENERAL STRUCTURE OF THE α-ADRENERGIC RECEPTORS

One of the intriguing discoveries resulting from the initial cloning of the β- and α-adrenergic receptors was that they were members of a much larger family of related proteins. This family was related by a common modus operandi: they all used G-proteins to transduce an initial external stimulus into a cellular response. Members of this family now include the adrenergic receptors, the muscarinic receptors, the opsins, the dopamine receptors, the chemoattractant receptor, the 5HT receptors, the neuropeptide receptors, and the glycoprotein hormone receptors. It is expected that this family will grow to include other receptors that are known to interact with G-proteins. The basic working model for the α-adrenergic receptors and for G-protein-coupled receptors as a whole is shown in Figure 4.3. The model is that of an integral membrane protein with seven membrane-spanning domains. Within these membrane-spanning domains there is a remarkable conservation of certain amino acid residues between all the members of the family. As will become evident, this homology increases greatly among the various subfamilies of receptors. In contrast, other regions of these proteins show little homology, even between closely related subtypes. The amino-terminus which is positioned on the outside of the cell membrane is one of these regions. Another is known as the third cytoplasmic loop (the i3 loop) which is also quite variable in length and which connects TM 5 and TM 6. In

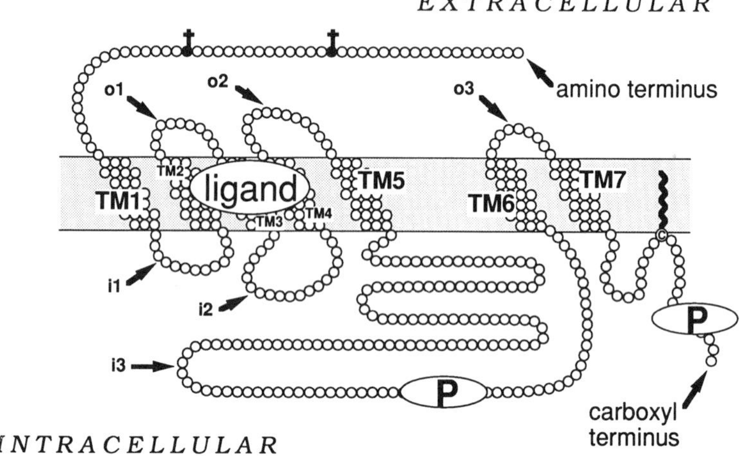

FIGURE 4.3. General model for the α-adrenergic receptors and for G-protein-coupled receptors as a whole. The chain of circles represents the individual amino acids that make up the receptor protein. The stippled area is the cell surface membrane. TM1–TM7 represent the putative membrane-spanning domains. o1–o3 and i1–i3 represent the extracellular and cytoplasmic loops that would be formed if this model is correct. "Ligand" indicates that the binding of ligands probably occurs in the transmembrane domains; "P" denotes potential sites of phosphorylation that may be important in the regulation of the receptor. The crosses are sites of potential N-linked glycosylation (note that the α_2-C2 lacks these sites). The bold swiggle attached to the cysteine in the carboxy-terminal region represents a potential site of fatty acylation.

general, the transmembrane domains consist of 20 to 25 predominantly hydrophobic amino acids, while the loops are of variable length and are predominantly hydrophilic. Potential posttranslational modifications include N-linked glycosylation, phosphorylation, and fatty acylation. The full extent of these modifications with respect to the α-adrenergic receptors remains to be determined.

3. Identification of Subtypes

3.1. Pharmacological Studies

3.1.1. α_2-ADRENERGIC RECEPTORS

As studies on the purification and cloning of the α_1- and α_2-adrenergic receptors progressed, pharmacological investigations were revealing further heterogeneity among these α-receptor subtypes. Some of the initial observations, made primarily through radioligand binding, were difficult to corroborate physiologically or biochemically, either because of the complexity of the systems or because the methods had not advanced sufficiently. For example, an early characterization of the binding of [^{3}H]rauwolscine demonstrated evidence of α_2-adrenergic receptor heterogeneity (Cheung et al., 1982). This study found that the affinity of prazosin for the human platelet α_2-receptor differed significantly from the value obtained in rat brain. Similarly, the K_i's of prazosin for neonatal rat lung and for other rodent species were consistently lower than that observed in pigs and humans (Feller and Bylund, 1984). Although the possibility of α_2-receptor heterogeneity was raised, it was not clear if this only represented species differences. In 1985, a classification of the α_2-adrenergic receptors into A and B subtypes was proposed largely on the basis of the differential affinity of prazosin and oxymetazoline for these subtypes (Bylund, 1985).

Further evidence of α_2-adrenergic receptor heterogeneity came from autoradiographical and functional studies. In mapping the location of α_2-receptors in rat brain, [^{3}H]idazoxan and [^{3}H]rauwolscine were found to label two populations of receptors (Boyajian et al., 1987). Both populations could bind [^{3}H]idazoxan with similar affinity, but one population had relatively low affinity for [^{3}H]rauwolscine and could not be labeled with this ligand. The latter group was referred to as α_2-R$_i$ for rauwolscine insensitive, while the former was called α_2-R$_s$ for rauwolscine sensitive. This classification has been criticized because the affinity of rauwolscine for the α_2-R$_i$ sites is too low to qualify these sites as being α_2-adrenergic. This issue has merit since two of the salient features of the α_2-adrenergic classification are high affinity for imidazolines, such as clonidine, and, comparatively high affinity for yohimbine (and rauwolscine) as compared with corynanthine.

However, the accuracy of the estimates may be unduly influenced by the vagaries of the [^{3}H]idazoxan binding itself, which can be dramatically influenced by the buffer composition. Another possibility is that the [^{3}H]idazoxan is binding to an imidazoline site that rauwolscine does not recognize. Thus, the relationship of the α_2-R$_s$ and α_2-R$_i$ to α_2-A and α_2-B subtypes is unclear and requires further study. (See Update, Sec. 6).

Functional studies using the α_2-selective antagonist, SKF 104078, suggest that some differences may exist between pre- and postsynaptic α_2-adrenergic receptors (Hieble et al., 1988). At postsynaptic α_2-receptors in dog saphenous vein, SKF 104078 blocks agonist-induced vasoconstriction with high affinity ($K_d \sim 80$ nM). At presynaptic α_2-receptors in the guinea pig atrium, however, SKF 104078 is virtually inactive ($K_d > 10$ μM). Other studies, however, using different models of the presynaptic α_2-receptor, report that SKF 104078 does have significant affinity for presynaptic α_2-receptors (Connaughton and Docherty, 1990). This suggests the possibility that there is heterogeneity among the presynaptic α_2-adrenergic receptors. The possible relationship between SKF 104078 insensitive receptors and the α_2-A and α_2-B subtypes has not been investigated. Table 4.1 summarizes the different nomenclatures that have been proposed for α_2-adrenergic receptor subtypes. A discussion of the classification based on molecular biological techniques follows in Section 3.2.1.

3.1.2. α_1-ADRENERGIC RECEPTORS

In contrast to the α_2-adrenergic receptors, some of the initial indications of α_1-adrenergic receptor heterogeneity came from functional studies (re-

TABLE 4.1. Proposed classifications of α_2-adrenergic receptor subtypes[a]

Basis of classification	Nomenclature	Prototypical tissue	Major characteristics
Radioligand binding/ biochemical[a]	α_2-A	Human platelet	Low affinity for prazosin
	α_2-B	Neonatal rat lung	High affinity for prazosin, nonglycosylated
Autoradiography[a]	α_2-Ri	Rat septum	Low affinity for rauwolscine
	α_2-R$_s$	Rat caudate-putamen	High affinity for rauwolscine
Functional assays[a]	α_2-Post	Dog saphenous vein	High affinity for SKF 104078
	α_2-Pre	Guinea pig atrium	Low affinity for SKF 104078
Molecular biology[b]	α_2-C10	Human platelet	Equivalent to the α_2-A
	α_2-C4	?CNS?	Possibly equivalent to the α_2-B identified pharmacologically in rat and human brain possibly related to α_2-R$_s$
	α_2-C2	?Peripheral?	Equivalent to the rat neonatal α_2-B

[a]See text Section 3.1.1 for discussion and references.
[b]See text Section 3.2.1 for discussion and references.

viewed: McGrath, 1982; Hieble et al., 1986; Minneman, 1988). As discussed previously, when [^{3}H]WB 4101 was introduced as a radioligand there was confusion about its selectivity and whether or not it was binding to multiple sites. In a reexamination of the binding of [^{3}H]WB 4101 and [^{3}H]prazosin to rat brain membranes, it was concluded that two α_1-adrenergic receptor subtypes were present (Morrow and Creese, 1986). These subtypes, named α_1-A and α_1-B, were also distinguished by the binding of chlorethylclonidine, an alkylating agent, and by evidence that they increased cytosolic calcium by different mechanisms (Han et al., 1987; Minneman et al., 1988). With respect to WB 4101 and chlorethylclonidine, the α_1-A subtype had higher affinity for WB 4101, and was resistant to inactivation by chlorethylclonidine as compared with the α_1-B subtype. As to increasing intracellular calcium, the α_1-A appears to be linked to a voltage-dependent calcium channel that is nifedipine sensitive, whereas the α_1-B increases calcium by stimulating phosphatidylinositol hydrolysis. Recently, a third α_1-adrenergic receptor has been identified by molecular biological studies (Schwinn et al., 1990). This "α_1-C" subtype has some of the features of the previously discussed α_1-A. However, it can be inactivated by chlorethylclonidine and it has not been identified in tissues where the α_1-A has been described functionally. The proposed classifications for α_1-adrenergic receptors are detailed in Table 4.2, and a more complete discussion of the molecular biological classification follows in Section 3.2.2.

TABLE 4.2. Proposed classifications of α_1-adrenergic receptor subtypes

Basis of classification	Nomenclature	Prototypical tissue (rat)	Major characteristics
Functional/radioligand binding[a]	α_1-A	Hippocampus, vas deferens	High affinity for WB 4101; CEC[b] insensitive; coupled to voltage-dependent Ca channel
	α_1-B	Liver/spleen	Low affinity for WB 4101; CEC[b] sensitive; coupled to phospholipase C
Molecular biology[c]	α_1-A	Hippocampus vas deferens/aorta[d]	Cloned from rat genomic DNA; probably equivalent to the α_1-A above
	α_1-B	DDT$_1$MF-2 cells liver/spleen/heart[d]	Equivalent to the α_1-B above
	α_1-C	?	Cloned from bovine brain cDNA; CEC[b] sensitive; coupled to phospholipase C

[a]See text Section 3.1.2 for discussion and references.
[b]Chlorethylclonidine.
[c]See text Section 3.2.2 for discussion and references.
[d]Based upon results of Northern blot analysis.

3.2. Cloning Studies

3.2.1. α_2-RECEPTOR SUBTYPES

One of the observations made following the cloning of the gene encoding the human platelet α_2-adrenergic receptor $(\alpha_2\text{-C10})$[3] was that Southern blot analysis provided evidence for the possible heterogeneity of α_2-receptors (Kobilka et al., 1987b). An example of this is shown in Figure 4.4 where human genomic DNA was hybridized with radiolabeled DNA containing the coding region of α_2-C10. Prior to hybridization, samples of genomic DNA had been cleaved separately with several restriction endonucleases (Bsu36I, PstI, HindIII, and BamHI); and the resulting digests were resolved by agarose gel electrophoresis and blotted on to nitrocellulose. Hybridization of the blot with the probe, followed by washing and autoradiography revealed several bands. The most intense band (C10) in each lane represents the DNA encoding the human platelet α_2-receptor itself. The less intense bands (C4, C2, a, b, etc.) represent DNAs that were related to the C10 band to the extent that they were recognized by the radiolabeled α_2-C10 probe. These results provided critical evidence and changed the debate on α_2-receptor heterogeneity from a question of "is there a subtype" to "how many subtypes are there?"

The first α_2-adrenergic receptor subtype to be positively identified by cloning was named α_2-C4 because it corresponded to the gene localized on human chromosome 4 (Regan et al., 1988). This clone was isolated from a human kidney cDNA library and its deduced amino acid sequence showed 44% overall homology with the corresponding sequence of α_2-C10 (the human platelet α_2). Southern blot analysis indicated that this clone was equivalent to the C4 bands shown in Figure 4.4. In contrast to its overall homology with α_2-C10, the α_2-C4 had 21% overall homology with the human β_2-adrenergic receptor and 17% homology with the human m4 muscarinic receptor. In the putative transmembrane-spanning domains, the homologies between α_2-C4 and α_2-C10 were higher: ranging from a low of 54% for TM 1 to highs of 88% for TM 3 and 7. When compared with the β_2-adrenergic receptor, homology in the transmembrane regions averaged 39%. Although α_2-C4 was clearly related to α_2-C10, expression and an examination of its radioligand binding properties were necessary to establish its identity as an α_2-adrenergic subtype.

α_2-C4 was expressed in mammalian COS-7 cells (Regan et al., 1988) using the methods described by Cullen (1987). With these methods, high but transient levels of expression of α_2-C4 were obtained 2–3 days after transfection. Membranes prepared from the COS-7 cells bound the α_2-

---

[3] α_2-C10, the name given to this gene, refers to its localization on human chromosome 10. Likewise, α_2-C4 and α_2-C2 refer to human α_2-adrenergic receptor subtypes whose genes are on human chromosomes 4 and 2, respectively.

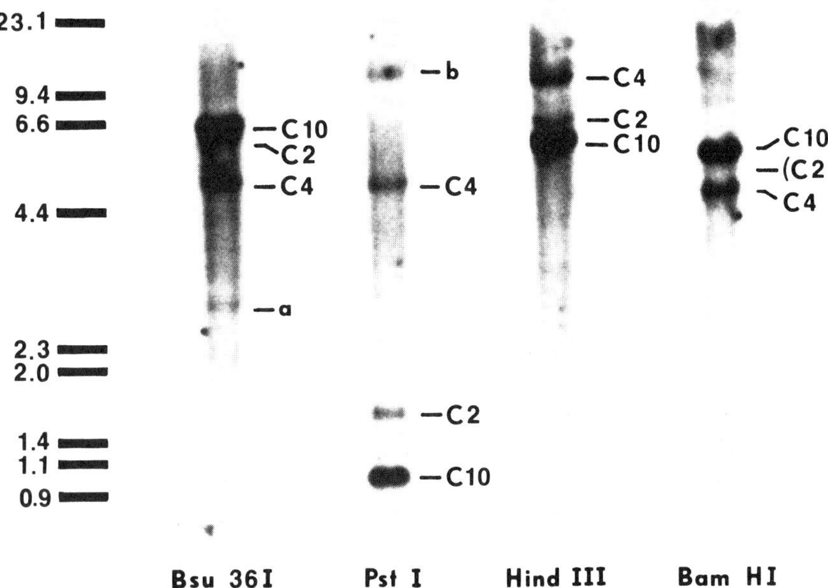

FIGURE 4.4. Southern blot of human genomic DNA probed with a ^{32}P-labeled α_2-C10 cDNA. Human genomic DNA (5 μg) was digested with the restriction enzymes indicated and the samples were electrophoresed on 0.8% agarose gels. The DNA was transferred to nitrocellulose and the blot was hybridized overnight with a nick-translated PstI fragment (0.95 kb) of α_2-C10 at 40°C in 50% formamide/0.8 M NaCl/2.5 × Denhardts/0.1% SDS/50 mM sodium phosphate (pH 6.5)/0.5 mM EDTA/400 μg/mL salmon sperm DNA. The blot was washed twice for 1 hr at 50°C in 2 × SSC/0.1% SDS and once for 1 hr at 56°C in the same buffer. The exposure was made for 5 days at − 80°C. The approximate positions of the molecular size standards are indicated on the left.

selective antagonist [3H]rauwolscine specifically and with high affinity, whereas binding was absent in the nontransfected control cells. Interestingly, the affinity of [3H]rauwolscine was higher for this newly cloned subtype than it was for α_2-C10 (see Table 4.3). This finding suggested a possible relationship of α_2-C4 with the α_2-R$_s$ subtype, wherein rauwolscine could discriminate between α_2-receptor subtypes (Table 4.1). In competition binding studies, prazosin showed higher affinity and oxymetazoline showed lower affinity for α_2-C4 as compared with α_2-C10. In addition, traditional α_2-agonists such as p-aminoclonidine and UK 14304 showed high affinity for this receptor. In many respects, the pharmacological characteristics of α_2-C4 were similar to the α_2-B subtype and it was tentatively identified as such; however, what most people would call the α_2-B is encoded by a different human gene. This gene, which is represented by the band labeled C2 in Figure 4.4, is α_2-C2.

Like the cloning of α_2-C4, α_2-C2 and what appears to be the rat homolog

TABLE 4.3. Ligand binding affinities (K_i's) for cloned α-adrenergic receptors expressed in COS-7 cells

	α_2 Subtypes (human)			α_1 Subtypes		
	α_2-C10[a] (platelet α_2-AR)	α-C4[a]	α_2-C2[b]	α_1-A[e] (rat)	α_1-B[c] (DDT$_1$MF-2 cell)	α_1-C[c] (bovine)
Antagonists						
Prazosin	1800	41	290(27)[d]	0.3	0.3	0.3
WB 4101	8	0.9	130	2.1	8	0.6
Phentolamine	10	33	9	110	160	5
Corynanthine	710	73	1000	250	640	78
Rauwolscine	3	0.4	11(2)[d]	?	3200	1400
Yohimbine	2	0.9	10	?	1300	320
SKF 104078	86	33	100	?	?	?
Agonists						
Epinephrine	1000	170	1800	550	4800	6600
Norepinephrine	2400	240	1300(370)[d]	100	9600	16000
Dopamine	4500	1000	?	?	>100000	>100000
Phenylephrine	1500	2900	?	1400	16000	15000
Methoxamine	?	?	?	110000	>100000	75000
p-Aminoclonidine	74	81	120	?	680	590
Oxymetazoline	11	62	1500(610)[d]	2100	290	27

[a]K_i's are in nanomolar. From Regan et al. (1988).
[b]From Lomasney et al. (1990).
[c]From Schwinn et al. (1990).
[d]Rat. From Zeng et al. (1990).
[e]From Lomasney et al. (1991).

of α_2-C2 were cloned on the basis of homology cross hybridization. In one approach, a clone (5A) encoding α_2-C2 was obtained using a probe based upon a DNA encoding the $5HT_{1A}$ receptor (Weinshank et al., 1990). A similar clone, named $RNG\alpha_2$ was isolated from a rat kidney cDNA library using a degenerate probe designed around the conserved amino acid sequence surrounding Asp^{113} in the β_2-adrenergic receptor (Zeng et al., 1990). The human α_2-C2 was also cloned with the help of the polymerase chain reaction (PCR), using primers based upon conserved sequences of α_2-C4 and α_2-C10 (Lomasney et al., 1990). The deduced amino acid sequence of the human α_2-C2 is shown in Figure 4.5, with the solid circles showing the amino acids that are conserved in both α_2-C10 and α_2-C4. When this sequence is compared to the rat $RNG\alpha_2$ (Zeng et al., 1990), the homology is much higher: 98% in the transmembrane region and 84% overall, with nearly all the divergence being accounted for by differences in the i3 loop.

Expression studies of the human α_2-C2 clone in mammalian cells show that it encodes an α_2-adrenergic receptor subtype (Lomasney et al., 1990; Weinshank et al., 1990). Thus, the expressed protein binds [^{3}H]yohimbine and the binding is competed for by appropriate α_2-adrenergic ligands (see

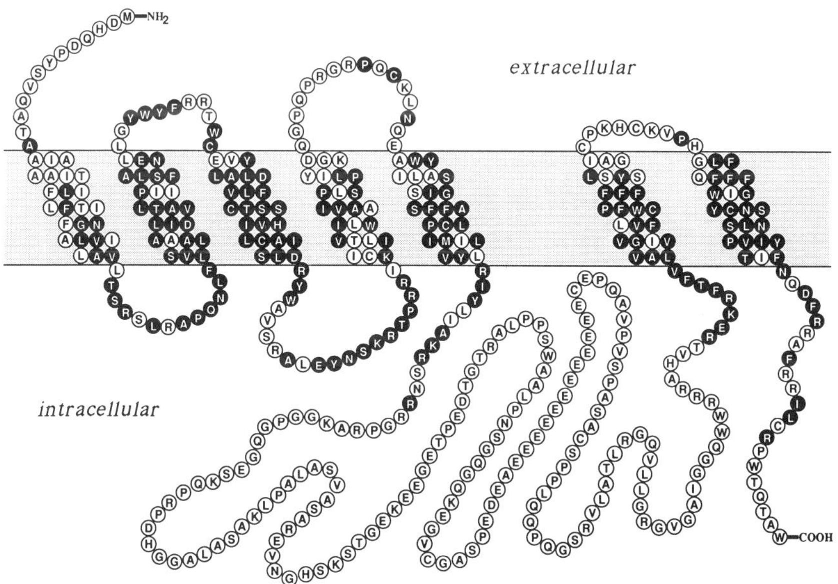

FIGURE 4.5. Deduced amino acid sequence and model of the human α_2-C2 adrenergic receptor. The cell surface membrane is indicated by shading. The circles represent individual amino acids (single-letter code). Amino acids that are conserved with both the human α_2-C10 and α_2-C4 are shown in black. Reprinted with permission of the National Academy of Science, from Lomasney et al. (1990).

Table 4.3). As compared with α_2-C10 (α_2-A), the α_2-C2 subtype has higher affinity for prazosin and significantly lower affinity for oxymetazoline. In this regard the pharmacological characteristics of α_2-C2 are similar to α_2-C4, which, in turn, is similar to the α_2-B as defined pharmacologically (Table 4.1). There is, however, convincing biochemical data that suggests that the α_2-C2 corresponds to the α_2-B.

In a comparison of the neonatal rat lung receptor (the prototypic α_2-B subtype; Bylund, 1985) with the human platelet α_2-receptor, it was determined that the former lacked N-linked glycosylation (Lanier et al., 1988), which is a characteristic of α_2-C10 and which appears to be present in α_2-C4 (Regan et al., 1988). Interestingly, both RNGα_2 and α_2-C2 lack consensus sites for N-linked glycosylation in their amino termini, which could explain why the neonatal rat lung α_2-receptor does not bind to wheat germ agglutinin and is unaffected by *N*-glycanase (Lanier et al., 1988). Northern blot analysis suggests that α_2-C2 is present in neonatal rat lung (Zeng et al., 1990) and adult rat kidney and liver but not brain (Lomasney et al., 1990). Interestingly, Northern blot analysis with α_2-C4 produced strong signals with rat brain RNA but, under the same conditions, no signal with RNA from peripheral tissues, including liver and kidney (Lorenz et al., 1990; see Table 4.4). These results imply that α_2-C4 is equivalent to the α_2-B subtype that has been ascribed to rat and human brain (Petrash and Bylund, 1986). Pharmacological data also support this hypothesis. Thus, in a direct comparison, α_2-receptors present in rat brain showed higher affinity for rauwolscine than did the α_2-receptors present in either neonatal rat lung or

TABLE 4.4. Northern blot analysis of α_2-adrenergic receptor subtypes in rat tissues[a]

	α_2-C10 (platelet α_2-AR) (3.8 kb)[b]	α_2-C4 (2.9 kb)[b]	α_2-C2 (4.1 kb)[b]
Cortex	+ + +	+ + +	0
Cerebellum	+ +	+ + +	0
Brainstem	+ + + +	+ + +	0
Hippocampus	+ + +	+ + +	0
Pituitary	+ +	0	0
Kidney	+ +	0	+ +
Liver	0	0	+ +
Lung	+	0	0
Aorta	+ +	0	0
Heart	0	0	0
Skeletal muscle	+ +	0	0
Spleen	+	0	0

[a]Derived from Lorenz et al. (1990). Probes were from the human α_2-adrenergic receptor subtypes. The data were analyzed in terms of the relative intensity of the bands observed after autoradiography. "0" means there was no clearly observable signal.
[b]Message size.

human platelets (Lanier et al., 1988). In addition, the affinity of oxymetazoline for these α_2-receptors gave the following order of potency: human platelet > rat brain cortex > neonatal rat lung (IC_{50}'s: 55, 600, and 6200 nM, respectively). This appears to be the same order of potency that has been found for oxymetazoline in competition binding studies done with the cloned receptors: that is, α_2-C10 > α_2-C4 > α_2-C2 (K_i's: 11, 62, and 1500 nM, respectively).

Further studies will be needed to establish the relationship of the cloned α_2-receptor subtypes with α_2-receptors that have been described pharmacologically or functionally. For example, the deduced molecular weight of the α_2-C2 is 54,000, whereas the M_r of the neonatal rat lung α_2-receptor was found by SDS-PAGE to be approximately 44,000 (Lanier et al., 1988). Likewise the α_2-C4 was cloned from human kidney cDNA (Regan et al., 1988), so it should be expressed in peripheral tissues as well as in the brain. It is possible that α_2-C4 is entirely a neuronal α_2 and is associated with the kidney by way of innervation. In this regard, it is interesting to consider the presynaptic α_2-adrenergic receptor, as defined functionally by SKF 104078 (Daly et al., 1988; Table 4.1). In competition binding studies with the α_2-subtypes cloned to date, SKF 104078 had moderately high affinity (K_i's of 86, 33, and 100 nM; Table 4.3). This compares favorably with the affinity of SKF 104078 for the dog saphenous vein or the human platelet α_2-receptors (K_i's of 76 and 78 nM, respectively) and it is in contrast to its affinity at guinea pig atrial α_2-receptors which is very low ($K_i > 10\ \mu M$, Daly et al., 1988). These data suggest that neither α_2-C10 nor α_2-C4 nor α_2-C2 encodes the α_2-receptor as represented by the presynaptic receptor in guinea pig atria.

3.2.2. α_1-RECEPTOR SUBTYPES

Functional evidence of α_1-adrenergic receptor heterogeneity has now been corroborated by the cloning of α_1-receptor subtypes. The first cloning of one of these subtypes was from a bovine brain cDNA library with a probe derived originally from the cloned α_1-receptor of DDT_1MF-2 cells (Schwinn et al., 1990). As for the α_2-adrenergic receptor subtypes, homology between the α_1-adrenergic receptor subtypes in the transmembrane domains was high. Between the DDT_1MF-2 α_1-receptor and the bovine α_1-receptor subtype, the homology was 72%, whereas between the bovine α_1 and the α_2-C10 it was 42% and between the bovine α_1 and the human β_1 it was 43%. It is notable that, from the point of view of the primary sequence, the α_1 and α_2-adrenergic receptors are no more related to each other than they are to the β-adrenergic receptors. This suggests that many of the conserved residues do not dictate the ligand binding characteristics but that they may be necessary to preserve some basic secondary or tertiary structure. Other evidence showing that these cloned α_1-receptors were unique came from somatic cell hybridization analysis. Thus, the genes encoding these receptors were localized, respectively, to human chromosomes 5 and 8, using the cloned DDT_1MF-2 α_1-receptor and the bovine brain α_1-receptor as probes.

With respect to the pharmacological classification of the α_1-receptors into the A and B subtypes, the cDNA encoding the DDT_1MF-2 α_1-receptor has been expressed, and binding studies show the characteristics of the α_1-B subtype. Specifically, like the α_1-B, the DDT_1MF-2 receptor has low affinity for WB 4101 and phentolamine as compared with the α_1-A subtype (Table 4.3). In addition, like the pharmacologically defined α_1-B subtype, the DDT_1MF-2 α_1-receptor is completely inactivated by chlorethylclonidine (Schwinn et al., 1990). Further evidence suggesting that the cloned DDT_1MF-2 α_1-receptor corresponds to the α_1-B comes from Northern blot analyses of its tissue distribution. Table 4.5 reveals that mRNA levels for the DDT_1MF-2 α_1-receptor are high in the liver, heart, and cerebral cortex of the rat, similar to the reported distribution of the α_1-B.

The relationship of the cloned bovine α_1-adrenergic receptor, on the other hand, to the previously defined α_1-A and α_1-B subtypes is less certain (Schwinn et al., 1990). For example, although the affinities of the cloned bovine α_1-receptor for WB 4101 and phentolamine are high, as expected of the α_1-A subtype, the cloned bovine α_1 is also inactivated by chlorethylclonidine which is contrary to the expected result. Furthermore, the mRNA tissue distribution of the cloned bovine α_1-receptor (Table 4.5) indicates

TABLE 4.5. Northern blot analysis of α_1-adrenergic receptor subtypes in rat tissues[a]

	α_1-A (rat brain) (3.0 kb)[c]	α_1-B (DDT$_1$MF-2 cell) (2.4 kb)[c]	α_1-C[b] (bovine)
Cerebral cortex	+ + +	+ + +	0
Cerebellum	+	+ +	0
Brainstem	+ +	+ + +	0
Hippocampus	+ + +	+	0
Pituitary	0	0	0
Kidney	+	+ +	0
Liver	0	+ + + +	0
Lung	+ +	+ +	0
Aorta	+ + +	0	0
Heart	+ +	+ + + +	0
Skeletal muscle	+	0	0
Spleen	+ +	+	0
Adipose tissue	+	0	0
Vas deferens	+ + + +	0	0

[a]Derived from Lomasney et al. (1991). The data were analyzed in terms of the relative intensity of the bands observed after autoradiography. "0" means there was no clearly observable signal.
[b]A message of 3–4.5 kb has been detected in rabbit liver.
[c]Message size.

that it is not present in tissues where the pharmacologically defined α_1-A has been characterized. By default, therefore, the cloned bovine α_1-adrenergic receptor can be defined as the α_1-C.

A third α_1-adrenergic receptor subtype has been cloned from a rat brain cDNA library (Lomasney et al., 1991). This clone, RA42, has 73% amino acid sequence homology in the transmembrane spanning domains with the DDT$_1$MF-2 α_1-B and 65% homology with the bovine α_1-C. Pharmacological studies of this α_1-receptor suggest that it is the previously defined α_1-A (Morrow and Creese, 1986). Thus, it has higher affinity than the α_1-B for the antagonist WB 4101 and higher affinity for the agonists phenylephrine and methoxamine. Compelling evidence that distinguishes RA42 from the bovine α_1-C is its tissue distribution. In contrast to the bovine α_1-C, which is difficult to locate in any tissue, RA42 is present in RNA from the cerebral cortex, hippocampus, and vas deferens: tissues previously defined as containing the α_1-A. The sensitivity of the expressed RA42 to inactivation by chlorethylclonidine is also consistent with its assignment as the rat α_1-A.

4. Second Messenger Systems

Traditionally, the second messenger systems of the α_1- and α_2-adrenergic receptors have been thought to be the stimulation of phosphatidylinositol turnover and the inhibition of adenylyl cyclase (reviewed, Bylund and U'Pritchard, 1983). Evidence from a variety of sources, however, indicates the possibility of multiple coupling mechanisms. This, combined with the increased number of α-receptor subtypes and the heterogeneity of the G-proteins, is clearly increasing the complexity of this system. Nevertheless, careful studies with natural and with recombinant expression systems are helping to elucidate these coupling mechanisms.

One example with respect to the α_1-adrenergic receptor subtypes, utilized native receptors present in the rat vas deferens and in the spleen (Han et al., 1987). In these tissues, the activation of α_1-adrenergic receptors increases cytosolic calcium which leads to contraction of the smooth muscle. It was found, however, that these receptors represented different subtypes and that they stimulated contraction by different mechanisms. In the vas deferens, the predominant subtype was the α_1-A, and the contractile response did not depend upon the activation of phosphatidylinositol hydrolysis. In the spleen, however, the major subtype was the α_1-B, and in this tissue contraction was dependent upon the activation of the inositol phospholipid pathway. Similar studies have been reported using the native α_1-receptors in rabbit aortic smooth muscle cells and in rabbit hepatocytes (Tsujimoto et al., 1989). In both of these systems, α_1-receptor activation can stimulate glycogen phosphorylase by increasing intracellular calcium. It appears, however, that both the receptors and the mechanisms of signal transduction that are responsible for this effect differ in the two tissues. In the rabbit aortic smooth muscle, the α_1-A subtype predominates, and increased intracellular calcium depends upon both intra- and extracellular sources. Following receptor activation, there is an initial influx of calcium by a voltage-sensitive calcium channel, followed by release from stores inside the cell. In rat hepatocytes, however, the receptors are of the α_1-B variety and stimulation of phosphorylase depends only upon the release of intracellular calcium caused by increased hydrolysis of phosphatidylinositol.

In virtually every system examined to date, the activation of α_2-adrenergic-receptors inhibits adenylyl cyclase. However, in many cases the inhibition of adenylyl cyclase does not correlate with the physiological effects of receptor activation (Isom and Limbird, 1988) and several possible second messengers have recently been implicated. These messengers include the Na^+/H^+ exchanger (Sweatt et al., 1986), potassium channels (North and Surprenant, 1985), voltage-sensitive calcium channels (Illes and Dorge, 1985), intracellular calcium release (Michel et al., 1989), and phosphatidylinositol hydrolysis (Cotecchia et al., 1990a). Studies with native receptors and with cloned receptors expressed in cultured cell lines suggest that the interactions of a given receptor subtype with a particular second messenger may be cell- or tissue-specific.

In human platelets, the α_2-adrenergic receptors appear to be coupled to Na^+/H^+ exchange as well as to the inhibition of adenylyl cyclase (Sweatt et al., 1986). In HT-29 cells, however, what is ostensibly the same α_2-adrenergic receptor subtype (α_2-A; Bylund et al., 1988) couples to the inhibition of adenylyl cyclase but not to Na^+/H^+ exchange (Cantiello and Lanier, 1989). Interestingly, the latter showed that certain α_2-adrenergic agonists interacted directly with the Na^+/H^+ exchanger but not in a receptor-specific manner. In human erythroleukemia cells, α_2-adrenergic receptor activation (α_2-A) inhibits adenylyl cyclase and increases cytosolic calcium through the release of intracellular calcium (Michel et al., 1989). Although mediated by a pertussis toxin-sensitive G-protein, this increase in cytosolic calcium was independent of the inhibition of cyclase and did not depend upon Na^+/H^+ exchange. Coupling of the α_2-adrenergic receptor to Na^+/H^+ exchange was not specifically examined in these cells.

The α_2-C10 and α_2-C4 adrenergic receptor subtypes have been cloned into cultured hamster lung fibroblasts (PS120 cells) and have been characterized with respect to the inhibition of adenylyl cyclase and stimulation of phosphatidylinositol hydrolysis (Cotecchia et al., 1990a). In these cells, which lack endogenous α_2-adrenergic receptors, activation of either α_2-C10 or α_2-C4 inhibited cyclase and weakly stimulated the hydrolysis of phosphatidylinositol. As shown in Figure 4.6, although both the α_2-C10 and α_2-C4 receptor subtypes inhibited adenylyl cyclase, α_2-C10 appeared to be more efficient. Thus, agonist activation of α_2-C10 gave a maximal inhibition of 70%, whereas α_2-C4 inhibited the enzyme approximately 30%. This

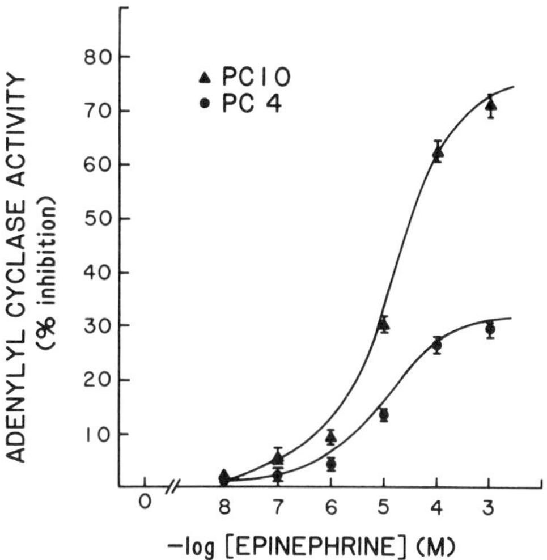

FIGURE 4.6. Inhibition of adenylyl cyclase by α_2-C10 (▲) and by α_2-C4 (●) in membranes prepared from stably transfected PS120 cells. Transfections and adenylyl cyclase assays were conducted as described (Cotecchia et al., 1990). Incubation of the membranes with epinephrine was for 15 min at 37°C. The receptor densities were 630 and 950 fmol/mg protein for α_2-C10 and α_2-C4, respectively. Reprinted with permission of the American Society for Biochemistry and Molecular Biology, from Cotecchia et al. (1990a).

apparent difference may be related to a more efficient interaction of α_2-C10 with inhibitory G-proteins. This has been examined by reconstitution of the cloned and partially purified α_2-C10 and α_2-C4 with purified recombinant G-proteins (Kurose et al., 1991). Using four different G-proteins, it was found that the order of potency for the interaction of both receptors was the same ($G_i3 > G_i1 > G_i2 > Go$) but that maximal stimulation of GTPase turnover was approximately 40% greater with α_2-C10 as compared with α_2-C4.

The α_2-C10 adrenergic receptor subtype has also been used to stably transfect cultured Chinese hamster ovary cells (CHO; Fraser et al., 1989). In these cells, receptor activation decreased intracellular cAMP but did not affect phosphatidylinositol hydrolysis. The decrease in cAMP was pertussis toxin-sensitive, which was also the case for the effects of both α_2-C10 and α_2-C4 on adenylyl cyclase and phosphatidylinositol hydrolysis in transfected PS120 cells (Cotecchia et al., 1990a). Surprisingly, in the CHO cells, while low concentrations of agonist decreased cAMP levels, high concentrations of agonist led to an increase in intracellular cAMP (Fraser et al., 1989). This effect, which was dependent on the level of receptor expression, was more pronounced in the presence of pertussis toxin. In the PS120 cells, the activation of α_2-C10 by high concentrations of agonist only slightly increased intracellular cAMP and this was only in the presence of pertussis toxin (Cotecchia et al., 1990a). As for the α_2-C4 subtype, cAMP concentrations were not increased either in the absence or presence of pertussis toxin. These observations suggest that, like other putative second messengers, the effect of increasing cAMP accumulation may be specific with respect to both the receptor subtype and the cell or tissue. Factors that contribute to this specificity could involve the types of G-proteins being expressed and/or the effector proteins themselves. To summarize, the α_2-C4 subtype has been shown to couple to the inhibition of adenylyl cyclase and to the stimulation of phosphatidylinositol turnover. The α_2-C10 subtype, or α_2-A, has been shown to couple to increases in intracellular calcium, inhibition of adenylyl cyclase, stimulation of phosphatidylinositol turnover, and increased cAMP accumulation. Some of these studies need to be investigated further to determine how they relate to physiological processes. In addition, it will have to be established if studies done with the pharmacologically defined α_2-A subtype actually represent the biochemically defined α_2-C10 subtype.

5. Structural Studies

The cloning of the adrenergic and other G-protein-coupled receptors has provided a major step toward understanding the structure and function of these proteins. The primary sequences are now known and understanding the secondary and tertiary structure, as well as the functional domains, has begun. Biochemical characteristics of the α-adrenergic receptors are listed in Table 4.6. The deduced primary sequences show that they consist potentially of 450 to 560 amino acids. This would yield proteins ranging in size from 50,000 to 62,000 Da: however, the full extent of posttranslational modifications are unknown at this time. It is known that many of the α-adrenergic receptors show N-linked glycosylation, although the α_2-C2 appears to be an exception. Fatty acylation by palmitic acid has been demonstrated for the β_2-adrenergic receptor (O'Dowd et al., 1989) but at present has not been examined for the α-adrenergic receptors. Other modifications, such as alternative splicing, which has been recently demonstrated for the dopamine receptors (Giros et al., 1989; Monsma et al., 1989), have not been documented in the α-receptors.

The construction of recombinant chimeric α_2/β_2-adrenergic receptors has shown the importance of i3 loop for coupling to adenylyl cyclase (Kobilka et al., 1988). In one such chimera, the α_2-C10 adrenergic receptor was modified by replacing the i3 loop, and portions of the adjacent membrane-spanning domains, with the corresponding region from the β_2-adrenergic receptor. In the presence of agonists, this chimera *stimulated* adenylyl cyclase with an α_2-adrenergic pharmacology (α_2-adrenergic receptors normally inhibit the enzyme). Similar studies have been done with α_1/β_2-

TABLE 4.6. Biochemical characteristics of the cloned α-adrenergic receptor subtypes

	α_2 Subtypes (human)			α_1 Subtypes		
	α_2-C10 (platelet α_2-AR)	α_2-C4	α_2-C2	α_1-A	α_1-B (DDT$_1$ MF-2 cell)	α_1-C (bovine)
Molecular size (M_r)						
SDS-PAGE	67000	75000	44000	?[a]	80000	?
Deduced	49500	50700	49500	61600	56600	51300
Amino acids						
Total	450	461	450	560	515	466
Amino terminal	34	52	13	93	48	27
Third intracell loop	158	151	179	74	71	68
Carboxy-terminal	21	22	21	161	164	137
Genes						
Human chromosome no.	10	4	2	5	5	8
Introns (coding seq)	0	?	0	?	Yes[b]	Yes
Glycosylation[c]	Yes (2)	Yes (2)	No (0)	2 Potential	Yes (4)	3 Potential
Phosphorylation[d]	β-ARK	?	?	?	PKC/PKA	?
Second message[e]	$-$AC/PI?/?	$-$AC/PI?/?	?	PI/Ca chan?	PI	PI

[a]?Means that no biochemical evidence is available for the expressed receptor.

[b]*Yes* means that at least one intron is present in the coding sequence.

[c]*Yes* means that independent biochemical evidence is available indicating that the expressed receptor is glycosylated. *No* means that independent biochemical evidence suggests that the expressed receptor is not glycosylated. The number in parentheses indicates the number of potential N-linked glycosylation sites present in the amino-terminal region of the receptor.

[d]Kinases that are known to be capable of phosphorylating the expressed receptor. β-ARK, β-adrenergic receptor kinase; PKC, protein kinase C (Ca dependent); PKA, protein kinase A (cAMP dependent).

[e]$-$AC, inhibition of adenylyl cyclase; PI, stimulation of phosphatidylinositol metabolism; Ca chan, opening of a voltage-dependent Ca channel; abbreviations followed by a ? means that evidence is available but the functional significance is still unknown; ? or /? means that there is no evidence to indicate what the second message is or that additional pathways may exist.

adrenergic receptor chimeras. Figure 4.7 shows results that were obtained
when the i3 loop of the β_2-receptor was replaced with the corresponding
loop of the α_1-B-adrenergic receptor (Cotecchia et al., 1990b). Normally,
stimulation of PI metabolism by the β_2-adrenergic receptor is negligible in
the presence of epinephrine; however, for this chimeric α_1/β_2-receptor, the
stimulation of PI metabolism was nearly indistinguishable from the native
α_1-receptor. These studies clearly establish the central role of the i3 loop in
coupling to second messenger systems.

Other chimeric receptors have shown the importance of the membrane-
spanning domains in ligand binding. For example, replacing the TM 7 of
the β_2-adrenergic receptor with α_2-receptor sequence yielded a chimeric
receptor that bound the α_2-selective antagonist [3H]yohimbine but did not
bind the β-selective antagonist [125I]cyanopindolol (CYP) (Kobilka et al.,
1988). Similarly, when TM 6 and 7 of the α_2-C10 were replaced with
β_2-receptor sequence, the resulting chimera bound [125I]CYP but not
[3H]yohimbine.

Photoaffinity labeling of purified human platelet α_2-adrenergic receptors
suggest that TM 4 is involved with ligand binding (Matsui et al., 1989).
These studies, which were based on peptide mapping, were consistent with
the hypothesis that the catechol moiety of epinephrine is directly involved
with binding to this region of the receptor. Site-directed mutagenesis studies
of the β_2-adrenergic receptor, on the other hand, suggest that the catechol

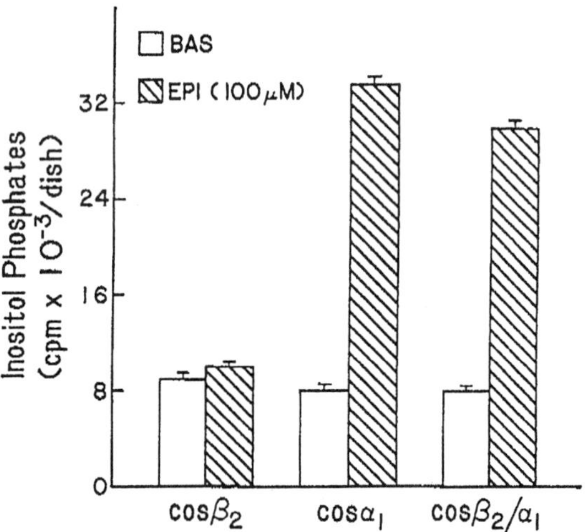

FIGURE 4.7. Stimulation of polyphosphoinositide metabolism by epinephrine in
COS-7 cells transiently transfected with DNA encoding either native β_2-, native
α_1-B-, or chimeric α_1-B/β_2-adrenergic receptors. The chimeric α_1-B/β_2-receptor
consisted of a human β_2-receptor in which the DNA encoding i3 loop was replaced
by the corresponding sequence from the α_1-B-adrenergic receptor. Reprinted with
permission of the National Academy of Science, from Cotecchia et al. (1990b).

hydroxyl groups interact with serine residues present in TM 5 rather than with serines present in TM 4 (Strader et al., 1989). These two studies are consistent with respect to the probable orientation of epinephrine when it binds to the adrenergic receptors. Thus, the amino group probably is binding to an acidic residue present in TM 3 while the catechol group interacts with TM 4 and/or TM 5. Studies with purified human platelet α_2-adrenergic receptors have implicated tyrosine (Nakata et al., 1986) and cysteine (Regan et al., 1986b) residues with either ligand binding or the formation of the ligand binding site. These residues, which are present in the sequence extending from the beginning of TM 4 to the end of TM 5, might be interesting candidates for mutagenesis studies.

Structural and functional studies of the α-adrenergic receptors are just beginning. Since they are also members of the larger class of G-protein-coupled receptors, it will be expected that some of the basic features will be shared. Many other features, however, such as pharmacological specificity, regulation, and coupling, are unique to the individual receptor subgroups. Understanding these unique features of the α-adrenergic receptors should yield important insight into these receptor proteins.

6. Update

Since the initial completion of this review, significant progress has been made in several areas with respect to our understanding of the molecular biology of α-adrenergic receptors. The following is a summary of this progress.

6.1. Cloning and Localization

In the area of cloning there has been considerable activity, especially as it relates to the identification of α_2-adrenergic receptor subtypes in the rat. Table 4.7 lists these studies along with information about which human subtype they might be equivalent to and their pharmacological classification. As can be seen, there is some confusion with respect to the pharmacological classification, spurred in part perhaps by the desire to find new subtypes but also by reported differences in apparent binding affinities. For example, although the deduced amino acid sequences of RG20 (Lanier et al., 1991) and cA-47 (Chalberg et al., 1990) are virtually identical, there are differences in their affinities for yohimbine. By competition curve analysis, RG20 has a K_i of 61 nM while cA-47 has a K_d of 2.1 nM as determined by the direct binding of [³H]yohimbine. Other studies (Harrison et al., 1991) verify the low affinity of yohimbine for RG20, and while there is no clear explanation for the high affinity of cA-47 for [³H]yohimbine, it is strange that 20 nM [³H]yohimbine was used for their competition curve studies

TABLE 4.7. Cloned α_2-adrenergic receptors

Nom de clone[a]	Type	Possible human equivalent[b]	Pharmacological classification[c]	Reference
Human platelet α_2	genomic	—	α_2-A	Kobilka et al. (1987b)
Human α_2-C2	genomic	—	α_2-B	Lomasney et al. (1990)
Human α_2-C4	kidney cDNA	—	α_2-B[d]	Regan et al. (1988)
Human α_2-B (5A)	genomic	α_2-C2	α_2-B	Weinshank et al. (1990)
Human α_2	genomic	α_2-C10	α_2-A	Fraser et al. (1989)
Rat RNGα_2	kidney cDNA	α_2-C2	α_2-B	Zeng et al. (1990)
Rat brain α_2 (cA-47)	brain cDNA	α_2-C10	α_2-A	Chalberg et al. (1990)
Rat RG20	genomic	α_2-C10	α_2-D	Lanier et al. (1991)
Rat α_2-C4	genomic/brain cDNA	α_2-C4	?[d]	Voigt et al. (1991)
Rat brain α_2-B	genomic/brain cDNA	α_2-C4	α_2-B[d]	Flordellis et al. (1991)
Rat RG10	genomic	α_2-C4	α_2-C	Lanier et al. (1991)
Porcine α_2	genomic	α_2-C10	α_2-A	Guyer et al. (1990)
Fish[e] α_2	genomic	α_2-C4	?	Svensson et al. (in preparation)

[a]As reported by referenced authors. Parentheses indicate an alternative name or clone number.
[b]As defined by chromosomal reference (see Section 3.2.1; also Regan et al., 1988).
[c]As reported by referenced authors (see Section 6.2; also Bylund 1988; Harrison et al., 1991). ? indicates that the authors were undecided or that it did not fit the existing classification.
[d]Now generally accepted as the α_2-C (see Section 6.2; also Harrison et al., 1991).
[e]Cuckoo wrasse *(Labrus ossifagus)*.

when the K_d was 2.1 nM. Nevertheless, these clones seem to be equivalent and appear to be the rat homologs of α_2-C10.

Similarly, the rat α_2-C4 (Voigt et al., 1991), the rat brain α_2-B (Flordellis et al., 1991) and RG10 (Lanier et al., 1991) seem to be the same and to be the rat homologs of α_2-C4. In all cases the pharmacology was thought to be "B-like" but on closer inspection was consistent with that of the OK cell α_2-C (Harrison et al., 1991). This is supported by Northern blot analysis which shows that OK cell RNA hybridizes preferentially to DNA encoding the α_2-C4 as opposed to the α_2-C10 or α_2-C4 (Lorenz et al., 1990). Northern blot analyses also continue to show a predominant localization of the α_2-C4 or its rat homolog in the CNS (Zeng and Lynch, 1991); however, α_2-C4 message has now been detected in rat heart (Voigt et al., 1991) and kidney (Voigt et al., 1991; Flordellis et al., 1991). An α_2-receptor that is associated with the sympathetic innervation of fish pigment cells (FPCα_2) has been cloned (Svensson et al., in preparation) and, based upon overall amino acid homology, is most closely related to the α_2-C4. Thus, it shows 56% homology with the α_2-C4, 46% homology with the α_2-C2 and 49% homology with the α_2-C10. This is much less than the homology that exists between the human α_2-C4 and its rat homolog. In the latter comparison there is 88% overall amino acid homology. When the rat α_2-C4 is compared to the human α_2-C2, the homology is 55% and when it is compared to the human α_2-C10, the homology is 58%. The presumed localization of the FPCα_2 appears to support the relationship of α_2-C4 with neuronal structures. With respect to its pharmacology, the FPCα_2 is interesting, having properties of both the α_2-A and α_2-C subtypes.

6.2. Functional Expression

The three presently known α_2-adrenergic receptors have been cloned from rat, expressed transiently in COS-1 cells, and their pharmacological properties studied (Harrison et al., 1991). This side-by-side comparison has also been done with the three human α_2-receptors expressed in COS-7 cells (Lomasney et al., 1990). In the former study, [^{3}H]rauwolscine was used as the radioligand while [^{3}H]yohimbine was used in the latter. The results of these studies are generally consistent with the assignment of the human α_2-C10/rat RG20 as the α_2-A, the human α_2-C4/RG10 as the α_2-C, and the human α_2-C2/rat RNGα_2 as the α_2-B. There are, however, some interesting differences between the rat and human α_2-receptors. For example, as judged by their K_i's, rauwolscine and yohimbine have significantly lower affinity for the rat α_2-A as compared with the human α_2-A (30–40 nM versus 2–7 nM). Also, corynanthine has moderate affinity for the rat α_2-B ($\sim$270 nM) but lower affinity for the human α_2-B ($\sim$1000 nM). Some of these species differences may help explain earlier results and they strengthen the possibility that the rauwolscine insensitive receptors identified in rat brain by autoradiography might represent the α_2-A subtype, while the rauwolscine sensitive receptors would represent the α_2-C subtype (Section 3.2.1; Boyajian et al., 1987).

α_2-Adrenergic receptor subtypes have been expressed stably in several cell lines and in some cases with fairly impressive results in terms of the level of expression. The porcine α_2-A has been subcloned into the plasmid pCMV4 and expressed in LLC-PK1 cells, a cell line derived from porcine kidney (Guyer et al., 1990). The cells were transfected using calcium phosphate precipitation, selected by resistance to neomycin, and a clonal line was obtained that bound [^{3}H]yohimbine with high affinity and with a receptor density of 109 pmol/mg protein. This is 500–600 times the receptor density that is present in human platelets, a tissue which is considered to be a good source of the α_2-A.

Using the human α_2-A, and the plasmid pBMT3X, a stably expressing α_2 cell line was obtained in mouse C127 cells (Bresnahan et al., 1990). Selection in this case was done on the basis of resistance to cadmium, and clones were obtained that had receptor densities in the range of 18–35 pmol/mg protein. Epinephrine could inhibit forskolin-stimulated cAMP formation, but at high concentrations the effect was reversed. However, since epinephrine also stimulated cAMP formation in nontransfected cells, it was not clear what this reversal meant. Guyer et al. (1990) did not report if reversal was observed with high concentrations of agonist in their pig cell line. In CHO cells transfected with the human α_2-A, epinephrine also produced a biphasic response with respect to forskolin-stimulated cAMP production (Section 4; Fraser et al., 1989). Using the same cell line this has been studied in more detail (Jones et al., 1991), and both the inhibition of forskolin-stimulated cAMP production by low concentrations of epinephrine and the potentiation by higher concentrations could be blocked with yohimbine. The inhibitory phase, but not the stimulatory phase, could also be blocked by pretreatment of the cells with pertussis toxin, suggesting that these effects are mediated by different second messenger systems.

The transcriptional regulation of α_2-adrenergic receptor expression has been found to be regulated by cAMP in HT-29 cells (Sakaue and Hoffman, 1991). These cells, which are derived from a human colonic adenocarcinoma, express endogenous α_2-adrenergic receptors that have been classified pharmacologically (Bylund et al., 1988) and by Northern blot analysis (Lorenz et al., 1990) to be of the α_2-A subtype. Exposing cells to the cAMP analog, dibutyryl-cAMP, increased α_2-A mRNA levels by approximately 5-fold. Similar results were also obtained when cAMP levels were increased with forskolin or VIP. In addition to the effects on message levels, dibutyryl-cAMP increased receptor density and the rate of α_2-A receptor gene transcription. Sakaue and Hoffman (1991) concluded that in HT-29 cells α_2-A receptor expression is regulated by cAMP, probably via cAMP-dependent protein kinase, and that 5′-flanking sequences of the α_2-A gene are important for this regulation. Sequences that are similar to known cAMP response elements have been noted in the 5′ untranslated region of the human α_2-A (Fraser et al., 1989), but whether or not these were responsible for the findings above was not examined.

6.3. Mutagenesis/Structural Studies

A series of mutations have been made in the human α_2-A that involved conserved aspartic acid and serine residues which may be involved in ligand binding to the receptor (Wang et al., 1991). These mutations, which are similar to mutations that have been made in the β-adrenergic receptor (Dixon et al., 1988), support the hypothesis that asp^{113} in TM 3 is binding to the amino group of catecholamines and that serine residues in TM 5 are binding to the catechol hydroxyl groups. If the hypothesis is correct, then Ser^{200} and Ser^{204} in the α_2-adrenergic receptor would be involved in binding to the meta and para hydroxyls, respectively. Wang et al. (1991) find evidence supporting the interaction of the para hydroxyl with Ser^{204} but not for the interaction of Ser^{200} with the meta hydroxyl.

An interesting mutation has been done wherein Phe^{412} in TM 7 of the human α_2-A has been changed to an Asn (Suryanarayana et al., 1991). This mutation caused a significant decrease in the affinity of the receptor for yohimbine and a dramatic increase in the affinity of the receptor for alprenolol and other related β-selective antagonists. The increase in affinity for alprenolol was 3,000-fold, going from a K_i of 6,600 nM in the wild-type α_2-receptor to 2.1 nM in the mutated receptor. The β-adrenergic and 5HT$_{-1A}$ receptors also bind alprenolol with high affinity and have an asparagine residue at an equivalent position, suggesting that certain antagonist interactions may be occurring at this position in TM 7. The affinity of agonists was generally decreased, but there was no clear pattern with respect to β- verus α-adrenergic specificity.

Ligand binding to α_2-adrenergic receptors is influenced by sodium and hydrogen ions acting via allosteric binding sites on the receptor. Proteolysis of purified porcine α_2-A adrenergic receptors has been used to try to localize where these sites might be (Wilson et al., 1990). Digestion of the receptor with trypsin, followed by repurification over wheat germ agglutinin-agarose, yields a hydrophobic core which is believed to consist of the amino terminus and transmembrane domains 1–5. This fragment binds [^{3}H]yohimbine and the binding is modulated by Na$^+$ and H$^+$. The authors propose that the allosteric effect occurs through interactions of monovalent ions with acidic amino acids in the fragment rather than with the i3 loop.

Although acidic amino acids in the i3 loop may not be involved in the allosteric effects of monovalent ions, they may be part of a consensus sequence required for phosphorylation of the α_2-adrenergic receptor by the β-adrenergic receptor kinase (β-ARK; Onorato et al., 1991). In vitro, β-ARK is capable of phosphorylating the human platelet α_2-A in an agonist-dependent fashion (Benovic et al., 1987). Interestingly, in contrast to the β-receptor where phosphorylation occurs in the carboxyl terminus, phosphorylation of the α_2-receptor probably takes place in the i3 loop since there are no serines or threonines in the carboxyl terminus of the α_2. To test the possibility that phosphorylation of the α_2-receptor takes place on serine residues present in an acidic region of the i3 loop, a peptide was synthesized corresponding to such a region of the human platelet α_2-A (Onorato et al., 1991). In the presence of β-ARK, the peptide served as a substrate for phosphorylation; however, phosphorylation did not occur when the acidic amino acids were changed to asparagine or glutamine. It was also shown that the requirement for acidic residues was on the amino terminal side of the serine(s) being phosphorylated.

With respect to the generally accepted model of G-protein-coupled receptors, with their seven transmembrane domains and loops (Figure 4.3), there is little direct evidence that actually supports this, especially as it concerns the α-adrenergic receptors. Using antibodies directed against the i3 loop of the human α_2-A, however, it has been found that the i3 loop has a cytoplasmic orientation (Schullery et al., in preparation). Thus, in cells transfected with the human α_2-A, indirect immunofluorescent labeling required permeabilization with detergent. This is consistent with the TM 7 spanning model and is supported by similar results obtained with antipeptide antibodies directed against selected regions of the β-adrenergic receptor (Wang et al., 1989).

Acknowledgments

The authors thank Jon Lomasney, Wulfing Lorenz, and Debra Schwinn for informative discussions concerning the cloning of the rat α_1-A-adrenergic receptor and for the Northern blot analyses of the α-adrenergic receptor subtypes. We thank Lee Limbird for information concerning the cloning of the porcine α_2-A, and Richard Weinshank, Paul Hartig, and Kevin Lynch for discussions concerning the cloning of the α_2-B adrenergic receptor. The authors extend their deepest appreciation to Bob Lefkowitz and Marc Caron who made much of this work possible. John W. Regan also acknowledges the support of Allergan Pharmaceuticals, the Pharmaceutical Manufacturers Association, the American Heart Association, and the National Institutes of Health.

References

Ahlquist RP (1948): A study of the adrenotropic receptors. *Am J Physiol* 153:586–600

Benovic JL, Regan JW, Matsui H, Mayor F, Cotecchia S, Lundberg LMFL, Caron MG, Lefkowitz RJ (1987): Agonist-dependent phosphorylation of the α_2-adrenergic receptor by the β-adrenergic receptor kinase. *J Biol Chem* 262:17251–17253

Berthelsen S, Pettinger WA (1977): A functional basis for classification of α-adrenergic receptors. *Life Sci* 21:595–606

Boyajian CL, Loughlin SE, Leslie FM (1987): Anatomical evidence for alpha-2 adrenoceptor heterogeneity: Differential autoradiographic distributions of [^{3}H]rauwolscine and [^{3}H]idazoxan in rat brain. *J Pharmacol Exp Ther* 241:1079–1091

Bresnahan MR, Flordellis CS, Vassilatis DK, Makrides SC, Zannis VI, Gavras H (1990): High level of expression of functional human platelet α_2-adrenergic receptors in a stable mouse C127 cell line. *Biochim Biophys Acta* 1052:439–445

Bylund DB, U'Prichard DC (1983): Characterization of alpha-1 and alpha-2 adrenergic receptors. *Int Rev Neurobiol* 24:343–431

Bylund DB (1985): Heterogeneity of alpha-2 adrenergic receptors. *Pharmacol Biochem Behav* 22:835–843

Bylund DB, Ray-Prenger C, Murphy TJ (1988): Alpha-2A and alpha-2B adrenergic receptor subtypes: Antagonist binding in tissues and cell lines containing only one subtype. *J Pharmacol Exp Ther* 245:600–607

Cantiello HF, Lanier SM (1989): α_2-Adrenergic receptors and the Na^+/H^+ exchanger in the intestinal epithelial cell line, HT-29. *J Biol Chem* 264:16000–16007

Cerione RA, Regan JW, Nakata H, Codina J, Benovic JL, Gierschik P, Somers RL, Spiegel AM, Birnbaumer L, Lefkowitz RJ, Caron MG (1986): Functional reconstitution of the α_2-adrenergic receptor with guanine nucleotide regulatory proteins in phospholipid vesicles. *J Biol Chem* 261:3901–3909

Chalberg SC, Duda T, Rhine JA, Sharma RK (1990): Molecular cloning, sequencing and expression of an α_2-adrenergic receptor complementary DNA from rat brain. *Mol Cell Biochem* 97:161–172

Cheung Y, Barnett DB, Nahorski SR (1982): [^{3}H]Rauwolscine and [^{3}H]yohimbine binding to rat cerebral and human platel membranes: Possible heterogeneity of α_2-adrenoceptors. *Eur J Pharmacol* 84:79–85

Connaughton S, Docherty JR (1990): No evidence for differences between pre- and postjunctional α_2 adrenoceptors in the periphery. *Br J Pharmacol* 99:97–102

Cotecchia S, Schwinn DA, Randall RR, Lefkowitz RJ, Caron MG, Kobilka BK (1988): Molecular cloning and expression of the cDNA for the hamster α_1-adrenergic receptor. *Proc Natl Acad Sci USA* 85:7159–7163

Cotecchia S, Kobilka BK, Daniel KW, Nolan RD, Lapetina EY, Caron MG, Lefkowitz RJ, Regan JW (1990a): Multiple second messenger pathways of α-adrenergic receptor subtypes expressed in eukaryotic cells. *J Biol Chem* 265:63–69

Cotecchia S, Exum S, Caron MG, Lefkowitz RJ (1990b): Regions of the α_1-adrenergic receptor involved in coupling to phosphatidylinositol hydrolysis and enhanced sensitivity of biological function. *Proc Natl Acad Sci USA* 87:2896–2900

Cullen BR (1987): Use of eukaryotic expression technology in the functional analysis of cloned genes. *Meth Enzymol* 152:684–704

Daly RN, Sulpizio AC, Levitt B, DeMarinis RM, Regan JW, Ruffolo RR, Hieble JP (1988): Evidence for heterogeneity between pre- and postjunctional α_2-adrenoceptors using 9-substituted 3 benzazepines. *J Pharmacol Exp Ther* 247:122–128

Dixon RAF, Kobilka BK, Strader DJ, Benovic JL, Dohlman HG, Frielle T,

Bolanowski MA, Bennett CD, Rands E, Diehl RE, Mumford RA, Slater EE, Sigal IS, Caron MG, Lefkowitz RJ, Strader CD (1986): Cloning of the gene and cDNA for mammalian β-adrenergic receptor and homology with rhodopsin. *Nature* 321:75–79

Dixon RAF, Sigal IS, Strader CD (1988): Structure-function analysis of the β-adrenergic receptor. *Cold Spring Harbor Symp Quant Biol* 53:487–497

Dohlman HG, Caron MG, Lefkowitz RJ (1987): A family of receptors coupled to guanine nucleotide regulatory proteins. *Biochemistry* 26:2657–2664

Dubocovich ML, Langer SZ (1974): Negative feed-back regulation of noradrenaline release by nerve stimulation in the perfused cat's spleen: Differences in potency of phenoxybenzamine in blocking the pre- and post-synaptic adrenergic receptors. *J Physiol (Lond)* 237:505–519

Feller DJ, Bylund DB (1984): Comparison of alpha-2 adrenergic receptors and their regulation in rodent and porcine species. *J Pharmacol Exp Ther* 228:275–281

Flordellis CS, Handy DE, Bresnahan MR, Zannis VI, Gavras H (1991): Cloning and expression of a rat brain α_{2B}-adrenergic receptor. *Proc Natl Acad Sci USA* 88:1019–1023

Fraser CM, Arakawa S, McCombie WR, Venter JC (1989): Cloning, sequence analysis, and permanent expression of a human α_2-adrenergic receptor in Chinese hamster ovary cells. *J Biol Chem* 264:11754–11761

Giros B, Sokoloff P, Martres M, Riou J, Emorine LJ, Schwartz J (1989): Alternative splicing directs the expression of two D2 dopamine receptor isoforms. *Nature* 342:923–926

Graham RM, Hess H, Homcy CJ (1982): Biophysical characterization of the purified α_1-adrenergic receptor and identification of the hormone binding subunit. *J Biol Chem* 257:15174–15181

Guyer CA, Horstman DA, Wilson AL, Clark JD, Cragoe EJ, Limbird LE (1990): Cloning, sequencing and expression of the gene encoding the porcine α_2-adrenergic receptor. *J Biol Chem* 265:17307–17317

Haga T, Haga K (1980): Characterization of alpha-adrenergic receptor subtypes in rat brain: Estimation of ability of adrenergic ligands to displace ^{3}H-dihydroergocryptine from the receptor subtypes. *Life Sci* 26:211–218

Han C, Abel PW, Minneman KP (1987): α_1-Adrenoceptor subtypes linked to different mechanisms for increasing intracellular Ca^{2+} in smooth muscle. *Nature* 329:333–335

Harrison JK, D'Angelo DD, Zeng D, Lynch KR (1991): Pharmacological characterization of rat α_2-adrenergic receptors. *Mol Pharmacol* 40:407–412

Hieble JP, Sulpizio AC, Nichols AJ, Willette RN, Ruffolo RR (1988): Pharmacologic characterization of SKF 104078, a novel alpha-2 adrenoceptor antagonist which discriminates between pre- and postjunctional alpha-2 adrenoceptors. *J Pharmacol Exp Ther* 247:645–652

Hieble JP, DeMarinis RM, Matthews WD (1986): Evidence for and against heterogeneity of alpha$_1$-adrenoceptors. *Life Sci* 38:1339–1350

Hoffman BB, Lefkowitz RJ (1980): [3]HWB4101—Caution about its role as an alpha-adrenergic subtype selective radioligand. *Biochem Pharmacol* 29:1537–1541

Illes P, Dorge L (1985): Mechanism of α_2-adrenergic inhibition of neuroeffector transmission in the mouse vas deferens. *Naunyn-Schmiedebergs Arch Pharmacol* 328:241–247

Isom LL, Limbird LE (1988): What happens next? In: *The Alpha-2 Adrenergic Receptors,* Limbird LE, ed. Clifton, N.J.; Humana Press, pp 323–363

Jones SB, Halenda SP, Bylund DB (1991): α_2-Adrenergic receptor stimulation of phospholipase A2 and of adenylate cyclase in transfected chinese hamster ovary cells is mediated by different mechanisms. *Mol Pharmacol* 39:239–245

Kobilka BK, MacGregor C, Daniel K, Kobilka TS, Caron MG, Lefkowitz RJ (1987a): Functional activity and regulation of human β_2-adrenergic receptors expressed in *Xenopus* oocytes. *J Biol Chem* 262:15796–15802

Kobilka BK, Matsui H, Kobilka TS, Yang-Feng TL, Francke U, Caron MG, Lefkowitz RJ, Regan JW (1987b): Cloning, sequencing and expression of the gene coding for the human platelet α_2-adrenergic receptor. *Science* 238:650–656

Kobilka BK, Kobilka TS, Daniel K, Regan JW, Caron MG, Lefkowitz RJ (1988): Chimeric α_2-β_2-adrenergic receptors: Delineation of domains involved in effector coupling and ligand binding specificity. *Science* 240:1310–1316

Kubo T, Fukuda K, Mikami A, Maeda A, Takahashi H, Mishina M, Haga T, Haga K, Ichiyama A, Kangawa K, Kojima M, Matsuo H, Hirose T, Numa S (1986): Cloning sequencing and expression of complementary DNA encoding the muscarinic acetylcholine receptor. *Nature* 323:411–416

Kunos G, Kan WH, Greguski R, Venter JC (1983): Selective affinity labeling and molecular characterization of hepatic α_1-adrenergic receptors with [^{3}H]phenoxybenzamine. *J Biol Chem* 258:326–332

Kurose H, Regan JW, Caron MG, Lefkowitz RJ (1991): Functional interactions of recombinant α_2 adrenergic receptor subtypes and G proteins in reconstituted phospholipid vesicles. *Biochemistry* 30:3335–3341

Lambert GA, Lang WJ, Friedman E, Meller E, Gershon S (1978): Pharmacological and biochemical properties of isomeric yohimbine alkaloids. *Eur J Pharmacol* 49:39–48

Lanier SM, Homcy CJ, Patenaude C, Graham RM (1988): Identification of structurally distinct α_2-adrenergic receptors. *J Biol Chem* 263:14491–14496

Lanier SM, Downing S, Duzic E, Homcy CJ (1991): Isolation of rat genomic clones encoding subtypes of the α_2-adrenergic receptor. *J Biol Chem* 266:10470–10478

Leeb-Lundberg LMF, Dickinson KEJ, Heald SL, Wikberg JES, DeBernardis JF, Winn M, Arendsen DL, Lekowitz RJ, Caron MG (1983): Covalent labeling of the cerebral cortex α_1-adrenergic receptor with a new high affinity radioiodinated photoaffinity probe. *Biochem Biophys Res Commun* 115:946–951

Leeb-Lundberg LMF, Dickinson KEJ, Heald SL, Wikberg JES, Hagen P, DeBernardis JF, Winn M, Arendsen DL, Lekowitz RJ, Caron MG (1984): Photoaffinity labeling of mammalian α_1-adrenergic receptors. *J Biol Chem* 259:2579–2587

Lomasney JW, Leeb-Lundberg LMF, Cotecchia S, Regan JW, DeBernardis JF, Caron MG, Lefkowitz RJ (1986): Mammalian α_1-adrenergic receptor: Purification and characterization of the native receptor ligand binding subunit. *J Biol Chem* 261:7710–7716

Lomasney JW, Lorenz W, Allen LF, King K, Regan JW, Yang-Feng TL, Caron MG, Lefkowitz RJ (1990): Expansion of the alpha$_2$-adrenergic receptor family: Cloning and characterization of a human alpha$_2$-adrenergic receptor subtype, the gene for which is located on chromosome 2. *Proc Natl Acad Sci USA* 87:5094–5098

Lomasney JW, Cotecchia S, Lorenz W, Leung WY, Schwinn DA, Yang-Feng TL, Brownstein M, Lefkowitz RJ, Caron MG (1991): Molecular cloning and expression of the cDNA for the α_{1A}-adrenergic receptor. *J Biol Chem* 266:6365–6369

Lorenz W, Lomasney JW, Collins S, Regan JW, Caron MG, Lefkowitz (1990): Expression of three α_2-adrenergic receptor subtypes in rat tissues: Implications for α_2 receptor classification. *Mol Pharmacol* 38:599–603

Matsui H, Lefkowitz RJ, Caron MG, Regan JW (1989): Localization of the fourth membrane spanning domain as a ligand binding site in the human platelet α_2-adrenergic receptor. *Biochemistry* 28:4125–4130

McGrath JC (1982): Evidence for more than one type of postjunctional α-adrenoceptor. *Biochem Pharmacol* 31:467–484

Michel MC, Brass LF, Williams A, Bokach GM, LaMorte VJ, Motulsky HJ (1989): α_2-Adrenergic receptor stimulation mobilizes intracellular Ca^{2+} in human erythroleukemia cells. *J Biol Chem* 264:4986–4991

Minneman KP (1988): α_1-Adrenergic receptor subtypes, inositol phosphates, and sources of cell Ca^{2+}. *Pharmacol Rev* 40:87–119

Minneman KP, Han C, Abel PW (1988): Comparison of α_1-adrenergic receptor subtypes distinguished by chlorethylclonidine and WB 4101. *Mol Pharmacol* 33:509–514

Monsma FJ, McVittie LD, Gerfen CR, Mahan LC, Sibley DR (1989): Multiple D2 dopamine receptors produced by alternative RNA splicing. *Nature* 342:926–929

Morrow AL, Creese I (1986): Characterization of α_1-adrenergic receptor subtypes in rat brain: A reevaluation of [^{3}H]WB4104 and [^{3}H]prazosin binding. *J Pharmacol Exp Ther* 29:321–330

Nakata H, Regan JW, Lefkowitz RJ (1986): Chemical modification of α_2-adrenoceptors: Possible role for tyrosine in the ligand binding site. *Biochem Pharmacol* 35:4089–4094

North RA, Surprenant A (1985): Inhibitory synaptic potentials resulting from α_2-adrenoceptor activation in guinea-pig submucous plexus neurones. *J Physiol (Lond)* 358:17–33

O'Dowd BF, Hnatowich M, Caron MG, Lefkowitz RJ, Bouvier M (1989): Palmitoylation of the human β_2-adrenergic receptor. *J Biol Chem* 264:7564–7569

Onorato JJ, Palczewski K, Regan JW, Caron MG, Lefkowitz RJ, Benovic JL (1991): Role of acidic amino acids in peptide substrates of the β-adrenergic receptor kinase and rhodopsin kinase. *Biochemistry* 30:5118–5125

Peralta EG, Winslow JW, Peterson GL, Smith DH, Ashkenazi A, Ramachandran J, Schimerlik MI, Capon DJ (1987): Primary structure and biochemical properties of an M2 muscarinic receptor. *Science* 236:600–605

Petrash AC, Bylund DB (1986): Alpha-2 adrenergic receptor subtypes indicated by [^{3}H]yohimbine binding in human brain. *Life Sci* 38:2129–2137

Regan JW (1988): Biochemistry of alpha-2 adrenergic receptors. In: *The Alpha-2 Adrenergic Receptors,* Limbird LE, ed. Clifton, N.J.: Humana Press, pp. 15–74

Regan JW, Matsui H (1990): α_2-Adrenergic receptor purification. In: *Receptor Biochemistry: A Practical Approach,* Hulme EC, ed. London: Oxford University Press

Regan JW, DeMarinis RM, Caron MG, Lefkowitz RJ (1984): Identification of the subunit-binding site of α_2-adrenergic receptors using [^{3}H]phenoxybenzamine. *J Biol Chem* 259:7864–7869

Regan JW, Raymond RJ, Lefkowitz RJ, DeMarinis RM (1986a): Photoaffinity labeling of human platelet and rabbit kidney α_2-adrenoceptors with ^{3}H-SKF102229. *Biochem Biophys Res Commun* 137:606–613

Regan JW, Nakata H, DeMarinis RM, Caron MG, Lefkowitz RJ (1986b):

Purification and characterization of the human platelet α_2-adrenergic receptor. *J Biol Chem* 261:3894–3900

Regan JW, Kobilka TS, Yang-Feng TL, Caron MG, Lefkowitz RJ, Kobilka BK (1988): Cloning and expression of a human kidney cDNA for an α_2-adrenergic receptor subtype. *Proc Natl Acad Sci USA* 85:6301–6305

Repaske MG, Nunnari JM, Limbird LE (1987): Purification of the alpha-2 adrenergic receptor from porcine brain using a yohimbine-agarose affinity matrix. *J Biol Chem* 262:12381–12386

Sakaue M, Hoffman BB (1991): cAMP regulates transcription of the α2A adrenergic receptor gene in HT-29 cells. *J Biol Chem* 266:5743–5749

Sawutz DG, Lanier SM, Warren CD, Graham RM (1987): Glycosylation of the mammalian α_1-adrenergic receptor by complex type N-linked oligosaccharides. *Mol Pharmacol* 32:565–571

Schullery D, Huang Y, Regan JW: Antibodies to the v–v1 loop of the α_2-C10 confirm its intracellular orientation. In preparation

Schwinn DA, Lomasney JW, Lorenz W, Szklut PJ, Fremeau RT, Yang-Feng TL, Caron MG, Lefkowitz RJ, Cotecchia S (1990): Molecular cloning and expression of the cDNA for a novel α_1-adrenergic receptor subtype. *J Biol Chem* 265:8183–8189

Starke K (1981): Alpha adrenoceptor subclassification. *Rev Physiol Biochem Pharmacol* 88:199–236

Strader CD, Candelore MR, Hill WS, Sigal IS, Dixon RAF (1989): Identification of two serine residues involved in agonist activation of the β-adrenergic receptor. *J Biol Chem* 264:13572–13578

Suryanarayana S, Daunt DA, Zastrow MV, Kobilka BK (1991): A point mutation in the seventh hydrophobic domain of the α_2 adrenergic receptor increases its affinity for a family of β receptor antagonists. *J Biol Chem* 266:15488–15492

Svensson S, Bailey TJ, Regan JW: Molecular cloning of a fish pigment cell α_2-adrenergic receptor. In preparation

Sweatt JD, Blair IA, Cragoe EJ, Limbird LE (1986): Inhibitors of Na^+/H^+ exchange block epinephrine- and ADP-induced stimulation of human platelet phospholipase C by blockade of arachidonic acid release at a prior step. *J Biol Chem* 261:8660–8666

Timmermans PBMWM, Hoefke W, Stahle H, van Zwieten PA (1980): Structure-activity relationships in clonidine-like imidazolidines and related compounds. *Prog Pharmacol* 3:21–62

Tsujimoto G, Tsujimoto A, Suzuki E, Hashimoto K (1989): Glycogen phosphory-lase activation by two different α_1-adrenergic receptor subtypes: Methoxamine selectively stimulates a putative α_1-adrenergic receptor subtype (α_{1a}) that couples with Ca^{2+} influx. *Mol Pharmacol* 36:166–176

U'Prichard DC, Greenberg DA, Snyder SH (1977): Binding characteristics of a radiolabeled agonist and antagonist at central nervous system alpha noradrenergic receptors. *Mol Pharmacol* 13:454–473

Voigt MM, McCune SK, Kanterman RY, Felder CC (1991): The rat α_2-C4 adrenergic receptor gene encodes a novel pharmacological subtype. *FEBS Lett* 278:45–50

Wang CD, Buck MA, Fraser CM (1991): Site-directed mutagenesis of α_{2A}-adrenergic receptors: Identification of amino acids involved in ligand binding and receptor activation by agonists. *Mol Pharmacol* 40:168–179

Wang HY, Lipfert L, Malbon CC, Bahouth S (1989): Site-directed anti-peptide antibodies define the topography of the β-adrenergic receptor. *J Biol Chem* 264:14424–14431

Weinshank RL, Zgombick JM, Macchi M, Adham N, Lichtblau H, Branchek TA, Hartig PR (1990): Cloning, expression and pharmacological characterization of a human α_{2B}-adrenergic receptor. *Mol Pharmacol* 38:681–688

Weitzell R, Tanaka T, Starke K (1979): Pre- and postsynaptic effects of yohimbine stereoisomers on noradrenergic transmission in the pulmonary artery of the rabbit. *Naunyn Schmiedebergs Arch Pharmacol* 308:127–136

Wilson AL, Guyer CA, Cragoe EJ, Limbird LE (1990): The hydrophobic tryptic core of the porcine α_2-adrenergic receptor retains allosteric modulation of binding by Na^+, H^+, and 5-amino-substituted amiloride analogs. *J Biol Chem* 265:17318–17322

Yarden Y, Rodriguez H, Wong SKF, Brandt DR, May DC, Burnier J, Harkins RN, Chen EY, Ramachandran J, Ullrich A, Ross EM (1986): The avian β-adrenergic receptor: Primary structure and membrane topology. *Proc Natl Acad Sci USA* 83:6795–6799

Zeng D, Harrison JK, D'Angelo DD, Barber CM, Tucker AL, Lu Z, Lynch KR (1990): Molecular characterization of a rat α_2B-adrenergic receptor. *Proc Natl Acad Sci USA* 87:3102–3106

Zeng D, Lynch KR (1991): Distribution of α_2-adrenergic receptor mRNAs in the rat CNS. *Mol Brain Res* 10:219–225

5

The 5-HT$_{1A}$ Receptor: From Molecular Characteristics to Clinical Correlates

JOHN R. RAYMOND, SALAH EL MESTIKAWY, AND ANNICK FARGIN

1. Introduction and Historical Perspective

The past decade has seen a remarkable growth in our understanding of the pharmacology and physiology of the various receptors for serotonin (5-hydroxytryptamine, 5-HT). Since 1979, when radioligand binding techniques were used to distinguish subtypes of 5-HT binding sites (Peroutka and Snyder, 1979), no less than ten subtypes of 5-HT receptors have been characterized (Richardson and Engel, 1986; Bradley et al., 1986; Mawe et al., 1986; Heuring and Peroutka, 1987; Dumuis et al., 1988b; Leonhardt et al., 1989; Conner and Monsour, 1990). However, the biochemical characterization of these receptors has been hampered by the lack of selective radioligands and/or cell lines expressing single well-characterized receptor subtypes.

Nowhere have the difficulties been more apparent than with the 5-HT$_{1A}$ receptor. Despite active development of radioligands (Gozlan et al., 1983, 1988a; Norman et al., 1985; Dompert et al., 1985; Moon and Taylor, 1985; Glaser and Traber, 1985; Ransom et al., 1986a; Cossery et al., 1987; Nelson et al., 1987; Herrick-Davis and Titeler, 1988; Gallagher and Wang, 1988), photoaffinity labels (Ransom et al., 1986b; Emerit et al., 1987; Gozlan et al., 1988b), and solubilization protocols (Asarch and Shih, 1987; Gozlan et al., 1987; El Mestikawy et al., 1988), attempts to purify the receptor were largely unsuccessful until recently (El Mestikawy et al., 1989). Fortunately, molecular biology techniques have proved to be powerful adjuncts to classic biochemical and pharmacological approaches to the characterization of this receptor subtype. The genes encoding this receptor have been cloned from human and rat by two different groups using low-stringency cross-hybridization (Kobilka et al., 1987b; Fargin et al., 1988; Albert et al., 1990).

Since this receptor is apparently well conserved among mammalian species (Hoyer, 1989), cloning and expression of the 5-HT$_{1A}$ receptor from other species such as hamster, guinea pig, and pig will undoubtedly soon follow.

The story of the initial cloning of the 5-HT$_{1A}$ receptor gene is fascinating, both from a point of historical perspective and from the remarkable insight which has been gained regarding the larger family of genes encoding G-protein-coupled receptors. This receptor was first cloned in 1987 by Brian Kobilka and colleagues. They had cloned first the hamster β_2-adrenergic receptor (β_2-AR) in collaboration with the group of Richard Dixon and Catherine Strader at Merck (Dixon et al., 1986), followed by the human β_2-AR (Kobilka et al., 1987a). They next sought to clone the human β_1-AR from a genomic library by low-stringency cross-hybridization using the full-length β_2-AR sequence as a probe. A single minor cross-hybridizing species termed G-21 was obtained. Like the β_2-AR gene, G-21 was intronless. This DNA was determined to encode a putative protein product with striking sequence and structural similarities to β_2-AR and rhodopsin (Figure 5.1), sharing particular sequence identity (41%) with the β_2-AR in the putative transmembrane domains. However, this DNA was found not to encode an adrenergic receptor. In fact, initial studies in *Xenopus* oocytes failed to reveal ligand binding or second messenger functions. Remarkably, the β_1-AR was subsequently cloned in a fortuitous fashion by using G-21 to probe a human placental library (Frielle et al., 1987).

In mid-1988, the G-21 DNA was transfected into COS-7 African green monkey kidney cells in order to screen for binding with adrenergic ligands. Although no specific binding was obtained with [^{3}H]epinephrine or [^{3}H]rauwolscine, specific binding with the β-AR partial agonist ligand [^{125}I]cyanopindolol was obtained in these cells by displacing with the β-AR antagonist propranolol (Fargin et al., 1988). However, saturation studies with [^{125}I]cyanopindolol revealed only a moderate affinity (≈ 10 nM), more consistent with a 5-HT$_{1A}$ receptor subtype (Hoyer et al., 1985a,b). Binding studies with the specific agonist 5-HT$_{1A}$ receptor ligand [^{3}H]8-dihydroxy-2-(di-*n*-propylamino)tetralin ([^{3}H]8-OH-DPAT) subsequently confirmed that clone G-21 encodes the human 5-HT$_{1A}$ receptor (Fargin et al., 1988; Raymond, 1989a). The identification of the protein encoded by G-21 using two "typical" β-AR ligands (cyanopindolol and propranolol) further illustrates the remarkable overlap in both the structure and functions of these members of the family of G-protein-coupled receptors.

2. Pharmacology

In 1979 Peroutka and Snyder used radioligand binding techniques to distinguish between two subtypes of 5-HT recognition sites in the central nervous system. According to their convention, 5-HT$_1$ binding sites are labeled by [^{3}H]5-HT, displaying nanomolar affinity for 5-HT. 5-HT$_2$

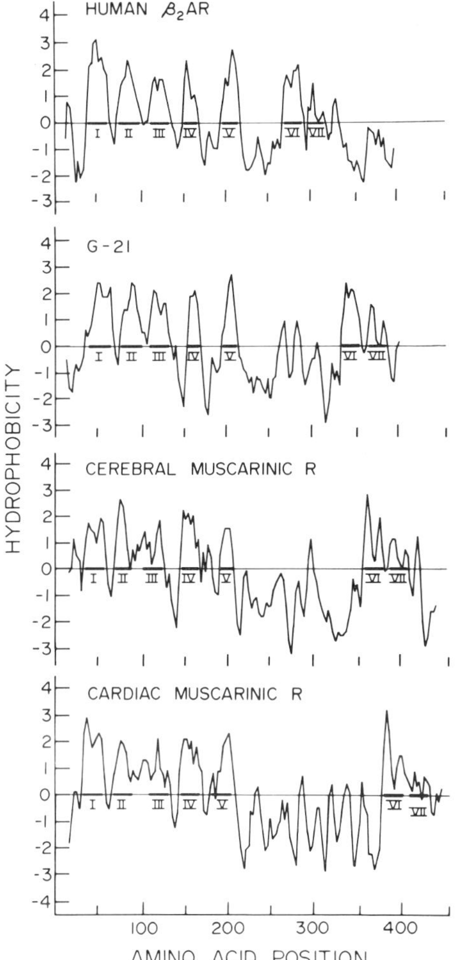

FIGURE 5.1. Hydropathicity plot of various G-protein-coupled receptors. The plots were constructed by the method of Kyte and Doolittle from the amino acid sequences of the respective receptors. The most negative values are assigned to highly charged regions which constitute hydrophilic domains, while positive values are assigned to uncharged residues. Each receptor has seven putative transmembrane domains (TM 1–7) (hydrophobic regions of at least twenty amino acids) denoted as in the figure by roman numbers I–VII. G-21 is the 5-HT$_{1A}$ receptor. (Figure provided by Drs. H. G. Dohlman, M. G. Caron, and R. J. Lefkowitz.)

binding sites are labeled by [^{3}H]spiperone (Leysen et al., 1979), displaying micromolar affinity for 5-HT. 5-HT$_1$ sites were subsequently divided into 5-HT$_{1A}$ and 5-HT$_{1B}$ sites based on the finding that spiperone inhibited [^{3}H]5-HT binding in a biphasic manner (Pedigo et al., 1981). The 5-HT$_{1A}$

receptor binding site was defined as that with nanomolar affinity for both 5-HT and spiperone. Since then, many other ligands have been utilized to radiolabel the 5-HT$_{1A}$ receptor (Figure 5.2). These include [^{3}H]8-OH-DPAT, [^{3}H]ipsapirone (TVX Q-7821), [^{3}H]WB 4101, [^{3}H]buspirone, [^{3}H]spiroxatrine, [^{3}H]1-[2-(4-aminophenyl)ethyl]-4-(3-trifluoromethyl-phenyl)piperazine ([^{3}H]PAPP), [^{3}H]5-methoxy-3-(di-n-propylamino)chroman ([^{3}H]5-MeO-DPAC), and [^{125}I]-labeled Bolton-Hunter-8-methoxy-2-[N-propyl-N-propylamino]tetralin ([^{125}I]BH-8-MeO-N-PAT).

Regardless of the labeled ligand used for binding studies, the 5-HT$_{1A}$ receptor displays a high affinity (<10 nM) for 8-OH-DPAT, 5-carboxyamidotryptamine (5-CT), ipsapirone (TVX Q-7821), spiroxatrine, PAPP, 5-HT, RU 24969, d-LSD, iodocyanopindolol, MDL 72823, metergoline, and lisuride. Other compounds with moderate affinity (10–100 nM) for the 5-HT$_{1A}$ receptor include bufotenine, methysergide, cyproheptadine, metitepine, pindolol, propranolol, (+)-butaclamol, buspirone, spiperone, urapidil, gepirone, ($-$)21009, and WB 4101 (Hoyer et al., 1985b; Mir et al., 1988; Leysen, 1988; Hoyer, 1989). Of particular note is the high affinity of the 5-HT$_{1A}$ receptor for potent β-AR blocking agents such as iodocyano-pindolol, pindolol, and ($-$)-21009, a feature shared with β_1- and β_2-adrenergic and 5-HT$_{1B}$ receptors. The cloned human 5-HT$_{1A}$ receptor has very similar binding properties to those previously described primarily in rat hippocampal membranes (Fargin et al., 1988; Raymond et al., 1989a).

Thus far, the most satisfactory and widely used radioligand is [^{3}H]8-OH-DPAT, which has many features that have made it the current best choice for binding studies of the 5-HT$_{1A}$ receptor. It has a high specific activity (110 μCi) for a tritiated ligand, is commercially available, possesses a very high degree of selectivity, and typically has low nonspecific binding. 8-OH-DPAT has an $\approx$ 3000-fold higher affinity for the 5-HT$_{1A}$ receptor than for the closely related 5-HT$_{1B}$ receptor (Middlemiss and Fozard, 1983). The other radioligands are much less selective. Ipsapirone (TVX Q-7821) and WB 4101 bind to α_1-adrenergic receptors, and buspirone to D2-dopamine receptor with high affinity (Leysen, 1988). Spiroxatrine binds with high affinity to D2-dopamine receptors and μ-opiate receptors (Niemegeers et al., 1964; Leysen et al., 1977; Leysen, 1988), and with moderate affinity to α_1-adrenergic receptors and 5-HT$_2$ receptors (Leysen, 1988).

Despite the advantages of [^{3}H]8-OH-DPAT discussed above, limitations to its usefulness exist. For example, the 8-OH-DPAT derivative [^{125}I]BH-8-MeO-N-PAT may be more useful for rapid autoradiographic mapping of 5-HT$_{1A}$ receptors at the microscopic level (Gozlan et al., 1988a) because of its higher specific activity ($\approx$ 1000 Ci/mmol). In addition, [^{3}H]8-OH-DPAT labels a striatal site called 5-HT$_{pre}$ (Gozlan et al., 1983; Hall et al., 1985), and a binding site on human platelets which may represent 5-HT uptake sites (Schoemaker and Langer, 1986; Inei and Meyerson, 1988). In that respect, the chroman derivative [^{3}H]5-MeO-DPAC labels 5-HT$_{1A}$ but not

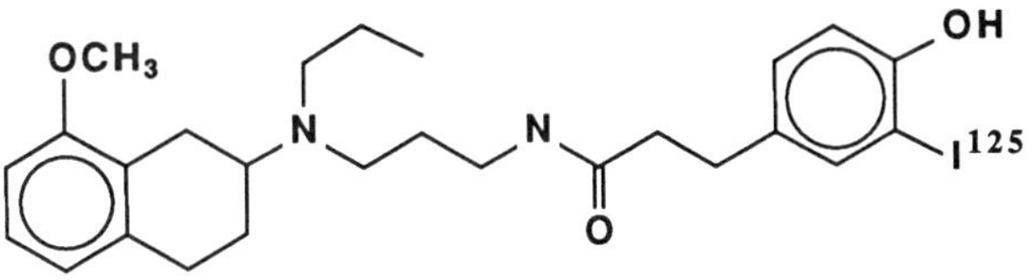

[^{125}I]-BH-8-MeO-N-PAT

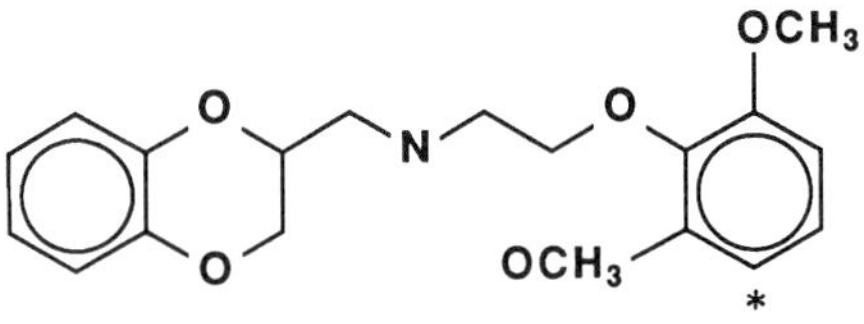

[^{3}H]-WB4101

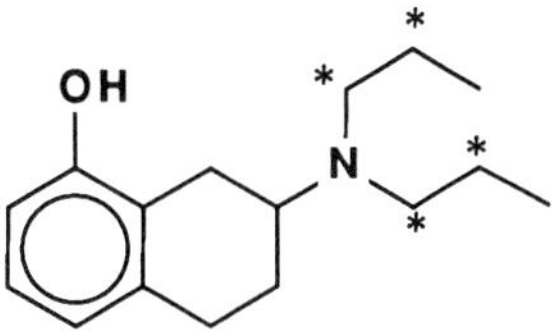

[^{3}H]-8-OH-DPAT

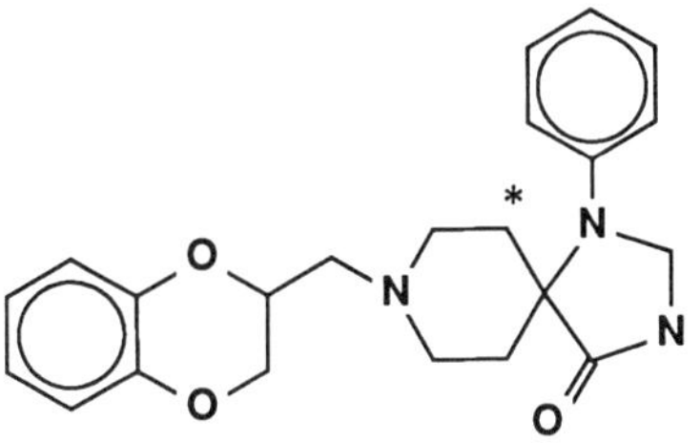

[^{3}H]-Spiroxatrine

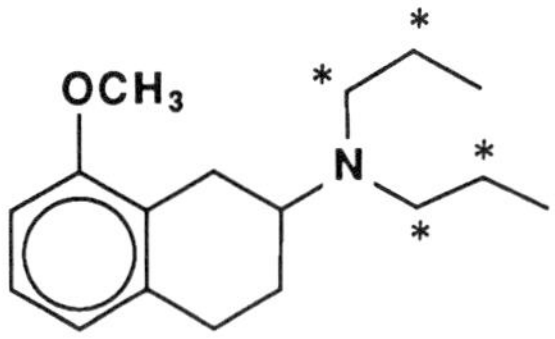

[^{3}H]-5-MeO-DPAC

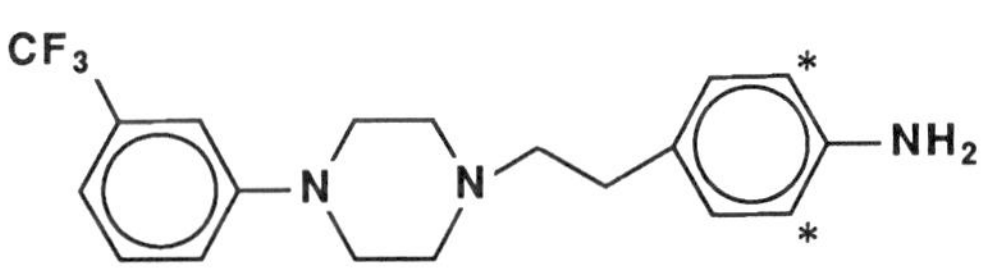

[^{3}H]-PAPP

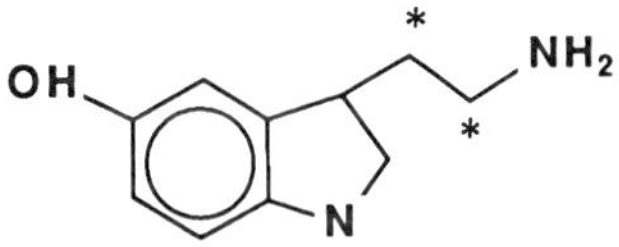

[^{3}H]-5-Hydroxytryptamine

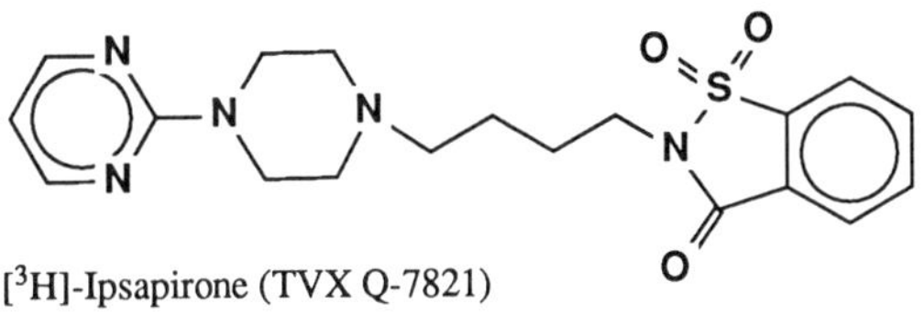

[^{3}H]-Ipsapirone (TVX Q-7821)

FIGURE 5.2. Comparison of the structures of some of the radioactive ligands which have been utilized to label the 5-HT_{1A} receptor. Asterisks denote the locations of tritium atoms (when known) as determined from manufacturers' product inserts or from primary references.

5-HT$_{pre}$ binding sites (Cossery et al., 1987). However, for most purposes, [^{3}H]8-OH-DPAT is still the ligand of choice for characterizing the 5-HT$_{1A}$ receptor.

One pharmacological problem that needs to be resolved for this receptor is the lack of specific antagonist compounds. Nonspecific agents such as β-blockers, spiperone, metitepine, and butaclamol have been the only available antagonists for the 5-HT$_{1A}$ receptor. However, recent work has resulted in the development of more specific partial agonists/antagonists for the 5-HT$_{1A}$ receptor such as buspirone, ipsapirone, MDL 73005, BMY 7378, NAN-190 (Yocca et al., 1987; Bockaert et al., 1987; Nelson et al., 1987; Dumuis et al., 1988a,b; Cornfield et al., 1989; Fitzgerald et al., 1989; Gartride et al., 1990; Hibert and Moser, 1990; Hjorth and Sharp, 1990; Moser et al., 1990; Przegalinksi et al., 1990; Rydelek-Fitzgerald et al., 1990; Sharp et al., 1990; Yocca, 1990; Zemlan et al., 1990).

3. Clinical Correlates

The study of 5-HT$_{1A}$ receptors is particularly pertinent for two broad clinical issues. These involve (1) the treatment and genesis of CNS disorders, particularly anxiety, sleep, depression, and migraine, and (2) regulation of vascular tone. The clinical correlates of 5-HT receptors and CNS function have been the subject of intense interest (Kennett et al., 1987; Peroutka, 1987a; Cooper and Abbott, 1988), and will not be discussed in detail here. It is important to note, however, that the 5-HT$_{1A}$ receptor agonist buspirone has been widely marketed as an anxiolytic. In addition, several behavioral (Lucki et al., 1983; Tricklebank et al., 1985; Hutson et al., 1986; Smith and Peroutka, 1986; Traber and Glaser, 1987; Carli and Samanin, 1988; Invernizzi et al., 1988) and physiological (Tricklebank, 1985; Goodwin and Green, 1985; Middlemiss et al., 1985; Kwong et al., 1986; Hutson et al., 1987; Hamon et al., 1988) correlates which may be mediated by 5-HT$_{1A}$ receptors have been described (see Dourish et al., 1987).

The potential role of 5-HT as a vasoregulatory agent has many clinical implications, particularly in the area of hypertension (see Saxena and Villalón, 1990, for review of 5-HT$_{1A}$ receptor antihypertensive mechanisms). For example, the antihypertensive urapidil, an α_1-adrenergic antagonist, may also lower blood pressure through its actions as a 5-HT$_{1A}$ receptor agonist. The structure of 5-HT was elucidated in 1948, after its discovery as a vasotonic serum factor (Page, 1952). 5-HT has subsequently been characterized as both a vasoconstrictor and as a vasorelaxant (Henning and Rubenson, 1971; McCall et al., 1987). When 5-HT is intravenously injected into intact animals, a classic triphasic blood pressure response is seen (Kalkman et al., 1984). An initial transient hypotensive effect may be secondary to activation of 5-HT$_3$ receptors. The subsequent pressor response is thought to be mediated by 5-HT$_2$ receptors and the third hypotensive phase by a 5-HT$_1$ subtype (Kalkman et al., 1984; Vanhoutte et al., 1984). Furthermore, vascular strips from rats with hypertension demonstrate an exaggerated responsiveness to 5-HT (Finch, 1974; Webb, 1982), perhaps through 5-HT$_2$ receptors. In contrast, the 5-HT$_1$ selective agonist 5-carboxamidotryptamine causes prolonged vasodilation and hypotension in both hyper- and normotensive animals (Saxena and Verdouw, 1985; Saxena and Lawang, 1985; Dalton et al., 1985). Of particular relevance is that several ligands with high affinity for the 5-HT$_{1A}$ receptor exert a centrally mediated hypotensive effect in rats and cats (reviewed in Schoeffter and Hoyer, 1988). These data strongly suggest that the 5-HT$_{1A}$ receptor mediates vasorelaxation. An exception may occur in the canine basilar artery, where several groups have proposed that the 5-HT$_{1A}$ receptor mediates vasoconstriction (Taylor et al., 1986; Peroutka et al., 1986). However, these data are controversial because (1) the efficacy of several potent 5-HT$_{1A}$ ligands was somewhat less than would be expected and (2) two other groups have attributed this effect to a 5-HT$_2$ receptor (Müller-Schwenitzer and Engel, 1983; Van Neuten et al., 1984). These data clearly show that the clinical role of the 5-HT$_{1A}$ receptor in vasoregulation still remains to be defined.

An additional role for the 5-HT$_{1A}$ receptor in immune function is also conceivable. The role of 5-HT as a potential immune modulator has long been suspected but is not well characterized (Devoino and Ilyutchenok, 1968; Lespinats et al., 1984; Sternberg et al., 1987). Specifically, the 5-HT$_{1A}$ receptor may play a function in the modulation of killer cell cytotoxicity (Hellstrand and Hermodsson, 1987). The intriguing potential involvement of the 5-HT$_{1A}$ receptor in immune function is further underscored by Northern blot analysis of human fetal RNA, which revealed high message levels in spleen, lymph nodes, and thymus (Kobilka et al., 1987b) (Figure 5.3). However, much study is needed in this area.

4. Anatomic Distribution in the Central Nervous System

The distribution of the 5-HT$_{1A}$ receptor has been delineated through a combination of techniques including radioligand autoradiography on brain slices and binding studies on membrane fractions. Radioligand binding studies utilizing membranes prepared from several brain areas have consistently shown high densities of 5-HT$_{1A}$ receptors in human, rat, and porcine hippocampus and cortex. These results were confirmed by autoradiography, revealing high densities in rat, mouse, and guinea pig hippocampal formations (particularly the dentate gyrus), the septal regions (particularly the lateral septal nucleus), the entorhinal cortices, and the dorsal raphe nuclei. In the human brain, the hippocampal formation is most enriched in

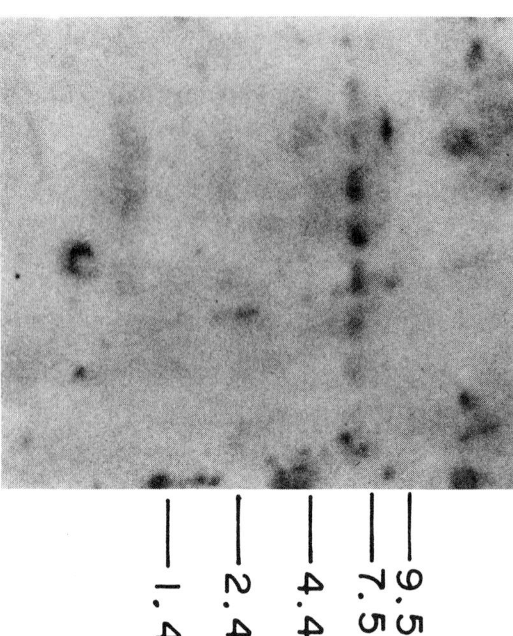

FIGURE 5.3. Northern blot analysis of G-21 transcripts of human fetal and placental total RNA. Each lane contains 15 μg of RNA, processed as previously described after electrophoresis through 1.2% agarose. From Kobilka et al. (1987b), reprinted with permission from *Nature* 329:75–79. Copyright © 1987 Macmillan Journals Limited.

5-HT_{1A} binding sites, particularly the CA_1 field, dentate gyrus, and subiculum. The external cortical areas, particularly layer II, and the raphe nuclei are also quite enriched in 5-HT_{1A} receptors (Marcinkiewicz et al., 1984; Hoyer et al., 1985a,b, 1986; Vergé et al., 1986; Pazos et al., 1986; Palacios et al., 1987). Studies have also demonstrated autoradiographic evidence of 5-HT_{1A} receptors in the spinal cord (Daval et al., 1987). Although autoradiography and ligand binding have been very useful in delineating brain regions with high levels of 5-HT_{1A} receptors, more sensitive techniques are desirable for cellular localization.

For example, antireceptor antibodies based on the putative amino acid sequence as determined from the nucleotide sequence of the 5-HT_{1A} receptor have allowed a detailed mapping of the cellular locations of rat brain 5-HT_{1A} receptors (El Mestikawy et al., 1990, 1991). This technique may be suitable for subcellular localization by electron microscopy in brain. In that regard, one of our groups has utilized affinity-purified IgG fractions raised against oligopeptide portions of the third intracellular [inner (i3)] loop of the human 5-HT_{1A} receptor to map the cellular and subcellular location of 5-HT_{1A} receptors in human and rat kidney (J. R. Raymond, J. Kim, R. E. Beach, and C. C. Tisher, unpublished observation) by light and electron microscopic techniques. These studies illustrate the importance of molecular biology techniques to the study of G-protein-coupled receptors. Moreover, the availability of the gene and cDNA for this receptor will be crucial for in situ hybridization studies.

The physiological correlates of the respective brain and spinal cord 5-HT$_{1A}$ binding sites have not yet been established. However, it has been shown that the 5-HT$_{1A}$ receptor inhibits neuronal firing of the dorsal raphe in several species (Aghajanian and Lakoski, 1984; Trulson and Aresteh, 1986; Sinton and Fallon, 1986; VanderMaelen et al., 1986; Wilkinson et al., 1987; Sprouse and Aghajanian, 1987). The 5-HT$_{1A}$ receptor may mediate neuronal function in the septum (Beck et al., 1985), the ventromedial hypothalamus (Newberry and Priestly, 1988), and the hippocampus in a similar fashion (Peroutka et al., 1987; Joëls et al., 1987; Zgombick et al., 1989). In the hippocampus, the 5-HT$_{1A}$ receptor hyperpolarize pyramidal cells of the CA$_1$ field by opening K$^+$ channels via a pertussis toxin-sensitive G-protein (Andrade and Nicoll, 1987a,b). Regarding the spinal cord 5-HT$_{1A}$ binding sites, descending projections from the nucleus raphe magnus to the dorsal horn of the spinal cord are involved in the control of ascending nociceptive messages (Oliveras et al., 1975; Daval et al., 1987).

5. Signal Transduction Mechanisms

The nature of the second messenger linkage(s) of the 5-HT$_{1A}$ receptor has been a point of considerable controversy (for review, see Hoyer, 1989), but cloning and functional expression of this receptor have shed some light on this controversy (Figure 5.4). As background, however, it is important to note that no naturally occurring continuous cell line constitutively expressing only the 5-HT$_{1A}$ receptor is known. Therefore, initial reports relied entirely on brain slices or heterogeneous membrane preparations probably containing multiple 5-HT receptor subtypes, which made interpretation of these studies difficult. Some have suggested that the 5-HT$_{1A}$ receptor is linked in a stimulatory fashion to adenylyl cyclase (Shenker et al., 1985; Markstein et al., 1986). Others have demonstrated inhibition of forskolin-stimulated adenylyl cyclase with a 5-HT$_{1A}$ receptor pharmacology (Weiss et al., 1986; De Vivo and Maayani, 1986; Bockaert et al., 1987; Makman et al., 1988; Schoeffter and Hoyer, 1988; Dumuis et al., 1988a; Cornfield et al., 1988). Electrophysiological studies have demonstrated an apparent 5-HT$_{1A}$ receptor-mediated stimulation of K$^+$ conductance (Andrade et al., 1986; Colino and Halliwell, 1987; Andrade and Nicoll 1987a,b). Finally, both stimulatory (Minchin et al., 1985) and inhibitory (Claustre et al., 1988) coupling of the 5-HT$_{1A}$ receptor to phospholipase C activity have been described.

How can these seemingly conflicting reports be reconciled? These data suggest either that different receptor subtypes are being studied, or that

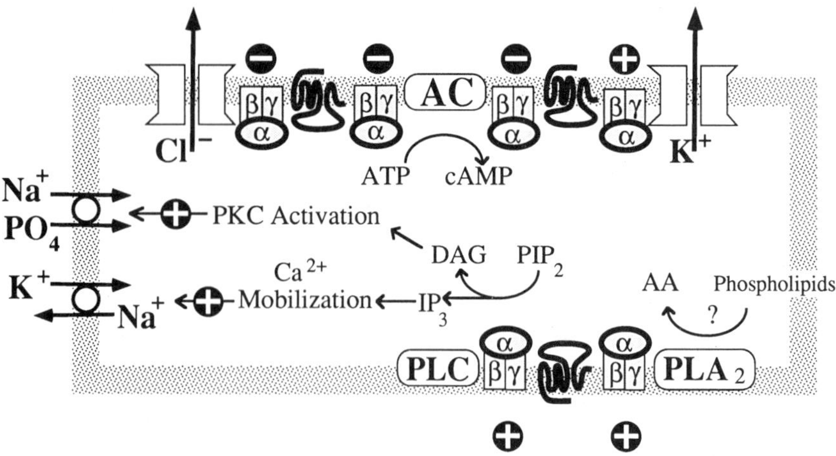

FIGURE 5.4. Coupling of second messengers and effectors to the 5-HT$_{1A}$ receptor in heterologous expression systems. Depicted are 5-HT$_{1A}$ receptor linkages with the inhibition of adenylyl cyclase (AC), stimulation of phosphatidylinositol 4,5-bisphosphate (PIP$_2$) hydrolysis, increased levels of intracellular Ca^{2+}, activation of protein kinase C (PKC), activation of K$^+$ channels, inhibition of Cl$^-$ channels, augmentation of arachidonic acid release, stimulation of the Na$^+$/K$^+$-ATPase and Na$^+$/PO$_4$ symporter. All linkages described are sensitive to pertussis toxin. α, β, and γ represent subunits of G-proteins. Other abbreviations used are IP$_3$ (inositol 1,4,5,-trisphosphate), DAG (diacylglycerol), AA (arachidonic acid), PLC (phospholipase C), and PLA$_2$ (phospholipase A$_2$). The exact nature of the coupling to arachidonic acid release is unknown; PLA$_2$ may be involved, but other effects such as on LDL uptake or modulation of enzymes such as diacylglycerol kinase by PKC have not yet been ruled out.

more than one potential second messenger can be linked to a single receptor subtype in a single or in multiple cell types. As stated previously, the primary obstacle to the resolution of this controversy lies in the heterogeneity of the tissue preparations. Paul Hartig (1989) has recently commented on this area, stating that "transfection of this [5-HT$_{1A}$ receptor] clone into a functional response system is eagerly awaited. . . ." Three groups have successfully accomplished transfection of the 5-HT$_{1A}$ receptor into mammalian expression systems, allowing a partial resolution of this issue (Fargin et al., 1989; Albert et al., 1990; Karschin et al., 1991).

The receptor has been expressed in four nonneuronal cell lines (COS-7 monkey kidney cells, HeLa human cervical carcinoma cells, CHO chinese hamster ovary cells, and rat atrial myocytes) and one neuronal cell line (GH$_4$C$_1$ cells). Using COS-7 cells for transient expression, Fargin et al. (1989) first showed that the human 5-HT$_{1A}$ receptor does not stimulate basal cAMP formation. Rather, it induces a clear ($>80\%$) inhibition of isoproterenol-stimulated cAMP accumulation in cells cotransfected with human β_2-AR. Using HeLa cell and CHO cell clones stably expressing various levels of human 5-HT$_{1A}$ receptor, our groups also demonstrated a negative coupling to adenylyl cyclase (IC$_{50}$ $\approx$ 20 nM for 5-HT) (Fargin et al., 1989; Raymond et al., 1992). The inhibition of cAMP synthesis was sensitive to pertussis toxin (IC$_{50}$ $\approx$ 10–20 ng/ml $\times$ 4 hr), implicating a Gi-like G-protein in this effect. No coupling to Gs could be demonstrated in these cells. Paul Albert, Olivier Civelli, and colleagues have demonstrated a similar inhibition of VIP-stimulated cAMP accumulation in GH$_4$ cells expressing the rat 5-HT$_{1A}$ receptor (Albert et al., 1990). These data clearly show a linkage between the 5-HT$_{1A}$ receptor and inhibition of adenylyl cyclase activity.

Regarding the possibility of coupling to multiple second messenger pathways, our group (Fargin et al., 1989, 1991; Raymond et al., 1989b, 1991) also demonstrated a 5-HT$_{1A}$ receptor-induced increase of accumulation of total inositol phosphates and inositol 1,4,5-triphosphate in the HeLa cell clones [also in CHO cells (Raymond et al., 1992)]. The coupling to phospholipase C appeared to be intramembranous (and not secondary to second messenger/kinase cross-talk) because activation of phospholipase C was demonstrated in washed plasma membranes (Fargin et al., 1991). This stimulation was pertussis toxin-sensitive (IC$_{50}$ $\approx$ 10–20 ng/ml $\times$ 4 hr). However, the half-maximal dosage of 5-HT for this effect was much higher (EC$_{50}$ $\approx$ 3 μM) than that associated with the inhibition of adenylyl cyclase. Although the stimulation of phosphatidylinositol hydrolysis in these cells was small, two subsequent effects could be measured (Raymond et al., 1989b). These effects were an inositol 1,4,5-triphosphate-mediated increase of intracellular Ca^{2+} (Middleton et al., 1990), and activation of protein kinase C as measured by the MARCKS (*m*yristoylated *a*lanine-*r*ich *C* *k*inase *s*ubstrate) protein phosphorylation assay (Blackshear, 1988). Moreover, the 5-HT$_{1A}$ receptor-induced protein kinase C activation was shown to be responsible for an increase in activity of a sodium-dependent phosphate transporter (Raymond et al., 1989b, 1991). Interestingly, the 5-HT$_{1A}$ receptor stimulates Na$^+$/K$^+$-ATPase activity in the same cells independent of effects of adenylyl cyclase or protein kinase C activities. This increase is most probably mediated via an increase in cytosolic Ca^{2+} levels (Middleton et al., 1990). Both transport effects are sensitive to pertussis toxin. Another example of the pluripotent nature of 5-HT$_{1A}$ receptor signal transduction can be demonstrated in CHO cells. In those cells, the 5-HT$_{1A}$ receptor augments arachidonic acid release initiated by the phospholipase A$_2$ activator melittin, the calcium ionophore A23187, and the agonists thrombin and ATP. However, 5-HT or 8-OH-DPAT *alone* does not increase arachidonic acid release (Raymond et al., 1992). This novel signal transduction mechanism may provide a mechanism by which copackaged or coreleased agonists can increase the effects of each other (for example, serotonin and ATP at neuronal junctions or serotonin and thrombin in the vasculature).

The 5-HT$_{1A}$ receptor may be involved in the regulation of other transport processes, such as an inhibition of electrically stimulated [^{3}H]acetylcholine release from guinea pig ileum (Kilbinger and Pfeuffer-Friedrich, 1985; Fozard and Kilbinger, 1985; Pfeuffer-Friedrich and Kilbinger, 1985) and an inhibition of [^{3}H]dopamine release (Ennis et al., 1988). The involvement of this receptor in other transport processes is just now being investigated. For example, Henry Lester's group recently used a novel vaccinia virus heterologous expression system to examine the effects of recombinant 5-HT$_{1A}$ receptors on cardiac K$^+$ channels (Karschin et al., 1991). In rat atrial myocytes, the receptor activated the same inwardly rectified K$^+$ channels as did endogenous muscarinic receptors. Our own studies in CHO cells have demonstrated that recombinant human 5-HT$_{1A}$ receptors inhibit a large conductance chloride channel through a pertussis toxin-sensitive G-protein (Mangel et al., submitted).

Of particular note regarding the signal transduction pathways linked to the 5-HT$_{1A}$ receptor, is a growing awareness that single receptor subtypes can be linked to distinct signal transduction pathways in the same cell (Limbird, 1988). The recent cloning of other G-protein-coupled receptors has allowed a detailed examination of various signaling pathways associated with recombinant receptors expressed in eukaryotic cell systems. Like the 5-HT$_{1A}$ receptor, various muscarinic receptor subtypes, and the α_2-AR's have each been clearly linked to more than one second messenger pathway in single cell type (Ashkenazi et al., 1987; Peralta et al., 1988; Cotecchia et al., 1990). Moreover, Liu and Albert (1991) have shown that signaling linkages of the 5-HT$_{1A}$ receptor depend on the host cell type.

The exact identity of the G-protein(s) which link the 5-HT$_{1A}$ receptor to various functional effects is uncertain. However, in the studies reported to date, all effects linked to the 5-HT$_{1A}$ receptor are sensitive to pertussis toxin (Andrade et al., 1986; Makman et al., 1988; Newberry and Priestly, 1988; Zgombick et al., 1989). In that regard, a recent study suggests that Gαi3 predominantly mediates both the inhibition of adenylyl cyclase and stimulation of phospholipase C in membranes derived from HeLa cells transfected to express the human 5-HT$_{1A}$ receptor (Fargin et al., 1992). Expression of the 5-HT$_{1A}$ receptor in other cell systems, perhaps with excesses or deficiencies of various other components of the signal transduction network, will be necessary to fully elucidate the complete nature of the second messenger linkages of this receptor.

6. Biochemistry

The biochemical characterization of the 5-HT$_{1A}$ receptor has been primarily carried out in rat hippocampal membrane preparations. The G-protein linkage of the 5-HT$_{1A}$ receptor was established by the demonstration that guanine nucleotides and magnesium modulate agonist binding to this receptor (Gozlan et al., 1983). Initial studies clearly demonstrated that like other members of the superfamily of G-protein-coupled receptors, the 5-HT$_{1A}$ receptor is a glycoprotein (El Mestikawy et al., 1988). However, attempts to purify this receptor were unsuccessful until recently (El Mestikawy et al., 1989), although a $\approx$ 1000-fold purification of a 5-HT$_1$-like receptor has been reported (Gallagher and Wang, 1988).[1] Successful solubilization of active 5-HT$_{1A}$ receptor has been achieved by two groups using the zwitterionic detergent CHAPS to solubilize from rat hippocampal membranes (El Mestikawy et al., 1988, 1989) and a mixture of digitonin and Nonidet P-40 to solubilize from bovine cortex (Asarch and Shih, 1987).

The molecular mass of the 5-HT$_{1A}$ receptor from rat hippocampus ($\approx$ 60 kDa) has been determined in membranes using the techniques of radiation inactivation (Gozlan et al., 1987), and by photoaffinity labeling with radioactive probes followed by sodium dodecyl sulfate–polyacrylamide gel electrophoresis (SDS-PAGE) (Ransom et al., 1986b; Emerit et al., 1987). Although these photoaffinity labels, [^{3}H]1-[2-(4-azidophenyl)ethyl]-4-(3-trifluoromethylphenyl)piperazine ([^{3}H]p-azido-PAPP) (Ransom et al., 1986b) and [^{3}H]8-methoxy-2-[N-n-propyl, N-3-(2-nitro-4-azidophenyl)-aminopropyl]tetralin (Emerit et al., 1987), have proved useful for the determination of the molecular mass of the rat hippocampal 5-HT$_{1A}$ receptor, their low specific activity (50–100 mCi/mmol) and limited commercial availability do not currently allow more widespread application. Subsequently, the expression of the human 5-HT$_{1A}$ receptor allowed another group to adapt the previously characterized D2-dopamine probe [^{125}I]N_3-NAPS (N-(p-azido-m [^{125}I]iodophenethyl)spiperone (Amlaiky and Caron, 1985) as a photoaffinity label for the human 5-HT$_{1A}$ receptor expressed in COS-7 and HeLa cells (Raymond et al., 1989a). In membranes prepared from these transfected cells, this ligand specifically labels a broad band of molecular mass $\approx$ 75 kDa which has the appropriate pharmacology of the 5-HT$_{1A}$ receptor. By sequentially photoaffinity labeling, solubilizing, and immunoprecipitating with an antipeptide antiserum (JWR21) raised against a 26 amino acid sequence predicted from the nucleotide sequence encoding the receptor (Fargin et al., 1988), the native human 5-HT$_{1A}$ receptor derived from frontal cortex and hippocampus was identified as single species migrating in a broad pattern at 62–64 kDa on 10% SDS-PAGE (Figure 5.5) (Raymond et al., 1989a).

The theoretical molecular mass obtained from the deduced amino acid sequence is 46 kDa. Because it has been previously demonstrated that the 5-HT$_{1A}$ receptor is glycosylated (El Mestikawy et al., 1989), the additional mass of the 5-HT$_{1A}$ receptor noted in transfected COS-7 and HeLa cells is likely due to N-linked glycosylation at three consensus sites on the amino

---

[1]Identification of the subtype of this purified receptor was not established, because no clear pharmacology was described in this study.

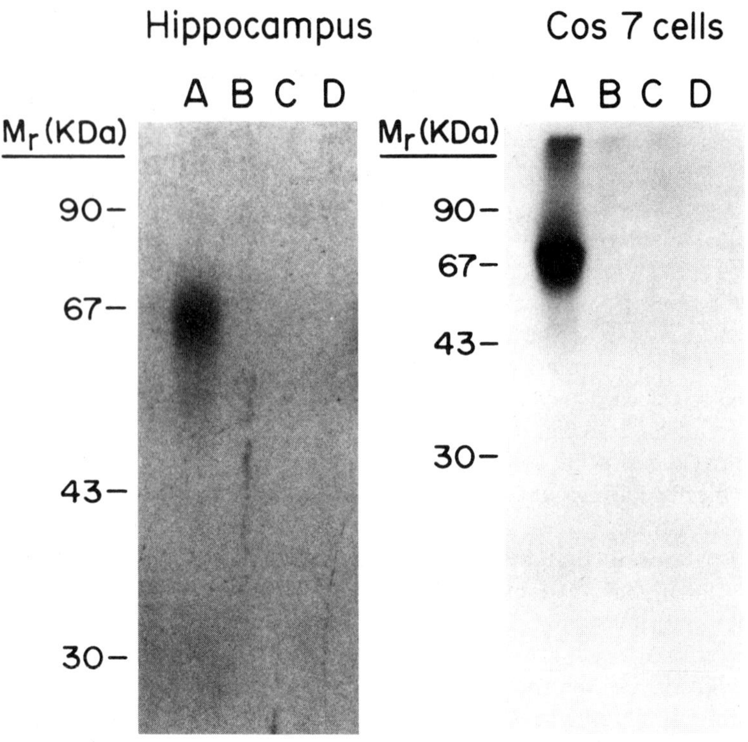

FIGURE 5.5. Photoaffinity labeling of human 5-HT$_{1A}$ receptors derived from hippocampus and transfected COS-7 cells transiently expressing $\approx$ 14 pmol/mg protein of the receptor. Procedures were performed exactly as described (Raymond et al., 1989a). Human hippocampal (200 μg) and COS-7 cell membranes (100 μg) were labeled with 700 pM [^{125}I]N$_3$-NAPS (N-(p-azido-m[^{125}I]iodophenethyl)spiperone (Amlaiky and Caron, 1985), solubilized, immunoprecipitated with JWR21, electrophoresed on 10% SDS-PAGE gels, dried, and exposed for 12 hr. Labeling was performed in the absence (A,C,D) or presence (B) of 10 μM 8-OH-DPAT. Lanes A and B represent immunoprecipitation with JWR21 alone, while C and D represent immunoprecipitation with JWR21 which was preblocked with 1 μM corresponding peptide. Preimmune sera from the same rabbit did not precipitate labeled receptor (not shown). The different molecular mass of the receptor expressed in these two tissues is presumably due to differential glycosylation.

terminus. The apparent differences in the sizes of the labeled bands in the various tissues may be due to differential glycosylation. It has been shown that the α_2-C4-receptor, the human β_2-receptor, and rhodopsin can migrate with larger molecular mass when expressed in transfected cell lines as opposed to native tissues (Karnik et al., 1988; Regan et al., 1988; Lanier et al., 1988). In the case of the human β_2-AR expressed in CHW cells, which migrates with a larger molecular mass than the native hamster β_2-AR, the larger molecular mass has been clearly shown to be due to different degrees of glycosylation (Bouvier et al., 1988).

7. Molecular Biology

A number of interesting observations can be derived from the primary amino acid sequence of the 5-HT$_{1A}$ receptor. This human protein, composed of 421 amino acids, is characterized by a core molecular weight of $\approx 46,000$ and an isoelectric point of 8.8. Since this receptor is highly analogous to the well-characterized opsins and β_2-adrenergic receptor, some predictions on the relationship of structure to functions can be generated.

7.1. Hydrophobic Transmembrane Domains

Hydropathicity analysis revealed that the 5-HT$_{1A}$ receptor contains seven hydrophobic stretches that could possibly form membrane-spanning α-helices. In addition, a high degree of homology with β_2-adrenergic receptor is present mostly within those domains. Those two major characteristics led to the proposal that the 5-HT$_{1A}$ receptor belongs to the large family of G-protein-coupled receptors. A logical prediction based on what is known about the α_2- and β_2-adrenergic receptors, porcine muscarinic receptor, and rhodopsin binding sites (see Sigal et al., 1988, or Raymond et al., 1989c, for review), is that the 5-HT$_{1A}$ receptor binding site is also located within these membrane-spanning domains. In this respect, it is interesting to note that the 5-HT$_{1A}$ receptor possesses two highly conserved acidic residues (Asp^{79} and Asp^{113} for the β_2-adrenergic receptor; Asp^{82} and Asp^{116} for the 5-HT$_{1A}$ receptor) located in transmembrane (TM) 2 and 3, which have both been proposed to act as a counterion for the amine group of the catecholamines (Strader et al., 1989). While the first acidic residue is conserved in all cloned G-protein-coupled receptors reported thus far, the second one is present in only the receptors which bind the bioamines epinephrine, norepinephrine, dopamine, acetylcholine, and serotonin. One of these two residues, presumably Asp^{116}, may serve as the counterion for serotonin binding to the 5-HT$_{1A}$ receptor. This hypothesis can be clarified by mutagenesis studies.

The 5-HT$_{1A}$ receptor displays considerable overall homology within the transmembrane domains (≈ 41–45%) to several receptors with which it shares similar binding characteristics like the β_1- and β_2-adrenergic receptors and D2-dopamine receptors (Bunzow et al., 1988), as compared to much lower levels of homology with receptors that have less obvious pharmacological overlap such as the α_2-C10-adrenergic and M1 muscarinic receptors (Table 5.1). Surprisingly, the 5-HT$_{1A}$ receptor shares less homology in the transmembrane regions with either the 5-HT$_{1C}$ or 5-HT$_2$ receptors than it does with the β_1- and β_2-adrenergic receptors and

TABLE 5.1. Comparison of amino acid identities of transmembrane regions of various cloned G-protein-coupled receptors[a]

| | Percentage homology with human 5-HT$_{1A}$ receptor | | | | | | | |
Receptor	TM 1	TM 2	TM 3	TM 4	TM 5	TM 6	TM 7	Overall
Rat 5-HT$_{1A}$	100	100	100	89	100	100	100	98
Human β_1	30	37	48	22	41	70	37	41
Human β_2	33	41	56	33	41	70	44	45
Rat D2	37	52	59	30	30	48	48	43
Human M1	22	26	37	33	37	48	44	35
Rat 5-HT$_{1C}$	26	37	59	19	33	37	41	36
Rat 5-HT$_2$	26	33	59	22	33	37	37	35

[a]The amino acid identities of the transmembrane regions of the various receptors compared to the human 5-HT$_{1A}$ receptor are expressed as percentage identical residues in each domain (TM 1–TM 7), and as the total identities of all of the transmembrane domains. Boxes indicate regions of particular homology with D2-dopamine receptor, and β_1- and β_2-adrenergic receptors, as discussed in the text.

D2-dopamine receptors. More careful examination of the transmembrane domains reveals a specific amino acid identity of $\approx 70\%$ in TM 6 between the 5-HT$_{1A}$ receptor and the β_1- and β_2-adrenergic receptors. This region may be important in conferring the ability of the 5-HT$_{1A}$ receptor to bind β-adrenergic antagonists, as has been demonstrated for the β_1- and β_2-adrenergic receptors (Kobilka et al., 1988; Frielle et al., 1988). In this region, only Met^{350} of the 5-HT$_{1A}$ receptor (Met^{279} of the β_2-adrenergic receptor) is highly specific for these three receptor subtypes. Again, mutations of this residue will clarify its importance in binding β-adrenergic antagonists.

7.2. Hydrophilic Domains

By analogy with the β_2-adrenergic receptor, and because of the presence of three consensus sequences for N-linked glycosylation on the amino terminus, the receptor is probably oriented in the plasma membrane with the amino-terminus facing the extracellular domain. The seven transmembrane hydrophobic regions are connected by hydrophilic sequences which are likely to form three inner (intracellular) and three outer (extracellular) loops. In the putative o2 domain, the 5-HT$_{1A}$ receptor possesses a cysteine residue (Cys^{186}) which may form a disulfide bond with Cys^{109} which is located at the limit between the o1 loop and the TM 3. By analogy with what has been shown for the β_2-adrenergic receptor (Fraser, 1989; Dohlman et al., 1990), such a disulfide bond may stabilize the receptor conformation and explain, in part, why reducing agents affect 5-HT$_{1A}$ binding properties (Gozlan et al., 1988b; Emerit et al., 1991).

In the intracellular domains, the 5-HT$_{1A}$ receptor shares the general features observed for nearly all of the cloned receptors which inhibit adenylyl cyclase, such as a long i3 loop (129 amino acids) and a short C-terminal tail (20 amino acids) (Table 5.2). This pattern is somewhat less clear with receptors which couple to phospholipase C (5-HT$_{1C}$, 5-HT$_2$, and M1 muscarinic receptors) or which stimulate adenylyl cyclase (β_1- and β_2-adrenergic receptors). The 5-HT$_{1A}$ receptor does not contain consensus sequences for protein kinase A within the intracellular domains. However, at least three potential sites for protein kinase C (PKC)-activated phosphorylation (K^{147}RTPR within the putative i2 loop, and R^{226}KTVR and R^{340}KTVK within the putative i3 loop). Since it has clearly been demonstrated that the 5-HT$_{1A}$ receptor can activate PKC (Raymond et al., 1989b), it is reasonable to hypothesize that these sequences may be involved in a homologous desensitization process. Blier and De Montigny (1990) have recently shown a 5-HT$_{1A}$ receptor desensitization in rat neurons, but the signal transduction pathway involved was not delineated. In that regard, phorbol esters have been shown to rapidly (within minutes) induce a phosphorylation of the 5-HT$_{1A}$ receptor expressed in CHO cells. This phosphorylation showed a physiological stoichiometry of ≈ 2 mol of phosphate per mole of receptor and was associated with a desensitization of the inhibition of adenylyl cyclase (Raymond, 1991). Further studies using mutant receptors with conservative substitutions of the putative threonine

TABLE 5.2. Characteristics of structure and functional linkages of cloned G-protein-coupled receptors^a

Receptor	Loop i3	C-Terminus	Pathway(s)
Human β_1	79	99	↑ AC
Human β_2	43	68	
Human α_2-C4	150	24	
Human α_2-C10	157	23	
Human M2	180	25	↓ AC, ↑ PLC
Human M3	184	25	
Human 5-HT$_{1A}$	129	20	
Rat D2	134	14*	
Hamster α_1	71	64	
Human M1	153	41	
Human M4	240	45	↑ PLC
Rat 5-HT$_{1C}$	77	87	
Rat 5-HT$_2$	67	88	

^{a}The number of amino acid residues in each domain of the respective receptors is indicated. AC, adenylyl cyclase; PLC, phospholipase C. The box emphasizes the association of certain structural features (long i3 and very short carboxyl-terminus) with receptors that inhibit adenylyl cyclase.

acceptor residues will be necessary to establish a causal link between phorbol ester-induced phosphorylation and desensitization of the 5-HT$_{1A}$ receptor.

A final feature of interest is the presence of two cysteine residues at the carboxyl-terminus of the 5-HT$_{1A}$ receptor (Cys^{416} and Cys^{419}) which might be palmitoylated, as has been reported for analogous residues in rhodopsin (Ovchinnikov et al., 1988) and the β_2-adrenergic receptor (O'Dowd et al., 1989). It has recently been proposed that this posttranslational modification may modulate the coupling efficiency of the β_2-adrenergic receptor to adenylyl cyclase (O'Dowd et al., 1989). The role of these residues in 5-HT$_{1A}$ receptor function is currently undefined.

7.3. Related Receptor Subtypes and Prospects for the Future

The cloning of the 5-HT$_{1A}$, 5-HT$_{1C}$, and 5-HT$_2$ receptors has bearing on the prospects for cloning other similar related subtypes. For example, the 5-HT$_{1B}$ and 5-HT$_{1D}$ receptors have binding characteristics which are very similar to the 5-HT$_{1A}$ receptor. Moreover, all three receptors are linked primarily to the inhibition of adenylyl cyclase (Leysen, 1988; Hoyer, 1989; Hamon et al., 1989). Based on these similarities, one would expect that these receptors share particular amino acid sequence homology within the transmembrane domains. Since 5-HT$_{1B}$ receptors share a high affinity for classic β-adrenergic receptor blockers with the 5-HT$_{1A}$ and β-adrenergic receptors, it is quite likely that the homology is particularly high in TM 6. It was postulated that one of the recently described unknown clones (RDC4) which displays a high overall homology (65%) with the 5-HT$_{1A}$ receptor is a 5-HT receptor subtype (Libert et al., 1989). RDC4 has subsequently been shown to be a 5-HT$_{1D}$ receptor (Hamblin and Metcalf, 1991; Zgombick et al., 1991).

The use of molecular biology techniques will undoubtedly shed light on the problem of the existence of multiple putative 5-HT$_{1A}$-like receptors, evidenced by different coupling properties to adenylyl cyclase, and from anatomical (Vergé et al., 1986; Weissmann-Nanopoulos et al., 1985), pharmacological and physiological (Colino and Halliwell, 1986; Hutson et al., 1986, 1987; Martin and Mason, 1987; Andrade and Nicoll, 1987a; Blier and De Montigny, 1990; Kennett et al., 1987; Dumuis et al., 1988a; Carli and Samanin, 1988; Invernizzi et al., 1988; Hamon et al., 1988), and behavioral (Smith and Peroutka, 1986) studies. These studies have been recently reviewed (Hamon et al., 1989). It is uncertain whether these differences are secondary to the existence of multiple genes for very similar receptors, to differential posttranslational processing of the receptors, or to differences in the cellular components (such as membrane lipids, G-proteins or effectors) in the various brain regions.

8. Conclusions

A combination of classic biochemical and cell biological techniques, coupled with molecular biology techniques, has allowed for the partial resolution of important issues regarding the 5-HT$_{1A}$ receptor. These approaches have also been instrumental for the recent development of useful tools such as cell lines expressing the 5-HT$_{1A}$ receptor, photoaffinity labels, and antibodies. This combination of approaches will undoubtedly lead to further elucidation of the nature of signal transduction systems associated with the 5-HT$_{1A}$ receptor, the development and characterization of specific antagonists for the 5-HT$_{1A}$ receptor, and the discovery of closely related receptor subtypes.

Acknowledgments

We would like to acknowledge Drs. Henri Gozlan, Michel Hamon, Bob Lefkowitz, and Marc Caron for their interest and guidance, Dr. John Regan for his assistance in preparing antiserum JWR21, Dr. Susan Senogles for suggesting [^{125}I]N$_3$-NAPS as a potential photoaffinity label for the 5-HT$_{1A}$ receptor, and Dr. Mark R. Hnatowich for assistance with sequence alignments for the various receptors. This work was supported by a Department of Veterans Affairs Merit Award and a Grant-in-Aid from the American Heart Association (to JRR) and a grant from Sandoz (AF).

References

Aghajanian GK, Lakoski JM (1984): Hyperpolarization of serotonergic neurons by serotonin and LSD: Studies in brain slices showing increased K$^+$-conductance. *Brain Res* 305:181–185

Albert P, Zhou QY, Van Tol HHM, Bunzow JR, Civelli O (1990): Cloning, mRNA distribution and functional expression of the 5-HT$_{1A}$ receptor gene. *J Biol Chem* 265:5825–5832

Amlaiky N, Caron MG (1985): Photoaffinity labeling of the D$_2$-dopamine receptor using a novel high affinity radioiodinated probe. *J Biol Chem* 260:1983–1986

Andrade R, Malenka RC, Nicoll R (1986): A G protein couples serotonin and GABA$_B$ receptors to the same channels in hippocampus. *Science* 234:1261–1265

Andrade R, Nicoll R (1987a): Novel anxiolytics discriminate between postsynaptic serotonin receptors of the rat hippocampus. *Naunyn-Schmiedebergs Arch Pharmacol* 336:5–10

Andrade R, Nicoll R (1987b): Pharmacologically distinct actions of serotonin on single pyramidal neurones of the rat hippocampus recorded in vitro. *J Physiol (Lond)* 394:99–124

Asarch KB, Shih JC (1987): Solubilization of serotonin$_{1a}$ and serotonin$_{1b}$ binding sites from bovine brain. *J Neurochem* 48:1494–1501

Ashkenazi A, Winslow JW, Peralta EG, Peterson GL, Schimerlik MI, Capon DJ,

Ramachandran J (1987): An M_2 muscarinic receptor subtype coupled to both adenylyl cyclase and phosphoinositide turnover. *Science* 238:672–675

Beck SG, Clarke WP, Goldfarb J (1985): Spiperone differentiates multiple 5-hydroxytryptamine responses in rat hippocampal slices in vitro. *Eur J Pharmacol* 116:195–197

Blackshear PB (1988): Approaches to the study of protein kinase C involvement in signal transduction. *Am J Med Sci* 31:231–240

Blier P, De Montigny C (1990): Electrophysiological investigations of the adaptive response of the 5-HT system to the administration of 5-HT$_{1A}$ receptor agonists. *J Cardiovasc Pharmacol* 15(suppl 7):S42–S48

Bockaert J, Dumuis A, Bouhelal R, Sebben M, Cory RN (1987): Piperazine derivatives, including the putative anxiolytic drugs buspirone and ipsapirone, are agonists at 5-HT$_{1A}$ receptors negatively coupled with adenylate cyclase in hippocampal neurons. *Naunyn-Schmiedebergs Arch Pharmacol* 335:588–592

Bouvier M, Hnatowich M, Collins S, Kobilka BK, DeBlasi A, Lefkowitz RJ, Caron MG (1988): Expression of a human cDNA encoding the β_2-adrenergic receptor in Chinese hamster fibroblasts (CHW): Functionality and regulation of the expressed receptors. *Mol Pharmacol* 33:133–139

Bradley PB, Engel G, Feniuk W, Fozard JR, Humphrey PPA, Middlemiss DN, Mylecherane EJ, Richardson BP, Saxena PR (1986): Nomenclature of functional receptors for 5-hydroxytryptamine. *Neuropharmacology* 25:563–576

Bunzow JR, Van Tol HHM, Grandy DK, Albert P, Salon J, Christie M, Machida CA, Neve KA, Civelli O (1988): Cloning and expression of a rat D$_2$-dopamine receptor cDNA. *Nature* 336:783–787

Carli M, Samanin R (1988): Potential anxiolytic properties of 8-hydroxy-2-(di-*n*-propylamino) tetralin, a selective serotonin$_{1A}$ receptor agonist. *Psychopharmacology* 94:84–91

Claustre Y, Bénavides J, Scatton B (1988): 5-HT$_{1A}$ receptor agonists inhibit carbachol-induced stimulation of phosphoinositide turnover in the rat hippocampus. *Eur J Pharmacol* 149:149–153

Colino A, Halliwell JV (1986): 8-OH-DPAT is a strong antagonist of 5-HT action in rat hippocampus. *Eur J Pharmacol* 130:151–152

Colino A, Halliwell JV (1987): Differential modulation of three separate K-conductances in hippocampal CA1 neurons by serotonin. *Nature* 328:73–77

Conner DA, Mansour TE (1990): Serotonin receptor-mediated activation of adenylate cyclase in the neuroblastoma NCB.20: A novel 5-hydroxytryptamine receptor. *Mol Pharmacol* 37:742–751

Cooper SJ, Abbott A (1988): Clinical psychopharmacology may benefit from new advances in 5-HT pharmacology. *Trends Pharmacol Sci* 9:269–271

Cornfield LJ, Nelson DL, Monroe PJ, Taylor EW, Nikam SS (1988): Use of forskolin stimulated adenylate cyclase in rat hippocampus as a screen for compounds that act through 5-HT$_{1A}$ receptors. *Proc West Pharmacol Soc* 31:265–267

Cornfield LJ, Nelson DL, Taylor EW, Martin AR (1989): MDL 73005: Partial agonist at the 5-HT$_{1A}$ receptor negatively linked to adenylate cyclase. *Eur J Pharmacol* 173:189–192

Cossery JM, Gozlan H, Spampinato U, Perdicakis C, Guillaumet G, Pichat L, Hamon M (1987): The selective labelling of central 5-HT$_{1A}$ receptor binding sites by [^{3}H]5-methoxy-3-(di-*n*-propylamino) chroman. *Eur J Pharmacol* 140:143–155

Cotecchia S, Kobilka BK, Daniel KW, Nolan RD, Lapetina EY, et al. (1990): Multiple second messenger pathways of α-adrenergic receptor subtypes expressed in eukaryotic cells. *J Biol Chem* 265:63–69

Dalton DW, Feniuk W, Humphrey PPA (1985): The mechanism of the hypotensive action of 5-carboxyamidotryptamine in conscious DOCA-salt hypertensive rats. *Br J Pharmacol* 86:737P

Daval G, Vergé D, Basbaum AI, Bourgoin S, Hamon M (1987): Autoradiographic evidence of serotonin1 binding sites on primary afferent fibres in the dorsal horn of the rat spinal cord. *Neurosci Lett* 83:71–76

De Vivo M, Maayani S (1986): Characterization of the 5-HT$_{1A}$ receptor-mediated inhibition of forskolin-stimulated adenylate cyclase activity in guinea pig and rat hippocampal membranes. *J Pharmacol Exp Ther* 238:248–253

Devoino LV, Ilyutchenok R Yu (1968): Influence of some drugs on the immune response. *Eur J Pharmacol* 4:449–456

Dixon RAF, Kobilka BK, Strader DJ, Benovic JL, Dohlman HG, Frielle T, Bolanowski MA, Bennett CD, Rands E, Diehl RE, Mumford RA, Slater EE, Sigal IS, Caron MG, Lefkowitz RJ, Strader CD (1986): Cloning of the gene and cDNA for mammalian β-adrenergic receptor and homology with rhodopsin. *Nature* 321:75–79

Dohlman HG, Caron MG, DeBlasi A, Frielle T, Lefkowitz RJ (1990): A role of extracellular disulfided bonded cysteines in the ligand binding function of the β_2 adrenergic receptor. *Biochemistry* 29:2335–2342

Dompert WU, Glaser T, Traber J (1985): [^{3}H]-TVX Q 7821: identification of 5-HT$_1$ binding sites as target for a novel putative anxiolytic. *Naunyn-Schmiedebergs Arch Pharmacol* 328:467–470

Dourish CT, Ahlenius A, Hutson PL, eds. (1987): *Brain 5-HT$_{1A}$*. Chichester, England: Ellis Horwood

Dumuis A, Sebben M, Bockaert J (1988a): Pharmacology of 5-hydroxytrypta-mine$_{1A}$ receptors which inhibit cAMP production in hippocampal and cortical neurons in primary culture. *Mol Pharmacol* 33:176–186

Dumuis A, Bouhelal R, Sebben M, Cory R, Bockaert J (1988b): A nonclassical 5-hydroxytryptamine receptor positively coupled with adenylate cyclase in the CNS. *Mol Pharmacol* 34:880–887

El Mestikawy S, Cognard C, Gozlan H, Hamon M (1988): Pharmacological and biochemical characterization of rat hippocampal 5-hydroxytryptamine$_{1A}$ recep-tors solubilized by 3[3-(cholamidopropyl)dimethylammonio)]1-propane sulfonate (CHAPS). *J Neurochem* 51:1031–1040

El Mestikawy S, Taussig D, Gozlan H, Emerit MB, Ponchant M, Hamon M (1989): Chromatographic analysis of the 5-HT$_{1A}$ receptor solubilized from the rat hippocampus. *J Neurochem* 53:1555–1566

El Mestikawy S, Riad M, Laporte A-M, Vergé D, Daval G, Gozlan H, Hamon M (1990): Production of specific anti-rat 5-HT$_{1A}$ receptor antibodies in rabbits injected with a synthetic peptide. *Neurosci Lett* 118:189–192

El Mestikawy S, Fargin A, Raymond JR, Gozlan H, Hnatowich M (1991): The 5-HT$_{1A}$ receptor: An overview of recent advances. *Neurochem Res* 16:1–10

Emerit MB, El Mestikawy S, Gozlan H, Cossery JM, Besselièvre R, Marquet A, Hamon M (1987): Identification of the 5-HT$_{1A}$ binding subunit in rat brain membranes using the photoaffinity probe [^{3}H]8-methoxy-2-[*N-n*-propyl, *N*-3-(2-nitro-4-azidophenyl)aminopropyl]tetralin. *J Neurochem* 49:373–380

Emerit MB, Miquel MC, Gozlan H, Hamon M (1991): The GTP-insensitive component of high affinity [^{3}H]-8-hydroxy-2(di-n-propylamino)tetralin binding in the rat hippocampus corresponds to an oxidized state of the 5-HT$_{1A}$ receptor *J Neurochem* 4956:1705–1716

Ennis C, Kemp JD, Cox B (1981): Characterization of inhibitory 5-HT receptors that modulate dopamine release in the striatum. *J Neurochem* 36:1515–1520

Fargin A, Raymond JR, Lohse MJ, Kobilka BK, Lefkowitz RJ, Caron MG (1988): The genomic clone G-21 which resembles a β-adrenergic receptor sequence encodes the human 5-HT$_{1A}$ receptor. *Nature* 335:358–360

Fargin A, Raymond JR, Regan JW, Cotecchia S, Lefkowitz RJ, Caron MG (1989): Second messenger linkages of the 5-HT$_{1A}$ receptor expressed in eukaryotic cells. *J Biol Chem* 264:14848–14852

Fargin A, Yamamoto K, Cotecchia S, Goldsmith PK, Spiegel AM, Lapetina EG, Caron MG, Lefkowitz RJ (1991): Dual coupling of the cloned 5-HT$_{1A}$ receptor to both adenylyl cyclase and phospholipase C is modulated by the same G$_i$ protein. *Cell Signalling* 3:547–557

Finch L (1974): Vascular reactivity in hypertensive rats after treatment with antihypertensive agents. *Life Sci* 15:1827–1836

Fitzgerald L, Titeler M, Glennon RA, Yocca F (1989): Tritiated NAN-190 and tritiated BMY-7378: Antagonist radioligand probes of the brain 5-HT$_{1A}$ receptor. *Neurosci Abstr* 15:422

Fozard JR, Kilbinger H (1985): 8-OH-DPAT inhibits transmitter release from guinea-pig enteric cholinergic neurones by activating 5-HT$_{1A}$ receptors. *Br J Pharmacol* 86:601P

Fraser CM (1989): Site-directed mutagenesis of β-adrenergic receptors. Identification of conserved cysteine residues that independently affect ligand binding and receptor activation. *J Biol Chem* 264:9266–9270

Frielle T, Collins S, Daniel KW, Caron MG, Lefkowitz RJ, Kobilka BK (1987): Cloning of the cDNA for the human β_1-adrenergic receptor. *Proc Natl Acad Sci USA* 84:7920–7924

Frielle T, Daniel K, Caron MG, Lefkowitz RJ (1988): Structural basis of β-adrenergic receptor subtype specificity studied with chimeric β_1- β_2-adrenergic receptors. *Proc Natl Acad Sci USA* 85:9494–9498

Gallagher TK, Wang HH (1988): Purification and reconstitution of serotonin receptors from bovine brain. *Proc Natl Acad Sci USA* 85:2378–2382

Gartride SE, Cowen PJ, Hjorth S (1990): Effecfts of MDL 73005 on central pre- and postsynaptic 5-HT$_{1A}$ receptor function in the rat in vivo. *Eur J Pharmacol* 191:391–400

Glaser T, Traber J (1985): Binding of the putative anxiolytic TVX Q-7821 to hippocampal 5-hydroxytryptamine (5-HT) recognition sites. *Naunyn-Schmiedebergs Arch Pharmacol* 329:211–215

Goodwin GM, Green AR (1985): A behavioural and biochemical study in mice and rats of putative selective agonists and antagonists for 5-HT$_1$ and 5-HT$_2$ receptors. *Br J Pharmacol* 84:743–753

Gozlan H, El Mestikawy S, Pichat S, Glowinski J, Hamon M (1983): Identification of presynaptic serotonin autoreceptors using a new ligand: ^{3}H-PAT. *Nature* 305:140–142

Gozlan H, Emerit MB, Hall MD, Nielsen M, Hamon M (1987): In situ molecular sizes of the various types of 5-HT binding sites in the rat brain. *Biochem Pharmacol* 35:1891–1897

Gozlan H, Ponchant M, Daval G, Verge G, Menard F, Vanhove A, Beaucort JP, Hamon M (1988a): ^{125}I-Bolton-Hunter-8-methoxy-2-[N-propyl-N-propylamino]-tetralin as a new selective radioligand of 5-HT$_{1A}$ sites in the rat brain. *J Pharmacol Exp Ther* 244:751–759

Gozlan H, Emerit MB, El Mestikawy S, Cossery JM, Marquet A, Besselevièvre, Hamon M (1988b): Photoaffinity labeling and solubilization of the central 5-HT$_{1A}$ receptor binding site. *J Recept Res* 7:195–221

Hall MD, El Mestikawy S, Emerit MB, Pichat L, Hamon M, Gozlan H (1985): [^{3}H]-8-OH-2-di-n-propylamino]tetralin binding to pre- and post-synaptic 5-hydroxytryptamine binding sites in various regions of the rat brain. *J Neurochem* 44:1685–1692

Hamon M, Fattaccini CM, Adrien J, Gallissot MC, Martin P, Gozlan H (1988): Alterations of central serotonin and dopamine turnover in rats treated with ipsapirone and other 5-HT$_{1A}$ agonists with potential anxiolytic properties. *J Pharmacol Exp Ther* 246:745–752

Hamon M, Emerit MB, El Mestikawy S, Gallissot MC, Gozlan H (1990): Regional differences in the transduction mechanisms of serotonin receptors in the mammalian brain. In: *Cardiovascular Pharmacology of 5-HT: Prospective Therapeutic Applications,* Saxena PR, Wallis D, Woute W, Gevan P eds. Dordrecht: Kluwer Acad Publ, pp. 41–59

Hartig PR (1989): Molecular biology of 5-HT receptors. *Trends Pharmacol Sci* 10:64–69

Hellstrand K, Hermodsson S (1987): Role of serotonin in the regulation of human natural killer cell cytotoxicity. *J Immunol* 139:869–875

Hellstrand K, Hermodsson S (1987): Enhancement of human natural killer cell cytotoxicity by serotonin: Role of non-T/CD16+ NK cells, accessory monocytes, and 5-HT$_{1A}$ receptors. *Cell Immunol* 127:199–214

Henning M, Rubenson A (1971): Effects of 5-hydroxytryptophan on arterial blood pressure, body temperature and tissue monoamines in the rat. *Acta Pharmacol Toxicol* 29:145–154

Herrick-Davis K, Titeler M (1988): [^{3}H]-Spiroxatrine, a 5-HT$_{1A}$ radioligand with agonist binding properties. *J Neurochem* 50:528–533

Heuring RE, Peroutka SJ (1987): Characterization of a novel [^{3}H]5-hydroxytryptamine binding site in bovine brain membranes. *J Neurosci* 7:894–903

Hibert M, Moser P (1990): MDL-72382 and MDL-73005, novel, potent and selective 5-HT$_{1A}$ receptor ligands with different pharmacological properties. *Drugs Future* 15:159–170

Hjorth S, Sharp T (1990) Mixed agonist-antagonist properties of NAN-190 at 5-HT$_{1A}$ receptors behavioral and in-vivo brain microdialysis studies. *Life Sci* 46:955–963

Hoyer D, Engel G, Kalkman HO (1985a): Characterization of the 5-HT$_{1B}$ recognition site in rat and pig brain membranes: Binding studies with (−)[^{125}I]iodocyanopindolol. *Eur J Pharmacol* 118:1–12

Hoyer D, Engel G, Kalkman HO (1985b): Molecular pharmacology of 5-HT$_1$ and 5-HT$_2$ recognition sites in rat and pig brain membranes: Radioligand binding studies with [^{3}H]5-HT, [^{3}H]8-OH-DPAT, (−)[^{125}I]iodocyanopindolol, and [^{3}H]mesulergine. *Eur J Pharmacol* 118:13–23

Hoyer D, Pazos A, Probst A, Palacios JM (1986): Serotonin receptors in human brain. I. Characterization and autoradiographic localization of 5-HT$_{1A}$ recognition sites. Apparent absence of 5-HT$_{1B}$ recognition sites. *Brain Res* 376:85–96

Hoyer D (1989): 5-Hydroxytryptamine receptors and effector coupling mechanisms in peripheral tissues. In: *Peripheral Actions of 5-HT,* Fozard JR, ed. London: Oxford University Press, pp 72–99

Hutson PH, Dourish CT, Curzon G (1986): Neurochemical and behavioural evidence for mediation of the hyperphagic action of 8-OH-DPAT by 5-HT cell body autoreceptors. *Eur J Pharmacol* 129:347–352

Hutson PH, Donohoe TP, Curzon G (1987): Hypothermia induced by the putative 5-HT$_{1A}$ receptor agonists LY 165163 and 8-OH-DPAT is not prevented by 5-HT depletion. *Eur J Pharmacol* 143:221–228

Inei JR, Meyerson LR (1988): The 5-HT$_{1A}$ receptor probe [^{3}H]-8-OH-DPAT labels the 5-HT transporter in human platelets. *Life Sci* 42:311–320

Invernizzi RW, Cervo L, Samanin R (1988): 8-Hydroxy-2-(di-*n*-propylamin) tetralin, a selective serotonin$_{1A}$ receptor agonist, blocks haloperidol-induced catalepsy by an action on raphe nuclei medianus and dorsalis. *Neuropharmacology* 27:515–518

Joëls M, Shinnick-Gallagher P, Gallagher JP (1987): Effect of serotonin and serotonin analogues on passive membrane properties of lateral septal neurons in vitro. *Brain Res* 417:99–107

Kalkman HO, Engel G, Hoyer D (1984): Three distinct subtypes of serotonergic receptors mediate the triphasic blood pressure response to serotonin in rats. *J Hypertension* 2(suppl 3):143–145

Karnik SS, Sakmar TP, Chen H-B, Khorana HG (1988). Cysteine residues 110 and 187 are essential for the formation of correct structure in bovine rhodopsin. *Proc Natl Acad Sci USA* 85:8459–8463

Karschin A, Ho BY, LaBarca C, Elroy-Stein O, Moss B, Davidson N, Lester H (1991): Heterologously expressed serotonin 1A receptors couple to muscarinic K^{+} channels in heart. *Proc Natl Acad Sci USA* 88:5694–5698

Kennett GA, Marcou M, Dourish CT, Curzon G (1987): Single administration of 5-HT$_{1A}$ agonists decreases 5-HT$_{1A}$ presynaptic, but not postsynaptic receptor-mediated responses: Relationship to antidepressant-like action. *Eur J Pharmacol* 138:53–60

Kilbinger H, Pfeuffer-Friedrich I (1985): Two types of receptors for 5-hydroxytryptamine on the cholinergic nerves of the guinea-pig myenteric plexus. *Br J Pharmacol* 85:529–539

Kobilka BK, Dixon RAF, Frielle T, Dohlman HG, Bolanowski MA, Sigal IS, Yang-Feng T, Francke U, Caron MG, Lefkowitz RJ (1987a): cDNA for the human β_2-adrenergic receptor: A protein with multiple membrane-spanning domains encoded by a gene whose chromosomal location is shared with that of the receptor for platelet-derived growth factor. *Proc Natl Acad Sci USA* 84:46–50

Kobilka BK, Frielle T, Collins S, Yang-Feng T, Kobilka TS, Francke U, Lefkowitz RJ, Caron MG (1987b): Identification of an intronless gene which encodes a potential member of the family of G protein-coupled receptors. *Nature* 329:75–79

Kobilka BK, Kobilka TS, Daniel KW, Regan JW, Caron MG, Lefkowitz RJ (1988): Chimeric α_2- β_2-adrenergic receptors: Delineation of domains involved in effector coupling and ligand binding specificity. *Science* 240:1310–1316

Kwong LL, Smith ER, Davidson JM, Peroutka SJ (1986): Differential interactions of "prosexual" drugs with 5-hydroxytryptamine$_{1A}$ and α_2-adrenergic receptors. *Behav Neurosci* 100:664–668

Kyte J, Doolittle RF (1982): A simple method for displaying the hydropathic character of a protein. *J Mol Biol* 157:105–132

Lanier SM, Homcy CJ, Patenaude C, and Graham RM (1988): Identification of structurally distinct α_2-adrenergic receptors. *J Biol Chem* 263:14491–14496

Leonhardt S, Herrick-Davis K, Titeler M (1989): Detection of a novel serotonin receptor subtype (5-HT1E) in human brain: Interaction with a GTP-binding protein. *J Neurochem* 53:465–471

Lespinats G, Bonnett M, Tlouzeau S, Burtin C (1984): Enhancement by serotonin of intra-tumour penetration of spleen cells. *Br J Cancer* 50:545–547

Leysen JE, Tollenaere JP, Koch MHJ, Laduron P (1977): Differentiation of opiate and neuroleptic receptor binding in rat brain. *Eur J Pharmacol* 43:253–267

Leysen JE, Niemegeers CJE, Tollenaere JP, Laduron PM (1979): Serotonergic component of neuroleptic receptors. *Nature* 272:169–171

Leysen JE (1989): The use of 5-HT receptor agonists and antagonists for the characterization of their respective sites. In: *Neuromethods, Neuropharmacology II: Drugs as Tools in Neurotransmitter Research,* Boulton AA, Baker GB, Jourio AV, eds. Clifton, N.J.: Humana Press, pp 299–350

Libert F, Parmentier M, Lefort A, Dinsart C, Van Sande J, Maenhaut C, Simons M-J, Dumont JE, Vassart, G (1989): Selective amplification and cloning of four new members of the G protein-coupled receptor family. *Science* 244:569–572

Limbird LE (1988): Receptors linked to inhibition of adenylate cyclase: Additional signaling mechanisms. *FASEB J* 2:2686–2695

Liu YF, Albert PR (1991): Cell-specific signaling of the 5-HT$_{1A}$ receptor. *J Biol Chem* 266:23689–23697

Lucki I, Nobler MS, Frazer A (1983): Differential actions of serotonin antagonists on two behavioral models of seroronin receptor activation in the rat. *J Pharmacol Exp Ther* 228:133–139

Makman MH, Dvorkin B, Crain SM (1988): Modulation of adenylate cyclase activity of mouse spinal cord-ganglion explants by opioids, serotonin and pertussis toxin. *Brain Res* 445:303–313

Mangel A, Raymond JR, Fitz JG (1992): Co-regulation of high conductance anion channels by GTP-binding proteins in CHO cells. Submitted

Marcinkiewicz M, Vergé D, Gozlan H, Pichat L, Hamon M (1984): Autoradiographic evidence for the heterogeneity of 5-HT$_1$ sites in the rat brain. *Brain Res* 291:159–163

Markstein R, Hoyer D, Engel G (1986): 5-HT$_{1A}$ receptors mediate stimulation of adenylate cyclase in rat hippocampus. *Naunyn-Schmiedebergs Arch Pharmacol* 333:335–341

Martin KF, Mason R (1987): Ipsapirone is a partial agonist at 5-hydroxytryptamine$_{1A}$ (5-HT$_{1A}$) receptors in the rat hippocampus: Electrophysiological evidence. *Eur J Pharmacol* 141:479–483

Mawe GM, Branchek TA, Gershon MD (1986): Peripheral neural serotonin receptors: Identification and characterization with specific antagonists and agonists. *Proc Natl Acad Sci USA* 83:9799–9803

McCall RB, Patel BN, Harris LT (1987): Effects of serotonin 1 and 2 receptor agonists and antagonists on blood pressure, heart rate and sympathetic nerve activity. *J Pharmacol Exp Ther* 242:1152–1159

Middlemiss DN, Neill J, Tricklebank MD (1985): Subtypes of the 5-HT receptor involved in hypothermia and forepaw treading produced by 8-OH-DPAT. *Br J Pharmacol* 85:251P

Middlemiss DN, Fozard JR (1983): 8-OH-2-(di-*n*-propylamino)tetralin discriminates between subtypes of 5-HT$_1$ recognition sites. *Eur J Pharmacol* 90:151–153

Middleton JP, Raymond JR, Whorton ARR, Dennis VW (1990): Short-term regulation of Na^+/K^+ adenosine triphosphatase by recombinant human serotonin 5-HT_{1A} receptor expressed in HeLa cells. *J Clin Invest* 86:1799–1805

Minchin MCW, Godfrey PP, McClue SJ, Young MM (1985): 8-OH-DPAT stimulates inositol phospholipid breakdown in rat cerebral cortical slices. *J Neurochem* 44(suppl):S49

Mir AK, Hibert M, Tricklebank MD, Middlemiss DN, Kidd EJ, Fozard JR (1988): MDL 72832: A potent and stereoselective ligand at central and peripheral 5-HT_{1A} receptors. *Eur J Pharmacol* 149:107–120

Moon SL, Taylor DP (1985): In vitro autoradiography of ^{3}H-buspirone and ^{3}H-2-deoxyglucose after buspirone administration. *Soc Neurosci Abstr* 11:114

Moser PC, Ticklebank MD, Middlemiss DN, Mir AK, Hibert MF and Fozard JR (1990): Characterization of MDL-73005 as a 5-HT_{1A} selective ligand and its effects in animal models of anxiety: Comparison with buspirone, 8-hydroxy-DPAT and diazepam. *Br J Pharmacol* 99:343–349

Müller-Schwenitzer E, Engel G (1983): Evidence for mediation by 5-HT_2 receptors of 5-hydroxytryptamine-induced contraction of canine basilar artery. *Naunyn-Schmiedebergs Arch Pharmacol* 327:18–22

Nelson DL, Monroe PJ, Lambert G, Yamamura HI (1987): [^{3}H]-Spiroxatrine labels serotonin$_{1A}$-like sites in the rat hippocampus. *Life Sci* 41:1567–1576

Newberry NR, Priestly T (1988): A 5-HT1-like receptor mediates a pertussis toxin-sensitive inhibition of rat ventromedial hypothalamic neurones in vitro. *Br J Pharmacol* 95:6–8

Niemegeers CJE, Verbruggen FJ, Van Neuten JM, Janssen PAJ (1964): Spiroxamide (R 5188): A new compound producing morphine-like and chlorpromazine-like effects in animals. *Int J Neuropharmacol* 2:349–354

Norman AB, Battaglia G, Creese I (1985): [^{3}H]WB4101 labels the 5-HT_{1A} receptor subtype in rat brain. *Mol Pharmacol* 28:487–494

O'Dowd BF, Hnatowich M, Caron MG, Lefkowitz RJ, Bouvier M (1989): Palmitoylation of the human β_2-adrenergic receptor. *J Biol Chem* 264:7564–7569

Oliveras JL, Redjemi F, Guibaud G, Besson JM (1975): Analgesia induced by electrical stimulation of the inferior centralis nucleus of the raphe in the cat. *Pain* 1:139–145

Ovchinnikov YA, Adulaev NG, Bogachuk AS (1988): Two adjacent cysteine residues in the C-terminal cytoplasmic fragment of bovine rhodopsin are palmitylated. *FEBS Lett* 230:1–5

Page IH (1952): The vascular action of natural serotonin, 5- and 7-hydroxytryptamine and tryptamine. *J Pharmacol Exp Ther* 105:58–73

Palacios JM, Pazos A, Hoyer D (1987): Characterization and mapping of 5-HT_{1A} sites in the brain of animals and man. In: *Brain 5-HT$_{1A}$ Receptors.* Dourish CT, Ahlenius A, Hutson PH, eds. Chichester: Ellis Horwood, pp 67–81

Pazos A, Probst A, Palacios JM (1986): Serotonin receptors in the human brain. III. Autoradiographic mapping serotonin-1 receptors. *Neuroscience* 21:97–122

Pedigo NW, Yamamura HI, Nelson DL (1981): Discrimination of multiple [^{3}H]-5-hydroxytryptamine binding sites by the neuroleptic spiperone in rat brain. *J Neurochem* 36:220–226

Peralta EG, Ashkenazi A, Winslow JW, Ramachandran JS, Capon DJ (1988): Differential regulation of PI hydrolysis and adenylyl cyclase by muscarinic receptor subtypes. *Nature* 334:434–437

Peroutka SJ, Snyder SH (1979): Multiple serotonin receptors: Differential binding of [^{3}H]-5-hydroxytryptamine, [^{3}H]-lysergic acid diethylamide and [^{3}H]-spiroperidol. *Mol Pharmacol* 16:687–699

Peroutka SJ, Huang S, Allen GS (1986): Canine basilar artery contractions mediated by 5-hydroxytryptamine$_{1A}$ receptors. *J Pharmacol Exp Ther* 237:901–906

Peroutka SJ (1987a): Serotonin receptors. In: *Psychopharmacology: The Third Generation of Progress,* Meltzer H, ed. New York: Raven Press, pp 303–311

Peroutka SJ, Mauk MD, Kocsis JD (1987b): Modulation of hippocampal neuronal activity by 5-hydroxytryptamine and 5-hydroxytryptamine$_{1A}$ selective drugs. *Neuropharmacology* 26:139–146

Pfeuffer-Friedrich I, Kilbinger H (1985): The effect of LSD in the guinea-pig ileum. Inhibition of acetylcholine release and stimulation of smooth muscle. *Naunyn-Schmiedebergs Arch Pharmacol* 331:311–315

Przeglinksi E, Ismaiel AM, Chojnacka-Wojcik B, Budziszewska B, Tatarcynska D, Blaszcynska E (1990): The behavioral, but not the hypothermic or cortisone response to 8-hydroxy-2-(di-*N*-propylamino)-tetralin, is antagonized by NAN-190 in the rat. *Neuropharmacology* 29:521–526

Ransom RW, Asarch KB, Shih J (1986a): [^{3}H]1-[2-(4-aminophenyl)ethyl]-4-(3-trifluoromethylphenyl)piperazine: A selective radioligand for 5-HT$_{1A}$ receptors in rat brain. *J Neurochem* 46:68–75

Ransom RW, Asarch KB, Shih J (1986b): Photoaffinity labeling of the 5-hydroxytryptamine$_{1A}$ receptor in rat hippocampus. *J Neurochem* 47:1066–1072

Raymond JR, Fargin A, Lohse M, Senogles S, Regan JW, Lefkowitz RJ, Caron MG (1989a): Identification of the ligand binding subunit of the human 5-HT$_{1A}$ receptor. *Mol Pharmacol* 36:15–21

Raymond JR, Fargin AF, Middleton JP, Graff JM, Haupt DM, Caron MG, Lefkowitz RJ, Dennis VW (1989b): The human 5-HT$_{1A}$ receptor expressed in HeLa cells stimulates sodium-dependent phosphate uptake *via* protein kinase C. *J Biol Chem* 264:21943–21950

Raymond JR, Hnatowich M, Lefkowitz RJ, Caron MG (1989c): Adrenergic receptors: Models for the regulation of signal transduction processes. *Hypertension* 15:119–131

Raymond JR, Albers FJ, Middleton JP, Lefkowitz RJ, Caron MG, Obeid LM, Dennis VW (1991a): 5-HT$_{1A}$ and histamine H$_1$ receptors expressed in HeLa cells stimulate phosphoinositide hydrolysis and phosphate uptake via distinct G protein pools. *J Biol Chem* 266:372–379

Raymond JR (1991): Protein kinase C induces phosphorylation and desensitization of the human 5-HT$_{1A}$ receptor *J Biol Chem* 266:14747–14753

Raymond JR, Albers FJ, Middleton JP (1992): Functional expression of human 5-HT$_{1A}$ receptors and differential coupling to second messengers in CHO cells. *Naunyn Schmiedebergs Arch Pharmacol.* (in press)

Regan JW, Kobilka TS, Yang-Feng TL, Caron MG, Lefkowitz RJ, Kobilka BK (1988): Cloning and expression of a human kidney cDNA for an α_2-adrenergic receptor subtype. *Proc Natl Acad Sci USA.* 85:6301–6305

Richardson BP, Engel G. (1986): The pharmacology and function of 5-HT$_3$ receptors. *Trends Neurosci* 7:424–428

Rydelek-Fitzgerald L, Teitler M, Fletcher PW, Ismaiel AM, Glennon RA (1990):

NAN-190, agonist and antagonist interactions with brain 5-HT$_{1A}$ receptors. *Brain Res* 532:191–196

Saxena PR, Verdouw PD (1985): 5-Carboxyamide tryptamine, a compound with high affinity for 5-HT1 binding sites, dilates arterioles and constricts arteriovenous anastomoses. *Br J Pharmacol* 84:533–544

Saxena PR, Lawang A (1985): A comparison of cardiovascular and smooth muscle effects of 5-HT and 5-CT, a selective agonist of 5-HT$_1$ receptors. *Arch Int Pharmacodyn Ther* 277:235–252

Saxena PR, Villalón CM (1990): Brain 5-HT$_{1A}$ receptor agonism: A novel mechanism for antihypertensive action. *Trends Pharmacol Sci* 11:95–96

Schoeffter P, Hoyer D (1988): Centrally acting hypotensive agents with affinity to 5-HT$_{1A}$ binding sites inhibit forskolin-stimulated adenylate cyclase activity in calf hippocampus. *Br J Pharmacol* 95:975–985

Schoemaker H, Langer SZ (1986): [^{3}H]-8-OH-DPAT labels the serotonin transporter in rat striatum. *Eur J Pharmacol* 124:371–373

Sharp T, Beckus LI, Hjorth S, Bramwell SR, Grahame-Smith DG (1990): Further investigation of the in vivo pharmacological properties of the putative 5-HT$_{1A}$ antagonist, BMY 7378. *Eur J Pharmacol* 176:331–340

Shenker A, Maayani S, Weinstein H, Green JP (1985): Two 5-HT receptors linked to adenylate cyclase in guinea pig hippocampus are discriminated by 5-carboxamidotryptamine and spiperone. *Eur J Pharmacol* 109:427–429

Sigal IS, Dixon RAF, Strader CD (1988): Molecular biology of adrenergic receptors. *ISI Atlas Sci: Pharmacol* 2:387–391

Sinton CM, Fallon SL (1986): Differences in responses of dorsal and median raphe serotonergic neurons to 5-HT1 receptor ligands. *Soc Neurosci Abstr* 12:1239

Smith LM, Peroutka SJ (1986): Differential effects of 5-hydroxytryptamine$_{1A}$ selective drugs on the 5-HT behavioral syndrome. *Pharmacol Biochem Behav* 24:1513–1519

Sprouse JS, Aghajanian GK (1987): Electrophysiological responses of serotonergic dorsal raphe neurons to 5-HT$_{1A}$ and 5-HT$_{1B}$ agonists. *Synapse* 1:3–9

Sternberg EM, Wedner HJ, Leung MK, Parker CW (1987): Effect of 5-HT and other monoamines on murine macrophages. *J Immunol* 138:4360–4365

Strader CD, Candelore MR, Hill WS, Sigal IS, Dixon RAF (1989): Identification of two serine residues involved in agonist activation of the β-adrenergic receptor. *J Biol Chem* 264:13572–13578

Taylor EW, Duckles SP, Nelson DL (1986): Dissociation constants of serotonin agonists in the canine basilar artery correlate to Ki values at the 5-HT$_{1A}$ binding site. *J Pharmacol Exp Ther* 236:118–125

Traber J, Glaser T (1987): 5-HT$_{1A}$ receptor-related anxiolytics. *Trends Pharmacol Sci* 8:432–437

Tricklebank MD, Forler C, Fozard JR (1985): The involvement of subtypes of the 5-HT$_1$ receptor and of the catecholaminergic systems in the behavioral responses to 8-hydroxy-2-(di-*n*-propylamino)tetralin in the rat. *Eur J Pharmacol* 106:271–282

Tricklebank MD (1985): The behavioural response of 5-HT receptor agonists and subtypes of the central 5-HT receptor. *Trends Pharmacol Sci* 6:403–407

Trulson ME, Aresteh K (1986): Buspirone decreases the action of 5-hydroxytryptamine-containing dorsal raphe neurons in vitro. *J Pharm Pharmacol* 38:380–382

VanderMaelen CP, Matheson GK, Wilderman RC, Patterson LA (1986): Inhibition of serotonergic dorsal raphe neurons by systematic and iontophoretic administration of buspirone, a non-benzodiazepine anxiolytic drug. *Eur J Pharmacol* 129:123–130

Vanhoutte PM, Cohen RA, Van Neuten JM (1984): Serotonin and arterial vessels. *J Cardiovasc Pharmacol* 6(suppl 6):S421–S428

Van Nueten JM, Leysen JE, De Clerck F, Vanhoutte PM (1984): Serotonergic receptor subtypes and vascular reactivity. *J Cardiovasc Pharmacol* 6(suppl 4):S564–S574

Vergé D, Daval G, Marcinkiewicz, Patey A, El Mestikawy S, Gozlan H, Hamon M (1986): Q: Quantitative autoradiography of multiple 5-HT1 receptor subtypes in the brain of control or 5,7-dihydroxytryptamine-treated rats. *J Neurosci* 6:3474–3482

Vergé D, Daval G, Patey A, Gozlan H, El Mestikawy S, Hamon M (1985): Presynaptic 5-HT autoreceptors on serotonergic cell bodies and/or dendrites but not terminals are of the 5-HT$_{1A}$ subtype. *Eur J Pharmacol* 113:463–464

Webb RC (1982): Increased vascular sensitivity to 5-HT and methysergide in hypertension in rats. *Clin Sci* 63:73s

Weissmann-Nanopoulos D, Mach E, Magre J, Demassey Y, Pujol JF (1985): Evidence for the localization of 5-HT$_{1A}$ binding sites on serotonin containing neurons in the raphe dorsalis and raphe centralis nuclei of the rat brain. *Neurochem Int* 7:1061–1072

Weiss S, Sebben M, Kemp DE, Bockaert J (1986): Serotonin 5-HT$_1$ receptors mediate inhibition of cyclic AMP production in neurons. *Eur J Pharmacol* 120:227–230

Wilkinson LO, Abercrombie ED, Rasmussen K, Jacobs BL (1987): Effect of buspirone on single unit activity in locus coeruleus and dorsal raphe nucleus in behaving cats. *Eur J Pharmacol* 136:123–127

Yocca FD, Hyslop DK, Smith DW, Maayani S (1987): BMY 7378, a buspirone analog with high affinity, selectivity, and low intrinsic activity at the 5-HT$_{1A}$ receptor in rat and guinea-pig hippocampal membranes. *Eur J Pharmacol* 137:293–294

Yocca FD (1990): Neurochemistry and neurophysiology of buspirone and gepirone: Interactions at presynaptic and postsynaptic 5-HT$_{1A}$ receptors. *J Clin Psychopharmacol* 10:6S–12S

Zemlan FP, Zieleniewski-Murphy A, Murphy RM, Behbahani MM (1990): BMY-7378: Partial agonist at spinal cord 5-HT$_{1A}$ receptors. *Neurochem Int* 16:515–522

Zgombick JM, Beck SG, Mahle CD, Craddock-Royal B, Maayani S (1989): Pertussis toxin-sensitive guanine nucleotide-binding proteins couple adenosine A1 and 5-hydroxytryptamine1A receptors to the same effector systems in rat hippocampus: Biochemical and electrophysiological studies. *Mol Pharmacol* 35:484–494

Zgombick JM, Weinshank RL, Macchi M, Schecter LE, Branche KT, Hartig P (1991): Expression and pharmacological characterization of a canine 5-HT$_{1D}$ receptor subtype. *Mol Pharmacol* 40:1036–1042

6

The Dopamine D1 Receptors

Hyman B. Niznik, Roger K. Sunahara,
Hubert H. M. Van Tol, Philip Seeman,
David M. Weiner, Tom M. Stormann,
Mark R. Brann, and Brian F. O'Dowd

Many psychomotor signs and symptoms are influenced by the activity of dopaminergic neurons and by drugs that selectively interact with neuronal dopamine receptors. These include rigidity in Parkinson's disease, hallucinations in schizophrenia and Alzheimer's disease, dyskinesia in Huntington's chorea, and spontaneous oral dyskinesia in the elderly (Seeman, 1987; Seeman et al., 1987; Seeman and Niznik, 1990). These receptors may also play a significant role in drug addiction and alcoholism (Blum et al., 1990). On the basis of pharmacological, biochemical, and physiological criteria, receptors for dopamine have been classified into two types, termed D1 and D2 (Kebabian and Calne, 1979; Seeman, 1980; Niznik and Jarvie, 1989). These are members of a large gene family of hormone/neurotransmitter receptors that exert their biological actions via signal transduction pathways that involve guanine nucleotide-binding proteins. While D2 dopamine receptors inhibit the activation of adenylyl cyclase and appear to couple to numerous other effector systems, D1 dopamine receptors, found in the brain, retina, and parathyroid gland, stimulate adenylyl cyclase and subsequently activate cAMP-dependent protein kinases (Niznik, 1987; Seeman and Niznik, 1988; Hess and Creese, 1987; Hemmings et al., 1987).

Although the exact functional correlate of D1 dopamine receptor stimulation is unknown, D1 receptors appear to regulate neuron growth and differentiation (Lankford et al., 1988), elicit some behavioral responses, and modulate the functional expression of dopamine D2 receptor-mediated events (Clark and White, 1987; Waddington, 1989; Waddington and O'Boyle, 1989). However, the molecular mechanisms mediating the observed interactions between D1 and D2 dopamine receptors in vitro and its disruption in some human neuropsychiatric diseased tissues (Seeman et al., 1989) have not been elucidated. To facilitate the study of these molecular events we have cloned the gene encoding the dopamine D1 receptor (Sunahara et al., 1990) and report here on the examination of its deduced protein structure, its pharmacological characteristics following expression in cultured cells, chromosomal localization of the gene, and the pattern of its expression within the mammalian brain.

1. Cloning Strategy

Many methods are available for the isolation of specific genes. The most common of these requires purification to homogeneity of the gene protein product, a feat not yet achieved for the D1 dopamine receptor, or the isolation of a related gene from a different species. To circumvent the difficulties posed by protein purification, we devised a method, based on the polymerase chain reaction (PCR), to clone the dopamine D1 receptor. Dopamine receptors belong to the large G-linked family of receptors, the cloned members of which display substantial amino acid sequence conservation, especially in regions proposed to be transmembrane (TM) domains (O'Dowd et al., 1989a). This sequence homology suggested that the genes encoding other members of this receptor family, such as the dopamine D1 receptor, might be isolated with the use of degenerate PCR oligonucleotide primers. Using one such degenerate oligonucleotide (with sequence based on the D2 dopamine receptor within TM 3 and the same domain of 14 other G-linked receptors) and an oligonucleotide primer homologous to sequences in the phage vector of a rat striatal λgt11 cDNA library, PCR amplication of the cDNA resulted in the isolation of a rat clone (G-36) of approximately 450 bp. This clone appeared to code for a new member of the G-linked receptor family since it shared considerable amino acid homology, particularly within putative transmembrane domains (45%), with the rat β_2-AR (O'Dowd et al., 1990). We speculated that G-36 coded for a dopamine receptor, probably the dopamine D1 receptor, based on the following observations. First, sequence analysis of G-36 revealed two conserved serine residues in TM 5, which are speculated to represent the catechol binding site conserved in all catecholaminergic receptors (Strader et al., 1989a). Second, Northern blot analysis revealed the presence of mRNA in the bovine parathyroid gland, a tissue in which only dopamine D1 receptors are found, but not in the pituitary, a gland that only contains dopamine receptors of the D2 type. Finally, unlike dopamine D2 receptors, dopamine D1 receptors are known not to be present on terminal axonal projections of dopaminergic cell bodies in the substantia nigra, and in situ hybridization of various rat (Weiner et al., 1991) and human brain (Mengod et al., 1991) regions revealed the presence of this mRNA in striatum but not in the substantia nigra.

Subsequent screening of a human striatal cDNA library (Clontech) under high-stringency conditions permitted the identification of a clone (RS1; 545 bp), which upon sequence analysis revealed significant overall amino acid homology (84%) to G-36. Although further attempts to obtain cDNA clones encoding the full coding sequence of RS1 failed, we reasoned that the gene encoding the putative receptor RS1 may be intronless, since this clone shared considerably more homology to the β-adrenergic receptors, which are intronless (Kobilka et al., 1987), than to D2 dopamine receptors (32%), which contain at least seven intron–exon junctions (Grandy et al., 1989; Dal Toso et al., 1989). Hence, G-36 was used to probe a λ EMBL 3 SP6/T7 human genomic library (Clontech) under high-stringency hybridization conditions, and from this screening three positive clones with inserts of approximately 14 kb in length were isolated. Sequence and restriction map analysis indicated that all of the clones were identical and contained a sequence identical to the cDNA clone RS1. Southern blot analysis of a HindIII and BglII endonuclease digestion of the 14-kb genomic insert, using G-36 as a probe, identified two hybridizing fragments of 2 and 2.5 kb in length, respectively. These fragments were subcloned into pSP73 and, upon sequencing, revealed 500 bp of overlapping sequence which were subsequently excised and the resulting fragments ligated (RS2; 4 kb). Sequencing the clone revealed the presence of a putative initiation methionine and a long open reading frame of 1476 nucleotides encoding a 446-amino acid protein. The protein has a predicted molecular mass of 49,261 Da and is similar to the electrophoretic mobility (M_r 46000) of the deglycosylated striatal and bovine parathyroid photoaffinity-labeled D1 receptor as determined by SDS-PAGE (Jarvie et al., 1989). As has been reported for both human β_2-AR (Kobilka et al., 1987) and dopamine D2 receptor (Bunzow et al., 1988), a small open reading frame is present in the 5'-untranslated sequence of human D1 receptor (an in-frame tetrapeptide is situated 153 bp upstream from the putative initiation methionine).

2. D1 Receptor Protein Structure

The deduced amino acid sequence of the human gene encoding the D1 receptor exhibits many structural similarities with the family of receptor molecules that has been predicted to span the membrane seven times (O'Dowd et al., 1989a). The single polypeptide chain of D1 receptor has seven hydrophobic regions that probably represent transmembrane domains (Figure 6.1). The proposed model constrains the polypeptide chain to be divided into four extracellular, seven transmembrane, and four intracellular regions, and predicts a glycosylated amino-terminus on the extracellular surface and an intracellular carboxy-terminus.

A comparison of the deduced amino acid sequence of the human D1 receptor with the dopamine D2 receptor reveals 115 amino acids strictly

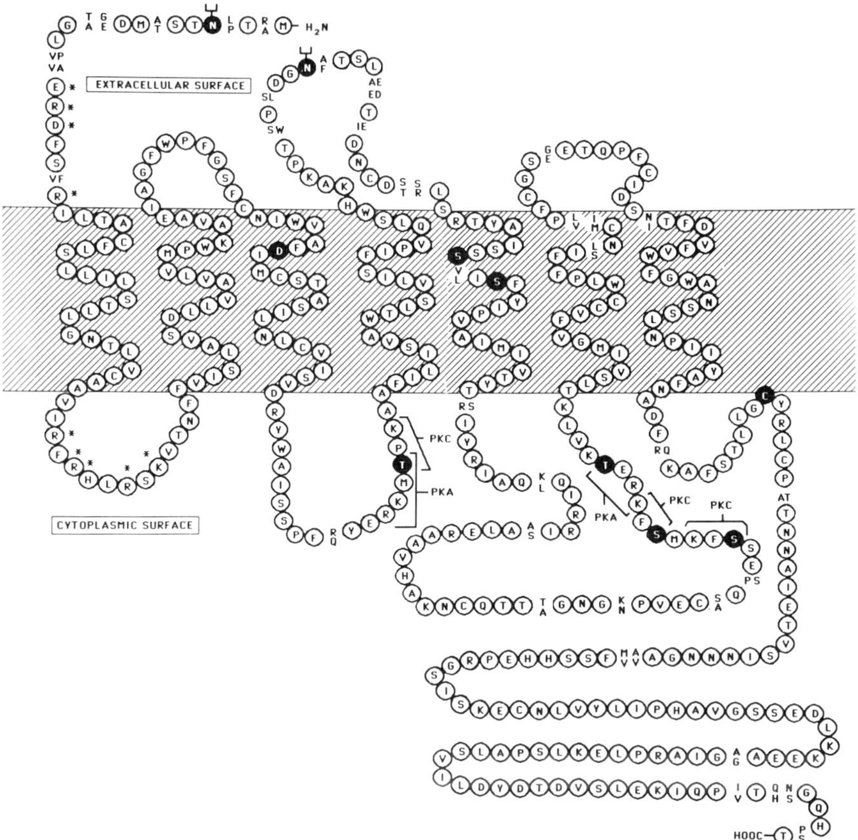

FIGURE 6.1. A model for the transmembrane topology of the human dopamine D1 receptor. The receptor is shown as containing seven transmembrane regions predicted from hydropathicity analysis of the deduced amino acid sequence. Asterisks (*) denote charged amino acids clustered adjacent to TM 1, which play a role in determining the topology of the receptor (Hartmann et al., 1989). Potential N-linked glycosylation sites on the extracellular surface, two overlapping putative PKA/PKC phosphorylation sites within the i2 and i3 loops as well as an additional PKC site in the i3 loop are highlighted. The C-terminal tail is shown anchored to the plasma membrane (at Cys^{348}; a cysteine residue in an equivalent position in β_2-AR is palmitoylated). The aspartate residue (Asp^{104}) shown highlighted in TM 3 and two Ser (200 and 203) residues in TM 5 are postulated to form part of the agonist binding site.

conserved (overall homology, 26%); with the D3 receptor (Sokoloff et al., 1990), 106 amino acids are identical (overall homology, 24%). The D2 and D3 receptors more closely resemble each other, containing 219 amino acids in strictly conserved positions, equivalent to an overall homology of 50%. Among these three dopaminergic receptors there are 91 amino acids which are identical (Figure 6.2) and 7 of these amino acids are found only in dopaminergic receptors (these are indicated in Figure 6.2 by letters within white circles). Most of the conserved residues are located either within the seven transmembrane regions or in small connecting loops (Figure 6.2). This observation correlates with the experimental evidence that residues in the transmembrane regions are responsible for forming the ligand binding domain. The highly conserved aspartic acid residue (Asp^{104} of D1 receptor) in TM 3 (Strader et al., 1987) and two conserved serine residues in TM 5

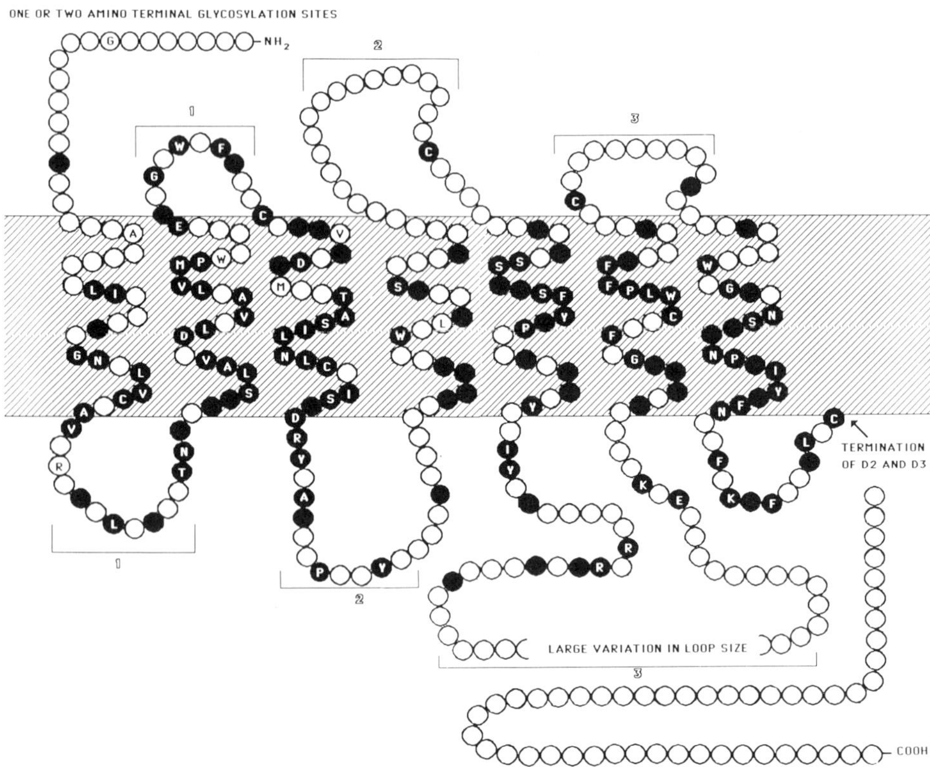

FIGURE 6.2. Distribution of conserved amino acids among the dopaminergic receptor family, D1, D2, and D3. Amino acids exclusive to dopaminergic receptors are highlighted by letters within white circles, amino acids identical in all dopamine receptors are indicated by letters in black circles, whereas conservative substitutions are indicated by black circles. Two of the connecting loops differ significantly in size in these three receptors. The o2 loop contains 20 amino acids residues in the human D1 receptor, compared to only 5 and 6 amino acids in the D2 and D3 receptors, respectively. The i3 loop contains 53 amino acid residues in the D1 receptor and 158 amino acid residues in the D2 and D3 receptors. Conservative amino acid substitutions are grouped as follows: S,T,P,A,G: M,I,L,V: F,Y,W: N,D,E,Q: R,K,H: C.

(Ser[200] and Ser[203] of D1 receptor) are implicated (Strader et al., 1989a) as being involved in binding of catecholaminergic agents (Figure 6.1). It is noteworthy that the positions of two of the seven amino acids found only in the dopaminergic receptors are to be found in TM 3 in close proximity to this conserved aspartic acid residue (Figure 6.3), perhaps forming part of the dopamine ligand binding pocket.

An overall comparison of the amino acid sequence of human D1 receptor with other members of the G-protein-linked receptor family, including adrenergic, serotonergic, and muscarinic receptors, reveals that 36 residues are identical and 72 conservatively substituted (Sunahara et al., 1990). Again, most of the conserved or identical residues are located either within the seven transmembrane regions or in several of the small connecting loops. The percentages of amino acids for several related receptors that are identical with those in human D1 receptor, within the transmembrane regions (175 residues) and overall for the entire protein, are as follows: human β_1-AR, 44 and 33%; human β_2-AR, 46 and 30%; human platelet α_2-AR, 40 and 24%; $5HT_{1A}$, 44 and 28%; and muscarinic m1 receptor, 31 and 24%.

The structural features that vary appreciably among receptors in this family are the sizes of the i3 (cytoplasmic) loop and the C-terminal tail. In this context the D1 receptor more closely resembles the β_2-AR (short i3 loop

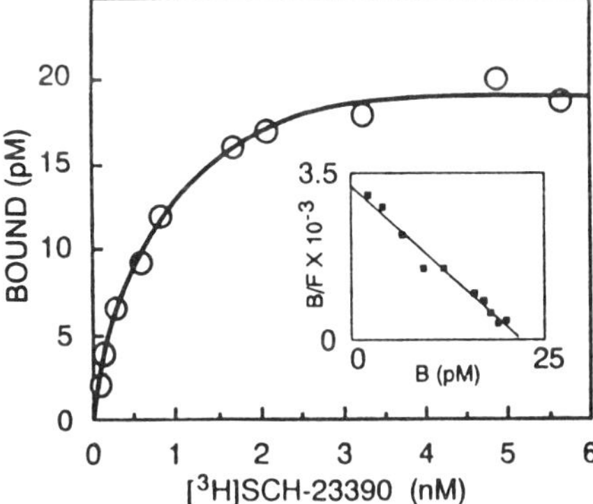

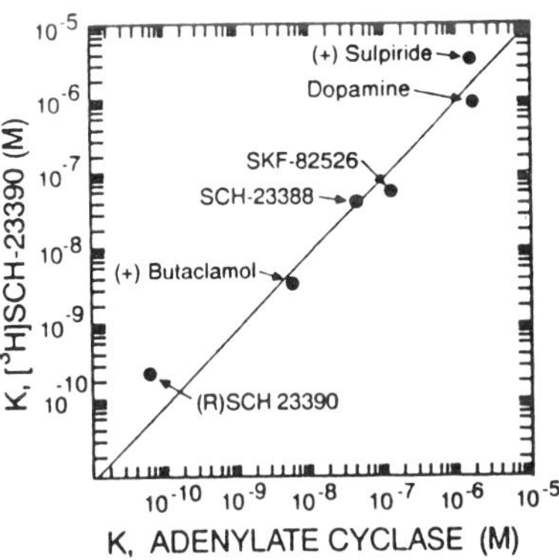

FIGURE 6.3. Binding of [³H]SCH-23390 to transformed COS-7 cell membranes expressing the human D1 receptor. *(Left)* Saturation isotherm of the specific binding of [³H]SCH-23390 to membranes from COS-7 cells expressing the human dopamine D1 receptor. The results shown are representative of three independent experiments each conducted in duplicate. *Inset:* Scatchard plot of the same data. Estimated B_{max} (889 fmol/mg protein) and K_d (656 ± 83 pM) values were obtained by LIGAND. *(Right)* Line of identity of K values for cAMP accumulation in bovine parathyroid gland (taken from Niznik et al., 1988a) and [³H]SCH-23390 binding to COS-7 cells of various dopaminergic agonists and antagonists.

and long C-terminal tail) than the D2 or D3 (long i3 loop and short C-terminal tail) receptor. Although the exact significance of this close structural similarity between the D1 receptor and β_2-AR is not clear, it is probably a consequence of the requirement of these receptors to interact with the common cytoplasmic transducing molecules (Gs proteins). It has previously been determined that amino acid sequences in the carboxy-terminus of the i3 loop of β_2-AR appear to be critical in the formation of the binding site for Gs, and this site as well as sites in the amino-terminus of the i3 contain residues that confer G-protein binding specificity (Gi, Gs, and Gp; see O'Dowd et al., 1988; Strader et al., 1989b). Close inspection of amino acid sequences in this critical region of the C-terminal tail of the i3 loop (see below) in D1 receptor and β_2-AR reveals that these receptors share seven of nine either identical amino acids or conserved substitutions.

$$K \quad E \quad H \quad K \quad A \quad L \quad K \quad T \quad L \; (\beta_2\text{-AR})$$

$$\underline{R} \quad \underline{E} \quad T \quad \underline{K} \quad V \quad L \quad \underline{K} \quad T \quad L \; (\text{D1 receptor})$$

Recent mutagenesis studies of the human β_2-AR concluded that the conserved charged distribution (underlined above) probably represents a common G-protein binding element and that this region also contains uncharged residues that confer G-protein binding specificity (O'Dowd et al., 1991a). The D1 receptor, in common with the human β_2-AR, contains putative glycosylation sites in both the amino-terminus and also in the extracellular loop between TM 4 and 5 (Figure 6.1). Each of the receptors (aligned in Figure 1 of Chapter 3) and also D1 receptor have conserved cysteine residues in loops o1 and o2. Data from site-specific mutagenesis of these conserved residues in β_2-AR are consistent with the notion that a disulfide bond may join these extracellular loops (Strader et al., 1989b). The effect of a disulfide bridge would be to divide the large extracellular loop of D1 receptor in two, possibly also constraining the receptor into a conformation whereby the TM 5 would be in close proximity to TM 3, forming the ligand binding pocket. In the topographical model of D1 receptor (Figure 6.1), the C-terminal tail is shown attached to the plasma membrane by a cysteine residue (Cys^{348}), while D2 and D3 receptors terminate at this point. This model of an i4 loop has been proposed for rhodopsin (Ovchinnikov et al., 1988) and β_2-AR (O'Dowd et al., 1989b) based on the observation that equivalent cysteine residues in these proteins are palmitoylated, whereby palmitic acid may anchor the C-terminal tail to the membrane.

Consensus sequences for phosphorylation by cAMP-dependent protein kinase (PKA) (Figure 6.1) appear twice in the sequence of D1 receptor, while three sites are present for phosphorylation by protein kinase C (PKC). Similarly with the β_2-AR PKA and PKC sites are located in the carboxy-terminus of the i3 loop, near the postulated receptor–Gs binding site, indicating that perhaps phosphate groups in this region could play a role in the affinity of such interactions.

Although no evidence exists to suggest that D1 receptor is a substrate for β-adrenergic receptor kinase (β-ARK), fifteen serine or threonine residues exist (distal to Cys^{348}) in the carboxy-terminus of D1 receptor, while there is also a small cluster of threonine residues in the i3 loop. In rhodopsin, β_2-AR and α_2-AR serine and threonine residues found in the carboxy-terminus and/or the i3 loop represent sites of phosphorylation by rhodopsin kinase or β-ARK (Benovic et al., 1989; O'Dowd et al., 1989b).

3. Expression of the D1 Receptor

The intronless gene encoding the dopamine D1 receptor was ligated into an expression vector (pBC12B) and we assessed the ability of the selective dopamine D1 receptor antagonist [³H]SCH-23390 to bind transfected COS-7 cell membranes with an appropriate pharmacological specificity for D1 receptors. As depicted in Figure 6.3a, COS-7 cells transfected with the gene bound [³H]SCH-23390 in a saturable manner with high affinity (K_d, 630 pM) and expressed quantities of receptor protein ranging from 510 to 890 fmol/mg protein. The observed K_d of [³H]SCH-23390 in membranes of transformed cells was equivalent to that seen in native human caudate membranes (K_d ~650 pM; Seeman, et al., 1987, 1989) while reported receptor densities are approximately two- to fourfold that seen in human brain.

Dopaminergic agonists and antagonists inhibited the binding of [³H]SCH-23390 in a concentration-dependent and stereoselective manner with an appropriate rank order of potency for the D1 dopamine receptor: SCH-23390 > (+)butaclamol > SCH-23388 > SKF-82526 > dopamine > noradrenaline $\gg$ serotonin. The selective dopamine D2 receptor agonist, LY-171555, and antagonist, eticlopride, were virtually without effect on [³H]SCH-23390 binding. Estimated K_i values for these compounds are listed in Table 6.1 along with respective values obtained in human brain membranes. Unlike that seen with native membrane preparations, agonist/[³H]SCH-23390 competition curves were uniphasic, consistent with the predominant presence of a non-guanine nucleotide-sensitive low-affinity form of the receptor (Niznik et al., 1988a). As is evident from Table 6.1, the K_i constants for these compounds correlate extremely well with those K_i values observed in native membranes. Moreover, as depicted in Figure 6.3b, K_i values of dopaminergic agonists and antagonists for the cloned human D1 receptor correlate well with both K_a and K_i values of these drugs for dopamine-stimulated adenylyl cyclase activity in dispersed cells of the bovine parathyroid gland (Niznik et al., 1989; Brown et al., 1989); the archetypical localization of the dopamine D1 receptor stimulating adenylyl cyclase (Kebabian and Calne, 1979). The cloned dopamine D1 receptor stimulates adenylyl cyclase (Dearry et al., 1990; Zhou et al., 1990).

TABLE 6.1. K_i values (nM) for [³H]SCH-23390 binding sites to transformed COS-7 cells and human brain membranes[a]

Drug	COS-7 transfected cells		Human caudate
	D1	D5	
(R)-SCH-23390	0.35	0.30	0.60
(+)-Butaclamol	3	27	2
α-Flupenthixol	4	8	7
Fluphenazine	22	14	20
SKF-82526	38	15	60
(S)-SCH-23388	41	35	43
Dopamine	3400	228	4000
Eticlopride	30000	19000	25322
Noradrenaline	28000	12000	41000
Serotonin	49000	3000	84000
LY-171555	100000	>20,000	100000

[a]SKF-82526 and dopamine did not recognize a high-affinity guanine-nucleotide-sensitive [³H]SCH-23390 binding component in transformed COS-7 cells. For comparative purposes, K_i values reported for agonist/[³H]SCH-23390 competition binding to native human caudate membranes represent the low-affinity non-guanine nucleotide-sensitive site.

4. mRNA Distribution

Northern blot analysis of RNA extracted from rat striatum and olfactory tubercles with the rat cDNA clone G-36, revealed the presence of an mRNA species of size 3.8 kb. Trace amounts were observed in the frontal cortex and hypothalamus. No detectable mRNA species were observed in rat olfactory bulbs, hippocampus, cerebellum, the anterior and neurointermediate lobes of the pituitary gland, and in rat kidney, liver, lung, and heart tissues. In the bovine parathyroid gland, two mRNA species of ~4.2 and 3.8 kb were observed and most probably represent transcripts from different polyadenylation sites (Zhou et al., 1990).

In situ hybridization histochemistry was used to map the distribution of D1 receptor mRNA expression in rat brain and to compare this expression to that of the dopamine D2 receptor mRNA (Weiner et al., 1991). These mRNAs showed largely overlapping yet distinct patterns of expression. The highest levels of expression of both mRNAs were observed in the caudate-putamen, nucleus accumbens, and olfactory tubercle (Figure 6.4). Within the caudate-putamen, approximately 50% of the medium-sized neurons expressed each of the mRNAs, whereas D2 receptor mRNA was observed in larger neurons. Dopamine D1 and D2 mRNAs were expressed in most cortical regions, with the highest levels in the prefrontal and entorhinal cortices. Within the neocortex, D1 mRNA was observed primarily in layer

6 and D2 mRNA in layers 4–5. In the amygdala, D1 receptor mRNA was observed in the intercalated nuclei and D2 mRNA in the central nucleus. Within the hypothalamus, D1 mRNA was observed in the suprachiasmatic nucleus and D2 mRNA in many of the dopaminergic cell groups. Within the septum, globus pallidus, superior and inferior colliculi, mamillary bodies, and substantia nigra only D2 receptor mRNA was detected while both D1 and D2 mRNAs were found in retina (Weiner and Brann, 1989; Weiner et al., 1991; Stormann et al., 1990). Overall, the distribution of D1 receptor mRNAs is in markedly good agreement with dopaminergic innervation patterns of the various brain regions and the distribution of [^{3}H]SCH-23390 binding sites.

The expression of both D1 and D2 receptor mRNAs in about one-half of the medium-sized interneurons of the caudate-putamen provides an neuroanatomical basis for the observed dopamine D1 and D2 receptor interaction seen in vitro using radioligand assays (Seeman et al., 1989). Interestingly enough, this observed D1–D2 receptor link is missing in approximately one-half of schizophrenic and Huntington's chorea diseased population.

Only dopamine D2 but not D1 receptor mRNA was detected in cholinergic neurons, suggesting that dopamine D1 selective drugs may have fewer effects on cholinergic systems than D2 selective compounds. If recent hypotheses concerning cholinergic systems in the pathophysiology of tardive dyskinesia prove correct, then one may predict from these data that chronic blockade of dopamine D1 receptors may not induce this disorder.

5. Presence of a Restriction Fragment-Length Polymorphism (RFLP)

Although a great deal of information is available regarding the biochemical characteristics of dopamine D1 receptors, the exact functional relationship between D1 receptors and neuropsychiatric diseases remains largely unknown. In order to explore the possible role that dopamine D1 receptors may play in human disease, a search for RFLPs recognized by the rat cDNA clone G-36 or of the human gene was carried out. Of the restriction enzymes tested (PstI, PvuII, BamHI, BglII, HindIII, MspI, TaqI, and EcoRI) on a panel of 5 unrelated individuals, only EcoRI generated an RFLP. EcoRI-digested DNA from 20 unrelated individuals identified a simple two-allele RFLP, with bands at 10.5 and 6.8 kb. Allelic frequencies were calculated for the A1 (10.5 kb) allele = 0.28, and the A2 (6.8 kb) allele = 0.72 (Sunahara et al., 1990). Mendelian inheritance was observed in several informative families. A longer probe involving the full human D1 receptor coding sequence recognized two additional hybridizing bands, using EcoRI. The band hybridizing at 6.0 kb represents the remaining part of the gene encoding the D1 receptor while the exact molecular nature of the 3.0-kb band is unknown and currently being investigated.

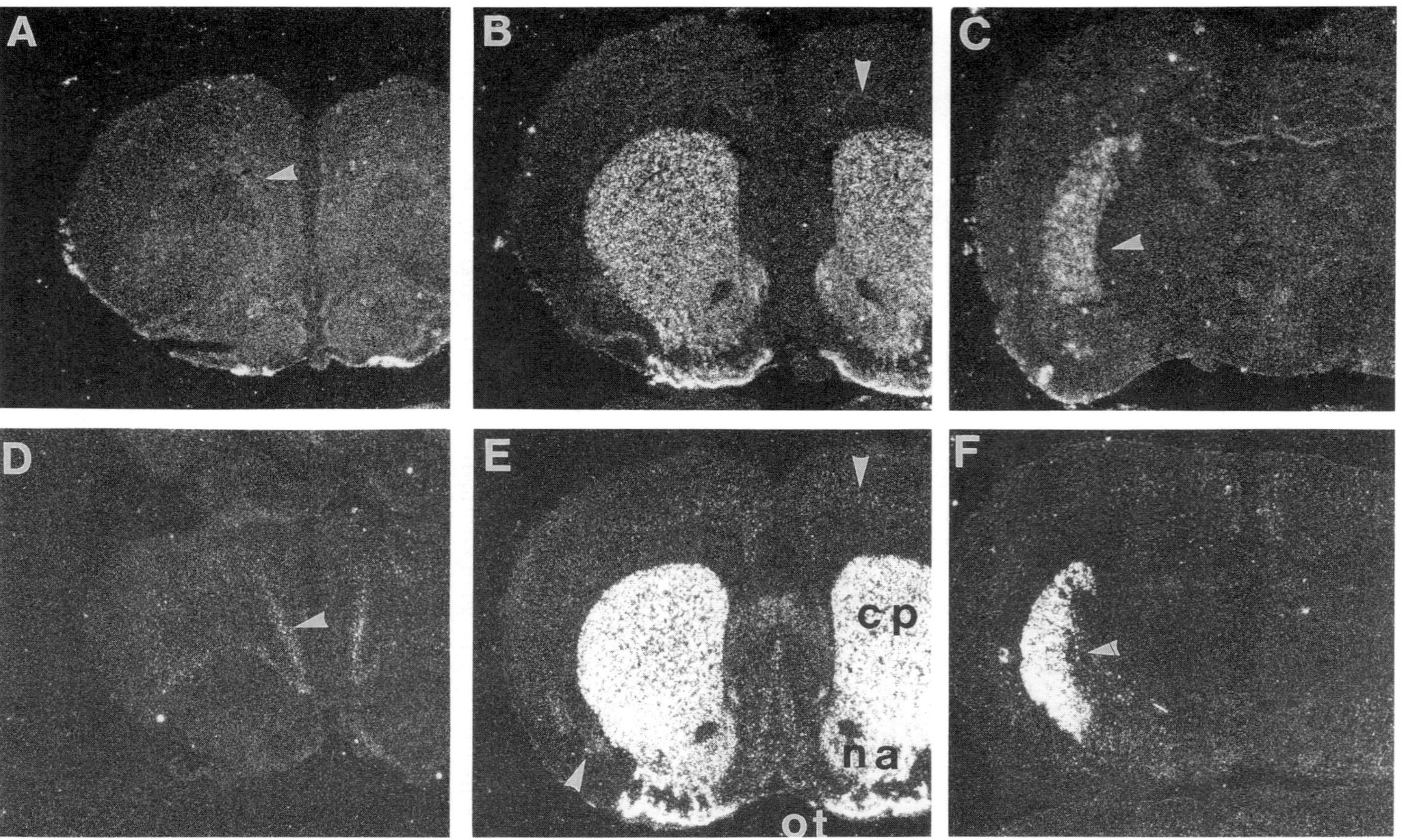

A
B
C
D
E
F
cp
na
ot

6. Chromosomal Location of D1 Receptor

Chromosomal localization of the D1 receptor gene (DRD1) was determined by Southern blot hybridization of DNA from a series of rodent–human hybrids. Chromosome 5 was the only human chromosome that segregated perfectly with the DRD1. All the other human chromosomes were discordant by two or more discordant hybrids. To further locate the DRD1 in a human genetic map, linkage analysis was carried out, using the EcoRI RFLP of DRD1. Significantly negative lod scores with 17 markers exclude DRD1 from most of the short arm (except for the telomeric region and a gap in the middle of the arm) and from the proximal one-third of the long arm of chromosome 5. Two markers in the more distal region of the long arm, namely GRL and D5S22, showed positive lod scores at varying recombination distances from DRD1. Obligate crossovers were revealed with each of these loci. The maximum lod score of 2.56 at theta = 0.15 between DRD1 and D5S22 is significant evidence of linkage, given the prior evidence for localization to chromosome 5. However, the one lod confidence interval is from 7 to 28% recombination, emphasizing the preliminary nature of these results. A multipoint analysis using both GRL and D5S22 gives a maximum lod score of 2.9 at 11 cM from GRL and 23 cM from D5S22. These preliminary data do not allow definitive ordering of these loci but do confirm that DRD1 maps to 5q31–34 possibly near the homologous genes for the β_2-AR and α_1-AR receptors (Yang-Feng et al., 1990). As the serotonergic receptor $5HT_{1A}$ has been mapped to the 5q11.2–q13 region (Yang-Feng et al., 1990), the DRD1 is now the fourth member of the G-protein-linked receptor family to be mapped to the long arm of chromosome 5. Further work is being carried out, using other markers in this region, to create a more detailed genetic map, and recent fluorescent in situ hybridization of the gene to human metaphase chromosomes has localized D1DR to 5q35.1 (Grandy et al., 1990).

Genetic linkage of markers on chromosome 5 to a dominant gene for the susceptibility to schizophrenia has been reported for various families with high incidence of this disease (Basset et al., 1988; Sherrington et al., 1988). Despite failures to replicate (Detera-Wadleigh et al., 1989; St. Clair et al.,

---

FIGURE 6.4. Expression of D1 (A–C) and D2 (D–F) dopamine receptor mRNAs in rat brain: All panels are coronal left hemisections in which regions expressing mRNA appear white in the photographs. Sections for D1 mRNA were from a different brain than was used for D2 mRNA, but were from anatomically similar planes. Images were obtained from 5-week exposures of sections against film. Arrowheads indicate layer 6 of anteromedial prefrontal cortex in A and D, layer 6 of supragenual cortex in B, and layers 4–5 in E, and the globus pallidus in C and F. na, nucleus accumbens; cp, caudate-putamen; ot, olfactory tubercle.

1989; Gilliam et al., 1989) and the exclusion of the GRL-D5S22 locus in one study (Kennedy et al., 1988), the DRD1 may still remain a candidate gene for schizophrenia and other neuropsychiatric disorders.

7. Dopamine D1 Receptor Variants: The D5 Receptor

Although the classic dopamine D1 receptor has been cloned, there is much experimental evidence (see Niznik, 1987; Andersen et al., 1990; Waddington and O'Boyle, 1989) to suggest the existence of at least two or more dopamine D1-like receptors that are expressed in brain and peripheral tissues. The evidence may be summarized as follows:

1. There is a general lack of correspondence between [^{3}H]SCH-23390 binding to dopamine D1 receptors and dopamine-stimulated adenylyl cyclase activity in terms of subcellular distribution, regional brain localization, and susceptibility to various membrane protein alkylating reagents.
2. Multiple pharmacological profiles of [^{3}H]SCH-23390 binding to dopamine D1 receptors have been observed. Thus, guanine nucleotide regulation of agonist interactions with [^{3}H]SCH-23390 binding sites are reportedly distinct in different brain regions; agonists potencies at D1 receptors in brain and bovine parathyroid gland differ by a factor of ~ 10, while antagonist potencies remain identical, and the observed affinity of [^{3}H]SCH-23390 and other specific dopamine D1 receptor ligands in brain and kidney differ again by at least tenfold (Felder et al., 1989).
3. While the cornerstone of dopamine D1 receptor classification rests on the observation of a dopamine-stimulated adenylyl cyclase, recent evidence suggests the existence of D1-like receptors in both the brain and periphery (kidney) which appear to activate inositol phosphate production and Ca^{2+} mobilization (Felder et al., 1989; Mahan et al., 1990) independent of adenylyl cyclase activity. Interestingly enough, D1 receptors identified in the retina are regulators of neuron growth and differentiation (Lankford et al., 1988), a characteristic of receptors coupled to the inositol phosphate cycle (Ashkenazi et al., 1989; Gutkind et al., 1991).
4. Finally, radiation inactivation/target size estimates of D1 receptors labeled by selective dopamine D1 receptor agonists and antagonists are not the same and photoaffinity labeling and partial proteolytic peptide mapping experiments reveal D1 receptor polypeptides displaying higher affinity for the endogenous agonist, dopamine, and distinct D1 receptor fragments found only in brain but not in the bovine parathyroid gland (Niznik et al., 1988b, 1989). Whether these D1 receptors are the same as those described above is unknown at present.

In this regard, we recently reported (Sunahara et al., 1991) the cloning of an intronless gene encoding a 477-amino acid protein sharing strong homology with the cloned D1 receptor. The expressed receptor, termed D5, binds drugs with a pharmacological profile similar to that of the cloned D1 receptor, but displays a tenfold higher affinity for the endogenous agonist, dopamine (Table 6.1). As with D1, the dopamine D5 receptor stimulates adenylyl cyclase activity. Northern blot and in situ hybridization analysis reveal a neuron-specific localization primarily within limbic regions with no mRNA detectable in kidney, liver, heart, and parathyroid gland. The existence of another dopamine D1-like receptor with these characteristics may represent an alternate pathway for dopamine-mediated events and regulation of dopamine D2, D3 (Sokoloff et al., 1990), and D4 (Van Tol et al., 1991) receptor activity.

We have attempted to search for other human genes homologous to the dopamine D5 receptor. Use of the PCR, with oligonucleotides based on the D5 receptor sequence, has allowed for the isolation of two additional intronless genes. The deduced amino acid sequences of these genes PG-1 and PG-2 shows an overall homology to the D5 receptor of 90%. However, these genes, PG-1 and PG-2, contain stop codons in their coding regions that would render these genes incapable of encoding functional receptors (O'Dowd et al., 1991b). In any event, it seems clear that there are many more dopamine receptors than were originally defined by pharmacological and functional criteria. In all likelihood, the G-linked superfamily of receptors will contain many members carrying the dopamine receptor trait.

8. Summary

Dopamine D1 and D2 receptors are targets of drug therapy in many psychomotor disorders, including Parkinson's disease and schizophrenia. We have cloned the human dopamine D1 receptor which resides on an intronless region on the long arm of chromosome 5, within the vicinity of two other members of the G-linked receptor family. Isolated from a human genomic library, and having 1476 nucleotides coding for 446 amino acids, the expressed protein encoded by the intronless gene binds drugs with affinities identical to the native human D1 receptor. A differential expression of dopamine D1 and D2 receptor mRNAs was observed in various brain regions which receive dopaminergic innervation. The presence of a D1 receptor gene restriction fragment-length polymorphism may be helpful for future disease linkage studies.

Acknowledgments

This work was supported in part by grants from the Addiction Research Foundation of Ontario (ARF), the National Alliance for Research on Schizophrenia and

Depression (NARSAD), the Ontario Mental Health Foundation, by an MRC predoctoral fellowship to R.K.S., National Institute of Mental Health, the Scottish Rite Schizophrenia Research Program, and by the Veteran's Administration National Center for PTSD, West Haven, CT. H.B.N. is a recipient of a NARSAD Young Investigator Award and H.B.N. and H.H.M.V.T. are Career Scientists of the Ontario Ministry of Health (HRPDP). D.M.W. is a recipient of a NIH/Howard Hughes Medical Institute Research Scholars Program.

References

Andersen PH, Gingrich JA, Bates MD, Dearry A, Falardeau P, Senogles SE, Caron MG (1990): Dopamine receptor subtypes: Beyond the D1/D2 classification. *Trends Pharmacol Sci* 11:231–236

Ashkenazi A, Ramachandran J, Capon DJ (1989): ACH analogue stimulate DNA synthesis in brain cells via specific muscarinic receptor subtypes. *Nature* 340:146–150

Basset AS, McGilvray BC, Jones BD, Pantzar JT (1988): Partial trisomy of chromosome 5 cosegregating with schizophrenia. *Lancet* i:799–801

Benovic JL, Deblasi A, Stone WC, Caron MG, Lefkowitz RJ (1989): β-Adrenergic receptor kinase: Primary structure delineates a multigene family. *Science* 246:235–246

Blum K, Noble EP, Sheridan PJ, Montgomery A, Ritchie T, Jagadeeswarran P, Nogami H, Briggs AH, Cohn JB (1990): Allelic association of human dopamine D2 receptor gene in alcoholism. *JAMA* 263:2055–2060

Brown EM, Chen CJ, Niznik HB, Fogel EL, Hawkins D (1989): Control of parathyroid function by dopamine: Characterization of D1 receptors and their interaction with other secretagogues in bovine parathyroid cells. In: *Peripheral Dopamine Pathophysiology,* Amenta F, ed. Boca Raton, Florida: CRC Press, pp 99–116

Bunzow JR, Van Tol HHM, Grandy DK, Albert P, Salon J, Christie M, Machida CA, Neve KA, Civelli O (1988): Cloning and expression of a rat D2 dopamine receptor cDNA. *Nature* 336:783–787

Clark D, White FJ (1987): D1 dopamine receptor—the search for a function. *Synapse* 1:347–388

Dal Toso R, Sommer B, Ewert M, Herb A, Pritchett DB, Bach A, Shivers BD, Seeberg PH (1989): The dopamine D2 receptor: Two molecular forms generated by alternative splicing. *EMBO J* 8(13):4025–4034

Dearry AG, Gingrich JA, Falardeau P, Fremeau RT, Bates MD, Caron MG (1990): Molecular cloning and expression of the gene for a human D1 dopamine receptor. *Nature* 347:72–76

Detera-Wadleigh DDD, Goldin LR, Sherrington R, Encio I, de Miiguel C, Berrettini W, Gurling H, Gershon ES (1989): Exclusion of linkage to 5q11-13 in families with schizophrenia and other psychiatric disorders. *Nature* 340:391–393

Felder RA, Felder CC, Eisner GM, Jose PA (1989): The dopamine receptor in adult and maturing kidney. *Am J Physiol* 257:F315–F327

Gilliam TC, Freimer NB, Kaufman CA, Powchik PP, Basset AS, Bengtsson U, Wasmuth JJ (1989): Deletion mapping of DNA markers to a region of chromosome 5 that co-segregates with schizophrenia. *Genomics* 5:940–944

Grandy DK, Marchionni MA, Makam H, Stofko RE, Alfano M, Frothingham L, Fischer JB, Burke-Howie KJ, Bunzow JR, Server AC, Civelli, O (1989): Cloning of the cDNA and gene for a human D2 dopamine receptor. *Proc Natl Acad Sci USA* 86:9762–9766

Grandy DK, Zhou QY, Allen L, Litt R, Magenis E, Civelli O, Litt M (1990): A human D1 dopamine receptor gene is located on chromosome 5 at q35.1 and identifies an *Eco* R1 RFLP. *Amer J Hum Genet* 47:828–834

Gutkind C, Novotny E, Brann MR, Robbins K (1991): Muscarinic receptors as agonist dependent oncogenes. *Proc Natl Acad Sci USA* 88:1459–1463

Hartmann E, Rapoport TA, Lodish HF (1989): Predicting the orientation of eukaryotic membrane spanning proteins. *Proc Natl Acad Sci USA* 86:5786–5790

Hemmings HC, Walaas SI, Oiumet CC, Greengard P (1987): Dopaminergic regulation of protein phosphorylation in the striatum: DARP-32. *Trends Neurosci* 10:77–82

Hess EJ, Creese I (1987): Biochemical characterization of dopamine receptors. In *Receptor Biochemistry and Methodology: Dopamine Receptors,* Creese I, Fraser CM, eds New York: Liss, Vol 8, pp 57–71

Jarvie KR, Booth G, Brown EM, Niznik HB (1989): Glycoprotein nature of dopamine D1 receptors. *Mol Pharmacol* 34:566–574

Kebabian JW, Calne DB (1979): Multiple receptors for dopamine. *Nature* 277:93–96

Kennedy JL, Giuffra LA, Moises HW, Cavalli-Sforza LL, Pakstis AJ, Kidd JR, Castiglione CM, Sjögren B, Wetterberg L, Kidd KK (1988): Evidence against linkage of schizophrenia to markers on chromosome 5 in a northern Swedish kindred. *Nature* 336:167–170

Kobilka BK, Frielle T, Dohlman HG, Bolanowski MA, Dixon RAF, Keller P, Caron MG, Lefkowitz RJ (1987): Delineation of the intronless nature of the genes for the human and hamster β2-adrenergic receptor and their putative promoter regions. *J Biol Chem* 262(15):7321–7327

Lankford KL, Demello FG, and Klein WL (1988): D1 type dopamine receptors inhibit growth cone motility in cultured retinal neurons: Evidence that neurotransmitters act as morphogenic regulators in the developing nervous system. *Proc Natl Acad Sci USA* 85:2839–2843

Mahan LC, Burch RM, Monsma FJ, Sibley DR (1990): Expression of striatal D1 dopamine receptors coupled to inositolphosphate production and calcium mobilization in *Xenopus* oocytes. *Proc Natl Acad Sci USA* 87:2196–2200

Mengod G, Vilaro MT, Niznik HB, Sunahara RK, Seeman P, O'Dowd BF, Palacios JM (1991): Visualization of a dopamine D1 receptor mRNA in human and rat brain. *Mol Brain Res* 10:185–191

Niznik HB (1987): Dopamine receptors; molecular structure and function. *Mol Cell Endocrinol* 54:1–22

Niznik HB, Fogel EL, Chen CJ, Congo D, Brown EM, Seeman P (1988a): Dopamine D1 receptors of the calf parathyroid gland; Identification and characterization. *Mol Pharmacol* 34:29–36

Niznik HB, Jarvie KR, Bzowej NH, Seeman P, Garlick RK, Miller JJ, Baindur N, Neumeyer JL (1988b): Photoaffinity labeling of dopamine D1 receptors. *Biochemistry* 27:7594–7599

Niznik HB, Jarvie KR (1989): Dopamine receptors. In: *Receptor Pharmacology and Function,* Williams M, Glennon RA, Timmermans PBMWM eds. New York: Dekker, pp 717–768

Niznik HB, Jarvie KR, Brown EM (1989): Dopamine D1 receptors of the calf parathyroid gland: Identification of a ligand binding subunit with lower apparent molecular weight but similar primary structure to neuronal D1 receptors. *Biochemistry* 28:6925–6930

O'Dowd BF, Hnatowich M, Regan JW, Leader WM, Caron MG, Lefkowitz RJ (1988): Site directed mutagenesis of the cytoplasmic domains of the human β_2-adrenergic receptor. *J Biol Chem* 263:15985–15992

O'Dowd BF, Lefkowitz RJ, Caron MG (1989a): Structure of the adrenergic and related receptors. *Annu Rev Neurosci* 12:67–83

O'Dowd BF, Hnatowich M, Caron MG, Lefkowitz RJ, Bouvier M (1989b): Palmitoylation of the human β_2-adrenergic receptor: Mutation of cys^{341} in the carboxy tail leads to an uncoupled nonpalmitylated form of the receptor. *J Biol Chem* 264(13):7564–7569

O'Dowd BF, Nguyen T, Tirpak A, Jarvie KR, Israel Y, Seeman P, Niznik HB (1990): Cloning of two additional catecholamine receptors from rat brain. *FEBS Lett* 262:8–12

O'Dowd BF, Hnatowich M, Lefkowitz RJ (1991a): Adrenergic and related G protein coupled receptors: Structure and function. *Encycl Hum Biol* 1:81–92

O'Dowd BF, Nguyen T, Jin H, Grupp L, Seeman P (1991b): Two human pseudogenes homologous to the D5 dopamine receptors. *Abstr Neurosci Soc* 17:598

Ovchinnikov YA, Abdulaev NG, Bogachuk AS (1988): Two adjacent cysteine residues in the c-terminal cytoplasmic fragment of bovine rhodopsin are palmitylated. *FEBS Lett* 230:1–5

Seeman P (1980): Dopamine receptors. *Pharmacol Rev* 32:229–313

Seeman P (1987): Dopamine receptors and the dopamine hypothesis of schizophrenia. *Synapse* 1:133–152

Seeman P, Bzowej NH, Guan HC, Bergeron C, Reynolds GP, Bird ED, Riederer P, Jellinger K, Tourtellotte WW (1987): Human brain D1 and D2 dopamine receptors in schizophrenia, Alzheimer's, Parkinson's and Huntington's diseases. *Neuropsychopharmacology* 1:5–15

Seeman P, Niznik HB (1988): Dopamine D1 receptor pharmacology. *ISI Atlas Sci: Pharmacol* 2:161–170

Seeman P, Niznik HB, Guan HC, Booth G, Ulpian C (1989): Link between D1 and D2 dopamine receptors is reduced in schizophrenia and Huntington diseased brain. *Proc Natl Acad Sci USA* 86:10156–10160

Seeman P, Niznik HB (1990): Dopamine receptors and transporters in Parkinson's disease and schizophrenia. *FASEB J* 4:2737–2744

Sherrington R, Brynjolfsson J, Petursson H, Potter M, Dudleston K, Barraclough B, Wasmuth J, Dobbs M, Gurling H (1988): Localization of a susceptibility locus for schizophrenia on chromosome 5. *Nature* 336:164–167

Sokoloff P, Giros B, Martes MP, Bouthenet ML, Schwartz JC (1990): Molecular cloning and characterization of a novel dopamine receptor (D3) as a target for neuroleptics. *Nature* 347:146–151

St. Clair D, Blackwood D, Muir W, Baille D, Hubbard A, Wright A, Evans HJ (1989): No linkage of chromosome 5q11-q-13 markers to schizophrenia in Scottish families. *Nature* 339:305–309

Stormann TM, Gdula P, Weiner DM, Brann MR (1990): Molecular cloning and expression of a dopamine D2 receptor from human retina. *Mol Pharmacol* 37:1–6

Strader CD, Sigal IS, Register RB, Candelore MR, Rands E, Dixon RAF (1987): Identification of residues required for ligand binding to the β-adrenergic receptor. *Proc Natl Acad Sci USA* 84:4384–4388

Strader CD, Candelore MR, Hill WS, Sigal IS, Dixon RAF (1989a): Identification of two serine residues involved in agonist activation of the β-adrenergic receptor. *J Biol Chem* 264:13572–13578

Strader CD, Sigal IS, Dixon RAF (1989b): Structural basis of β-adrenergic receptor function. *FASEB J* 3:1825–1832

Sunahara RK, Niznik HB, Weiner DM, Stormann TM, Brann MR, Kennedy JL, Gelerntner JL, Rozmahel R, Yang Y, Israel Y, Seeman P, O'Dowd BF (1990): Human dopamine D1 receptor encoded by an intronless gene on chromosome 5. *Nature* 347:80–83

Sunahara RK, Guan H-C, O'Dowd BF, Seeman P, Laurier LG, Ng G, George SR, Torchia J, Van Tol HHM, Niznik HB (1991): Cloning a human dopamine receptor gene (D5) with higher affinity for dopamine than D1. *Nature* 350:614–619

Van Tol HHM, Bunzow JR, Guan HG, Sunahara RK, Seeman P, Niznik HB, Civelli O (1991): Cloning a human dopamine D4 receptor gene with high affinity for the antipsychotic clozapine. *Nature* 350:610–614

Waddington JL (1989): Functional interactions between D1 and D2 dopamine receptor systems. *J Psychopharmacol* 3:54–63

Waddington JL, O'Boyle KM (1989): Drug acting on dopamine receptors: A conceptual reevaluation five years after the first selective D1 antagonist. *Pharmacol Ther* 43:1–52

Weiner DM, Brann MR (1989): The distribution of dopamine D2 receptor mRNA in rat brain. *FEBS Lett* 253:207–213

Weiner DM, Levey AI, Sunahara RK, Niznik HB, O'Dowd BF, Brann MR (1991): Dopamine D1 and D2 receptor mRNA expression in rat brain. *Proc Natl Acad Sci USA* 88:1859–1863

Yang-Feng TL, Fhong XF, Cotecchia S, Frielle T, Caron MG, Lefkowitz RJ (1990): Chromosomal organization of adrenergic receptor genes. *Proc Natl Acad Sci USA* 87:1516–1520

Zhou QY, Grandy DK, Thambi L, Kushner JA, Van Tol HHM, Cone R, Pribnow D, Salon J, Bunzow JR, Civelli O (1990): Cloning and the expression of human and rat D1 dopamine receptors. *Nature* 347:76–80

7

The Dopamine D2 Receptor

Olivier Civelli, James Bunzow, Paul Albert, Hubert H. M. Van Tol, and David Grandy

1. Introduction

Dopamine is the dominant catecholamine neurotransmitter in the mammalian brain. It is found throughout the entire central nervous system, but is predominant in the nigrostriatal, mesolimbic, and tuberoinfundibular tracts (Creese et al., 1983). Dopamine exerts its effects through binding to two types of receptor, the D1 and D2 receptors (Kebabian and Calne, 1979). Binding of dopamine to its receptors induces several second messenger systems, most importantly affecting cAMP levels (Vallar and Meldolesi, 1989). Activation of the D1 receptor stimulates adenylyl cyclase activity which results in an increase in intracellular cAMP levels while binding of dopamine to the D2 receptor inhibits the cyclase activity (Caron et al., 1978).

The importance of the dopamine system is evidenced by the number of physiological activities it modulates, which include control of movement, maintenance of emotional stability, and regulation of prolactin secretion (Hess and Creese, 1987). Interestingly, although the density of D1 receptors is higher than that of D2 receptors in the brain, it is the D2 receptor that has been specifically implicated in the pathophysiology of most of the dopamine-associated disorders. In this respect, it is noteworthy that many drugs used in treating mental disorders have high affinity for the D2 receptor site, making this receptor the focus of numerous studies (Seeman and Lee, 1975). It has been proposed that the D2 dopamine receptor is involved in the etiology of schizophrenia (Seeman, 1987), Parkinson's disease, and Tourette's syndrome.

In spite of progress made in understanding the pharmacology and physiology of the D2 receptor, its biochemical characterization has been difficult. Only recently has the D2 dopamine receptor been purified to homogeneity (Senogles et al., 1988). We discuss here our studies on the molecular cloning and expression of the rat brain D2 dopamine receptor.

2. Cloning of the Rat D2 Receptor

With the cloning of the β_2-adrenergic receptor in 1986 (Dixon et al., 1986), it was observed that its sequence was very similar to rhodopsin, the retinal receptor. Since both of these receptors are transduced by G-proteins, a new concept evolved which proposed that G-protein-coupled receptors might share certain sequence similarities with each other and thus be part of a large gene family.

Indeed, with the recent cloning of several other G-protein-coupled receptors (Stevens, 1987; Hall, 1987; Dolhman et al., 1987; Masu et al., 1987; Julius et al., 1988), it is clear that all of these receptors are evolutionarily related. In particular, it was shown that G-protein-coupled receptors share three structural characteristics: seven hydrophobic domains; about two dozen conserved amino acid residues, some of which might be targets of posttranslational modifications; and a significant degree of sequence similarity at both the peptide and nucleotide levels. It should be noted that not all of the G-protein-coupled receptors will necessarily share these characteristics, but, to date, all the proteins having these structural characteristics are G-protein-coupled receptors.

We embarked on the cloning of G-protein-coupled receptors by taking advantage of their sequence homology. We used the β_2-adrenergic receptor coding sequence as a hybridization probe to screen a rat genomic library under nonstringency hybridization conditions. We were able to isolate numerous clones; one, RGB-2, was studied in detail (Bunzow et al., 1988).

The RGB-2 clone contained a DNA fragment that hybridized with the β_2-adrenergic probe. This fragment could be translated into a protein which shared 48% amino acid identity with transmembrane (TM) 6 and 7 domains of the β_2-adrenergic receptor. However, as demonstrated later, this fragment also contained a 3' intron splice site and 400 bp of intronic sequence. The presence of at least one intron in its coding sequence was a finding in contrast with what was found for the adrenergic and muscarinic receptor genes.

The RGB-2 genomic fragment was found to hybridize to a 2.8-kb rat brain mRNA and was subsequently used to screen a rat brain cDNA library. Two clones, in one-half million, were positive. One, containing a 2.5-kb insert, was sequenced and its corresponding peptide sequence determined. This clone encodes a 415-amino acid protein with all the expected characteristics of a G-protein-coupled receptor: it has seven hydrophobic domains; the 21 amino acid residues conserved among all cloned G-protein-coupled receptors and potential glycosylation and phosphorylation sites; and a significant degree of sequence similarity with the other receptors in this gene family. Therefore, it appeared that RGB-2 could be a G-protein-coupled receptor.

To determine the ligand specificity of RGB-2, we first analyzed the tissue distribution of its mRNA in order to relate RGB-2's expression to the distribution of a known receptor. Northern blot analysis showed that RGB-2 mRNA sequences are expressed throughout the rat brain with the highest levels found in the striatum. We also determined that RGB-2 mRNA is present in high levels in the intermediate lobe of the pituitary and at much reduced levels in the anterior lobe. These data suggested that RGB-2 could encode a dopamine D2 receptor. To investigate this possibility, the RGB-2 cDNA was expressed in an heterologous system and the binding characteristics of the resulting protein were determined.

3. Expression of the D2 Receptor

The full-length RGB-2 cDNA was cloned into a plasmid containing the metallothionein promoter and this construct was cotransfected with pRSVneo (a selectable marker conferring resistance to the antibiotic neomycin) into mouse Ltk$^-$ cells. This fibroblast cell line does not express endogenous mouse RGB-2 mRNA sequences. Stable transfectants (expressing RGB-2 were isolated and membranes from one of these clones, L-RGB2Zem-1 (also called LZR-1 by Neve et al., 1989) were prepared and analyzed for their ability to bind dopamine ligands (Bunzow et al., 1988).

L-RGB2Zem-1 membranes bound D2-dopamine agonists and antagonists with the same pharmacological profile as do rat striatal membranes. These studies used the antagonist [^{3}H]spiperone whose binding was shown to be saturable (950 fmol/mg protein) and of high affinity (48 pM). [^{3}H]Spiperone binding to L-RGB2Zem-1 membranes was displaced by several antagonists with the stereospecificity expected of a D2 receptor and with the same K_i's as determined in rat striatal membranes. Finally, as determined using compounds which detect more than one receptor type, the transformed Ltk$^-$ cells expressed only one type of receptor. Therefore, we had demonstrated that RGB-2 encodes a protein that possesses the D2 receptor binding characteristics. The next step was to show that this D2 receptor was functional, that is, couples to a second messenger system.

4. Second Messenger Coupling and Induction

D2 receptors are present on lactotroph cells of the anterior pituitary where they regulate prolactin secretion. The somatomammotroph cell line GH$_4$C$_1$

is derived from a rat pituitary tumor and is known to secrete prolactin. This cell line, however, does not bind dopamine and therefore represents an excellent cell system in which to study exogenously expressed D2 receptor activity.

GH_4C_1 cells were transfected with the RGB-2 cDNA metallothionein construction and several stably transformed cells were cloned and raised. One, GH_4ZR_7, was found to express high levels of RGB-2 mRNA. Membranes prepared from GH_4ZR_7 cells bound [^{3}H]spiperone with saturable kinetics—a K_d of 96 pM and a B_{max} of 2.3 pmol/mg protein (Albert et al., 1989).

To analyze second messenger coupling, we first showed that dopamine inhibits [^{3}H]spiperone binding with an IC_{50} of 49 μM and a Hill coefficient of 0.69. A Hill coefficient less than unity is suggestive of high- and low-affinity binding sites. Addition of GTP and NaCl increased the IC_{50} value of dopamine twofold (109 μM) with a Hill coefficient approaching unity (0.93). Thus, the presence of GTP and NaCl converted the dopamine receptor from a population of high- and low-affinity receptors to the low-affinity state. These results indicated that the cloned dopamine receptor interacts with G-proteins.

We next analyzed the effects of dopamine binding on the levels of intra- and extracellular cAMP. Since the D2 receptor is expected to inhibit cAMP levels, VIP (vasoactive intestinal peptide) was used to first stimulate endogenous cAMP production. Dopamine inhibited both basal and VIP-stimulated cAMP levels in media from GH_4ZR_7 cells. Furthermore, intracellular cAMP levels were inhibited, albeit at a less pronounced level, probably due to the lower recovery of intracellular cAMP. The stereospecificity of these inhibitions was demonstrated using isomers of sulpiride: the active enantiomer ($-$)-sulpiride blocked the inhibition while ($+$)-sulpiride had no effect.

To demonstrate that the changes in cAMP levels were the result of an inhibition of adenylyl cyclase, dopamine was added to membranes of VIP- or forskolin-stimulated GH_4ZR_7 cells and adenylyl cyclase activity was measured. Dopamine inhibited the activity by 45%. This inhibition was stereoselective since the agonist quinpirole was active while its enantiomer LY181990 did not have any significant effect. Moreover, since receptors that couple to inhibitory G-proteins are known to be sensitive to pertussis toxin, adenylyl cyclase activity was measured in membranes prepared from GH_4ZR_7 cells pretreated with pertussis toxin and stimulated with forskolin or VIP. Pertussis toxin was able to uncouple dopamine-mediated inhibition of adenylyl cyclase and dopamine-stimulated cAMP accumulation. Therefore, the D2 receptor expressed in GH_4ZR_7 cells is capable of inhibiting adenylyl cyclase activity through a pertussis toxin-sensitive mechanism.

Finally, the inhibition of prolactin (PRL) secretion by dopamine was assayed in GH_4ZR_7 cells. VIP and thyrotropin-releasing hormone (TRH) are known to enhance PRL release by a cAMP-dependent and a cAMP-independent mechanism, respectively. Dopamine was able to inhibit PRL secretion stimulated by both hormones. These inhibitions were reversed by the active antagonist $(-)$-sulpiride but not $(+)$-sulpiride. Therefore, we have demonstrated that the RGB-2 cDNA encodes a D2 dopamine receptor which is functional, since it can couple to inhibitory G-protein and since this coupling results in an inhibition of adenylyl cyclase activity as measured by the drop in cAMP levels and in an inhibition of PRL secretion.

5. Model of the Rat Brain D2 Dopamine Receptor

The dopamine and the adrenergic receptors bind catecholamines. From the studies that have been done on the ligand binding site of the adrenergic receptors, it has become clear that the binding of adrenergic ligands involve the hydrophobic core of the receptors. In addition, several particular amino acid residues of the adrenergic receptors have been directly implicated in the ligand recognition. Interestingly, the same residues are found in the D2 dopamine receptors. On the basis of the studies that have been done on the adrenergic receptors, we propose the model shown in Figure 7.1 for the rat brain D2 dopamine receptor.

In this model, the receptor polypeptide spans the membrane seven times. Its amino-terminus is extracellular and contains the sequence of three potential glycosylation sites. Its carboxy-terminus is intracellular and terminates in Cys^{415} which might be palmitoylated and might anchor the receptor to the membrane (O'Dowd et al., 1989). The third cytoplasmic loop (i3) is large and contains at least two protein kinase A phosphorylation sites. In TM 2 and 7, Asp^{80} and Asn^{390} might be involved in agonist binding while in TM 3, Asp^{114} might be important for antagonist binding (Strader et al., 1987, 1988). Cys^{107} and Cys^{182} might form a disulfide bond which could affect ligand binding (Dixon et al., 1987). Although this model is hypothetical, it can direct our attempts at defining the domains of the receptor implicated in its functions.

6. Gene Localization of the Human D2 Receptor

As mentioned above, several human neurological diseases have been associated with imbalances in the dopaminergic system and, in some cases, D2-receptor involvement has been proposed. For example, several lines of evidence have linked the D2 receptor to schizophrenia. First, a linear relationship exists between the affinity of different neuroleptic drugs for the D2 receptor and their therapeutic dosage (Seeman et al., 1976). Also, the brains of some schizophrenics have been reported to possess an unusual density of D2 receptors in their striatum (Seeman, 1987). In addition, the

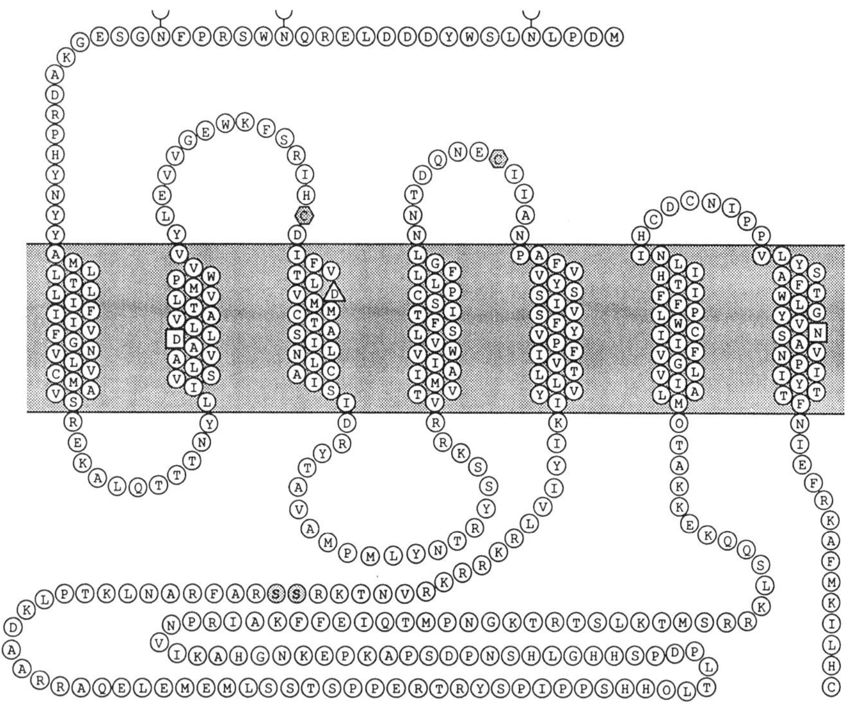

FIGURE 7.1. Model of the rat D2 dopamine receptor.

dopamine D2 receptor has been proposed to be involved in Parkinson's disease and Tourette's syndrome. In order to study the relationship between human diseases and the D2 receptor, we have cloned the human gene.

A human genomic library was screened with the rat D2 cDNA. Several clones were detected and one, λHD2G1, was found which contained a 1.6-kb BamHI fragment which was sequenced and found to code for the last 64 amino acids of the D2 receptor. It should be stressed that the D2 receptor gene has introns and that its entire sequence spans more than 30 kb.

The λHD2G1 DNA was hybridized to human metaphase chromosomes and to DNA prepared from a panel of rodent–human hybrids which have selectively lost human chromosomes (Grandy et al., 1989). All the data were consistent with localizing the D2 dopamine receptor to the long arm of chromosome 11 in the region q22–q23. Moreover, when the 1.6-kb BamHI fragment from λHD2G1 was used to probe human genomic DNA digested with different restriction enzymes, only single fragments were detected. This result, plus the fact that the total size of hybridizing bands in the hybrid panel analysis was consistent with the size of the λHD2G1 phage DNA under the hybridization conditions used, indicated the human genome contains only one D2 receptor gene. It should be mentioned that although our studies were done under stringent hybridization conditions, these conditions would have detected genes whose sequences are closely related (75% identity).

The presence of one D2 dopamine receptor gene is of pharmacological importance. The dopamine receptors have been classified into at least two subtypes, D1 and D2. The D2 receptors themselves have been further subdivided into auto- and postsynaptic receptors (Carlsson, 1975). The autoreceptors have a D2-like pharmacological profile but are found on neurons that synthesize dopamine. In these neurons, upon binding of dopamine, the autoreceptors regulate dopamine production through a feedback mechanism.

Our data, therefore, indicates that either the diverse D2 and D1 receptors are encoded by different genes whose sequences are divergent, at least more than the sequences of the muscarinic receptors, or that these dopamine receptors are encoded by the same unique gene and that the pharmacological differences are the results of either alternative posttranscriptional events or diverse posttranslational modifications. In view of the current knowledge about the differences between the D1 and D2 receptors, it seems that they may be coded for by different genes. On the other hand, some recent experiments indicate that the D2 autoreceptors are similar in sequence to the postsynaptic D2 receptor and that these two receptors are products of the same gene (Meador-Woodruff et al., 1989).

The human genomic clone, λHD2G1, has been used in several laboratories to investigate the linkage of the D2 receptor gene and several human genetic disorders. Thus far, the studies have focused on families afflicted with schizophrenia (Moises et al., 1989; Byerley et al., 1989); other kindreds afflicted with Tourette's syndrome (Gelernter et al., 1989; Isenberg et al., 1989); and with a high incidence of manic depression (Byerley et al., 1989). The D2 receptor was not directly linked to disease in any of these cases. It should be mentioned, however, that more families are presently being studied.

7. Conclusions and Perspectives

We have demonstrated that the G-protein-coupled receptor that we have cloned has the pharmacological profile and the biological function expected of the D2 dopamine receptor. Furthermore, we have localized the D2 dopamine receptor gene to the long arm of chromosome 11 in humans and have shown it to be represented as a single copy in the genome.

This work has established the foundations on which further studies of the D2 receptor can be based. For example, the D2 receptor has been shown to induce not only inhibition of adenylyl cyclase but also other second messenger systems. Cell lines transfected with the cloned receptor will be of great interest when analyzing the coupling of the D2 receptor to other second messenger pathways. In addition, the human D2 gene is the focus of many attempts to relate this gene to specific genetic diseases. Our inability to find, in the genomic analysis, closely related D2 sequences will not prevent us from searching for new receptors that would share a lower but significant degree of sequence similarity. This implies applying the cloning approach presented here and using the D2 receptor instead of the β_2-adrenergic receptor as probe. This search will be directed at the cloning of the D1 receptor. The success of this search, however, can only be guessed at since two receptors which bind the same ligand might not share a high degree of sequence similarity, as evidenced by the sequences of the serotonin 1a and 1c receptors. In this respect it is interesting that, evolutionarily, the cloned receptor most homologous to the D2 receptor is the α_2-adrenergic receptor (R. Doolittle, personal communication, 1989).

Finally, several remarks are warranted regarding our cloning strategy. Although this strategy has allowed us to clone several G-protein-coupled receptors, there are pitfalls inherent to this approach. First, it does not solely result in the isolation of G-protein-coupled receptors. Other gene fragments that share some random sequence similarities with parts of the probe are also detected. The major problem associated with this strategy is, however, our inability to direct it toward the cloning of a particular receptor instead of the large number of genes generally detected. One solution to this problem is to apply this strategy to the cloning of receptors from pharmacologically well-characterized cell lines and thus restrict the diversity of possible receptor clones. Increasing the stringency of the hybridization conditions can also be used to lower the number of receptors detected. On the other hand, the fact that this strategy leads to the isolation of numerous random receptor clones, can be viewed as an advantage. The challenge resides in sorting out the clones of interest. This is undoubtedly where new approaches will have to be developed. It would be very powerful, for example, to find a simple way to screen for all the known receptors and subject cloned receptor to this screening. Our search could be directed at finding new receptors for which endogenous ligands are presently unknown. The new receptor could then be synthesized *in vitro* and used as a matrix to isolate its natural ligand. This, of course, implies a large effort, but finding a new receptor and a new neurotransmitter or neuropeptide would make it worthwhile.

References

Albert PR, Neve KA, Bunzow JR, Civelli O (1990): Biological activity of the rat D2 dopamine receptor cDNA expressed in GH_4C_1 rat pituitary cells. *J Biol Chem* 265:2078–2104

Bunzow JR, Van Tol HHM, Grandy DK, Albert P, Salon J, Christie M, Machida

CA, Neve KA, Civelli O (1988): Cloning and expression of a rat D2 dopamine receptor cDNA. *Nature* 336:783–787

Byerley W, Mellon C, Holik J, Lubbers A, Leppert M, O'Connell P, Reimherr F, Grosser B, Wender P, Bunzow J, Grandy D, Civelli O, Lalouel JM, White R, Litt M (1989): Molecular genetics studies using the D2 dopamine receptor. *APA New Res Abstr*

Carlsson A (1975): Dopaminergic autoreceptors. In: *Chemical Tools in Catecholamine Research II: Regulation of Catecholamine Turnover,* Almgren O, Carlsson A, Engel J, eds. Amsterdam: North-Holland Publishing Company, pp 219–225

Caron MG, Beaulieu M, Raymond V, Gagne B, Drouin J, Lefkowitz J, Labrie F (1978): Dopaminergic receptors in the anterior pituitary gland. *J Biol Chem* 253:2244–2253

Creese I, Sibley DR, Hamblin MW, Leff SE (1983): The classification of dopamine receptors: Relationship to radioligand binding. *Annu Rev Neurosci* 6:43–71

Dixon RAF, Koblika BK, Strader DJ, Benovic JL, Dolhman HG, Frielle T, Bolanowski M, Bennett C, Rands E, Diehl R, Mumford R, Slater E, Sigal I, Caron M, Lefkowitz R, Strader C (1986): Cloning of the gene and cDNA for mammalian β-adrenergic receptor and homology with rhodopsin. *Nature* 321:75–79

Dixon RAF, Sigal IS, Candelore MR, Register RB, Scattergood W, Rands E, Strader CD (1987): Structural features required for ligand binding to the β-adrenergic receptor. *EMBO J* 6:3269–3275

Dohlman HG, Caron MG, Lefkowitz RJ (1987): A family of receptors coupled to guanine nucleotide regulatory proteins. *Biochemistry* 26:2657–2664

Gelernter J, Pakstis AJ, Chappell P, Kurlan R, Grandy DK, Bunzow J, Retief AE, Litt M, Civelli O, Kidd KK (1989): Tourette syndrome is not linked to D2 dopamine receptor. *Am Psych Assoc Abstr*

Grandy DK, Litt M, Allen L, Bunzow JR, Marchionni M, Makam H, Reed L, Magenis RE, Civelli O (1989): The human dopamine D2 receptor gene is located on chromosome 11 at q22-q23 and identifies a TaqI RFLP. *Am J Hum Genet* 45:778–785

Hall ZA (1987): Three of a kind: The β-adrenergic receptor, the muscarinic acetylcholine receptor, and rhodopsin. *Trends Neuro sci* 10:99–100

Hess EJ, Creese I (1987): Biochemical characterization of dopamine receptors. In: *Receptor Biochemistry and Methodology,* Creese I, Fraser CM, eds. New York: Liss, p 27

Isenberg KE, Burgess AK, Litt M, Grandy D, Civelli O, Devor EJ (1989): Genetic linkage study of a D2-dopamine receptor and flanking probes spanning 11q22-q23 in Tourette syndrome. *Am Soc Hum Genet Abstr*

Julius D, MacDermott AB, Axel R, Jessell TM (1988): Molecular characterization of a functional cDNA encoding the serotonin 1c receptor. *Science* 241:558–564

Kebabian JW, Calne DB (1979): Multiple receptors for dopamine. *Nature* 277:93–96

Masu Y, Nakayama K, Tamaki H, Harada Y, Kuno M, Nakanishi S (1987): cDNA cloning of bovine substance-K receptor through oocyte expression system. *Nature* 329:836–838

Meador-Woodruff JH, Mansour A, Bunzow JR, Van Tol HHM, Watson SJ, Civelli O (1989): Distribution of D2 dopamine receptor mRNA in rat brain. *Proc Natl Acad Sci USA* 86:7625–7628

Moises HW, Gelernter J, Grandy DK, Giuffra LA, Kidd JR, Pakstis AJ, Bunzow J, Sjögren B, Wettenberg L, Kennedy JL, Litt M, Civelli O, Kidd KK,

Cavalli-Sforza LL (1989): Exclusion of the D2-dopamine receptor gene as candidate gene for schizophrenia in a large pedigree from Sweden. *Psych genet Conf* Abstr

Neve KA, Henningsen RA, Bunzow JR, Civelli O (1989): Functional characterization of a rat dopamine D2 receptor cDNA expressed in a mammalian cell line. *Mol Pharmacol* 36:446–451

O'Dowd BF, Hnatowich M, Caron MG, Lefkowitz RJ, Bouvier, M (1989): Palmitoylation of the human β_2-adrenergic receptor. *J Biol Chem* 264:7564–7569

Seeman P (1987): Dopamine receptors and the dopamine hypothesis of schizophrenia. *Synapse* 1:133–152

Seeman P, Lee T (1975): Antipsychotic drugs: Direct correlation between clinical potency and presynaptic action on dopamine neurons. *Science* 188:1217–1219

Seeman P, Lee T, Chau-Wong M, Wong K (1976): Antipsychotic drug doses and neuroleptic/dopamine receptors. *Nature* 261:717–719

Senogles SE, Amlaiky N, Falardeau P, Caron MG (1988): Purification and characterization of the D2-dopamine receptor from bovine anterior pituitary. *J Biol Chem* 263:18996–19002

Stevens CF (1987): Channel families in the brain. *Nature* 328:198–199

Strader CD, Sigal IS, Candelore MR, Rands E, Hill WS, Dixon RAF (1988): Conserved aspartic acid residues 79 and 113 of the β-adrenergic receptor have different roles in receptor function. *J Biol Chem* 263:10267–10271

Strader CD, Sigal IS, Register RB, Candelore MR, Rands E, Dixon RAF (1987): Identification of residues required for ligand binding to the β-adrenergic receptor. *Proc Natl Acad Sci USA* 84:4384–4388

Vallar L, Meldolesi J (1989): Mechanisms of signal transduction at the dopamine D2 receptor. *Trends Pharmacol Sci* 10:74–77

8

Muscarinic Acetylcholine Receptors

S. V. Penelope Jones, Allan I. Levey, David M. Weiner,
John Ellis, Elizabeth Novotny, Shua-Hua Yu,
Frank Dorje, Jurgen Wess, and Mark R. Brann

1. Introduction

In 1914 Dale discovered two types of response to acetylcholine, one mimicked by muscarine and one by nicotine (Dale 1914; Dale and Ewin, 1914). This led to the subsequent discovery of nicotinic and muscarinic acetylcholine receptors. In addition to their pharmacological differences, muscarinic and nicotinic receptors can be differentiated by the mechanism and speed by which their cellular signals are transduced. Nicotinic receptors have a central pore through which sodium and potassium ions pass, resulting in depolarization of the cell membrane. Acetylcholine activates nicotinic receptors by opening the channel, and thus the response is as fast as the channel opening rate (ms). Muscarinic responses are more diverse, both hyperpolarizing and depolarizing cells by a variety of mechanisms. Muscarinic responses are also slower (on the order of 100's of milliseconds to seconds), due to their interaction with GTP-binding proteins (G-proteins) through which the cellular response is transduced. The slowest signals involve second messengers activated via the G-proteins. Examples include inhibition of adenylyl cyclase (reducing cAMP levels), stimulation of phosphatidylinositol (PI) hydrolysis (raising inositol tris phosphate which stimulates release of calcium from cytosolic stores), and arachidonic acid metabolism. These second messengers, in turn, stimulate a wide variety of responses. For example, mobilized calcium opens calcium-dependent po-tassium channels, and the activities of protein kinase A and C are dependent on cAMP and PI metabolism, respectively. On the other hand, there are muscarinic responses that are independent of second messengers. The activation of an inwardly rectifying potassium channel in the heart provides an example of a channel coupling directly with a G-protein.

In addition to this functional diversity, muscarinic receptors can be pharmacologically differentiated. For example, muscarinic receptors in cerebral cortex, salivary glands, and heart have high, moderate and low affinities for the antagonist pirenzepine (PZP), respectively. On the other hand, the antagonist AF-DX 116 has higher affinity for the heart than the other receptors. These and other studies led to the division of muscarinic receptors into three pharmacological subtypes (M1, e.g., cerebral cortex; M2, e.g., heart; M3, e.g., glands). (For general muscarinic receptor reviews, see Brann, 1989; Brann et al., 1990; Wolfe, 1989; Levine and Birdsall, 1989; Hulme et al., 1990; Nathanson, 1987.) Study of muscarinic receptor subtypes has recently intensified because of the growing list of potential clinical applications of selective compounds. For example, AF-DX 116 is used to treat bradycardia and, based on the selective loss of acetylcholine in brains of Alzheimer's patients, M1 selective agonists have been proposed as a therapy for this disorder (Whitehouse et al., 1982; Sunderland et al., 1988; Quirion et al., 1989).

Because of the pharmacological differences between muscarinic receptors located in heart and brain (Hammer et al., 1980), these tissues were initially targeted for purification efforts. Muscarinic receptors were purified to homogeneity from porcine brain (Haga and Haga, 1983) and atria (Peterson et al., 1984). Amino acid sequences were obtained from fragments of these pure proteins, and the corresponding cDNAs were cloned (Kubo et al., 1986a,b; Peralta et al., 1987b). Comparison of the two sequences indicates that atrial and brain muscarinic receptors are highly related but distinct proteins and that they are homologous with all members of the G-protein-coupled receptor superfamily (see other chapters in this volume). By screening various libraries at moderate stringency, species homologues of these two receptors, as well as three closely related receptors, were identified in rat and human (Bonner et al., 1987, 1988). Based on the pharmacology of the receptors encoded by these five related genes, they have all been shown to encode muscarinic acetylcholine receptors (Bonner et al., 1987, 1988). The cloning of all five of the receptor subtypes has been confirmed (Peralta et al., 1987a; Liao et al., 1989). Recently, non-mammalian homologues of these receptors have been cloned (Shapiro et al., 1989; Tietje et al., 1990). In this chapter we will use the consensus nomenclature as recently proposed (Levine and Birdsall, 1989): pharmacologically defined receptor subtypes are designated M1–M3 and molecular subtypes are designated m1–m5.

2. Pharmacological Properties

The pharmacological properties of cloned muscarinic receptor subtypes can be defined by the expression of the genes in cultured cells. Two model systems have been exploited for this purpose. The first is to synthesize mRNA in vitro from cloned DNA and inject it into *Xenopus* oocytes. These cells are easy to inject and are convenient for a limited set of electrophysiological recordings. Unfortunately, this approach has many limitations. *Xenopus* oocytes endogenously express low levels of muscarinic receptors, so that there are problems with background responses. The cells can not be propagated after injection, so that each injection represents a separate experiment (very labor intensive for scaling up to binding assays) and considerable variability exists between oocytes. The second strategy is to stably or transiently express the cloned DNA in cultured mammalian cells. Fortunately, this approach avoids the problems encountered with oocytes. In particular, mammalian cell lines can be selected which do not endogenously express muscarinic receptors and the transformed cells can be propagated so that unlimited amounts of receptor can be obtained. The affinities of ligands for the receptors can be readily obtained from binding assays, and agonist efficacy can be evaluated using functional responses.

For the pharmacological data obtained with cloned receptors to be physiologically relevant, the receptors must have similar properties when expressed endogenously in tissues and in transfected cell lines. Fortunately, at least with respect to antagonist binding, the available data indicate a virtual independence on cell type for the pharmacological properties of muscarinic receptors. For example, in our laboratories we have observed no differences in the affinities of several muscarinic antagonists for muscarinic receptors expressed by COS-7, CHO-K1, A9 L, RBL 2H3, and NIH 3T3 cells. Our antagonist binding data is also comparable to that observed in *Xenopus* oocytes (Akiba et al., 1988) and, where correlations have been possible (see Section 5) the pharmacology of a particular muscarinic receptor in a given tissue is very well correlated with that of the corresponding cloned receptor subtype. It should be emphasized that, while cell type seems to have little effect on antagonist binding, assay buffers (e.g., ionic strength) and other binding conditions can have significant effects (Hulme et al., 1990).

Table 8.1 illustrates the affinities of several muscarinic antagonists for the m1–m5 receptors. While none of the tested compounds has a marked selectivity for one muscarinic receptor over all others, each of the subtypes has a distinct pharmacological profile. Based on the similarities of their subtype selectivities, some of the drugs can be grouped into families. For example, *R*-trihexyphenidyl and PZP are more potent at m1 and m4 receptors than at the other receptor subtypes. On the other hand, methoctramine, himbacine, and the AF-DX derivatives are more potent at m2 and m4 receptors than at the other receptor subtypes. The latter compounds have the added property of having a very low affinity for m5 receptors. These data suggest that certain chemical classes of compounds may recognize distinct structural determinants. A structural feature shared by m1 and m4 receptors may be important for the binding of PZP, while a

TABLE 8.1. Affinities of muscarinic antagonists for cloned muscarinic receptors[a]

Antagonist	m1	m2	m3	m4	m5
(R)-Trihexyphenidyl	9.4	8.2	8.6	9.1	8.3
(S)-Trihexyphenidyl	6.9	6.3	5.9	6.6	6.2
(R)-Hexbutinol	8.7	7.7	8.7	8.5	8.3
(S)-Hexbutinol	7.6	7.1	7.3	7.7	7.1
Pirenzepine	8.2	6.7	6.9	7.4	7.0
UH-AH 37	8.7	7.4	8.2	8.3	8.3
HHSiD	7.4	7.0	8.0	—	7.2
p-F-HHSiD	7.6	6.9	7.8	7.5	7.0
HHD	8.0	7.2	7.8	—	7.1
Hexocyclium	8.6	8.1	8.9	8.3	8.4
Sila-hexocyclium	8.7	7.9	8.9	8.5	8.7
4-DAMP	9.2	8.4	9.3	8.9	9.0
Methoctramine	7.3	7.9	6.7	7.5	6.9
Himbacine	7.0	8.0	7.0	8.0	6.3
AF-DX 116	5.9	6.9	6.1	6.5	5.6
AF-DX 250	6.4	7.3	6.2	6.8	5.5
AF-DX 384	7.5	8.2	7.2	8.0	6.3
AQ-RA 741	7.5	8.4	7.2	8.2	6.1
NMS	10.3	10.1	10.3	10.6	10.0
Atropine	9.7	9.3	9.8	—	9.7

[a]All data are pK_i values derived from inhibition of [^{3}H]NMS binding to recombinant receptors expressed in CHO-K1 cells (Buckley et al., 1989; Dorje et al., 1991b; Wess et al., 1991c). With the exception of m1 data for HHSID, HHD, and hexocyclium, where the rat gene was used, all data were derived using human genes.

determinant shared by m2 and m4 receptors may have a greater contribution to the binding of himbacine and AF-DX 116 (see Section 4).

In addition to the competitive actions of the above muscarinic antagonists, a growing number of compounds appear to bind to a second (allosteric) site on muscarinic receptors. For example, the inhibition of functional responses and of receptor binding by gallamine does not follow the mass action predictions of competitive interactions; furthermore, gallamine alters the rates of dissociation of competitive ligands (Clark and Mitchelson, 1976; Stockton et al., 1983; Kenakin and Boselli, 1989). Some or all of these deviations from competitive behavior have also been reported for a pharmacologically diverse group of compounds that includes tetrahydroaminoacridine, methoctramine, verapamil, quinidine, and *d*-tubocurarine (Birdsall et al., 1987; Mitchelson, 1988; Henis et al., 1989). Muscarinic receptor subtypes are differentially susceptible to allosteric regulation; gallamine slows the rate of dissociation of [^{3}H]*N*-methylscopolamine (NMS) from all subtypes, with an order of potency of m2 > m4 > m1 > m3 > m5 (Ellis et al., 1991). The half-times for dissociation of [^{3}H]NMS themselves also vary considerably across subtypes, ranging from less than 5 min to more than 1 hr (m2 < m1 < m3 < m4 < m5). This order of dissociation rates for NMS from the cloned subtypes agrees well with studies by Waelbroeck et al. (1990), who have used differences in dissociation rates to help define four pharmacological subtypes in rat forebrain. Muscarinic allosteric effects do not seem to require any accessory proteins, as they persist after receptor solubilization and purification (Poyner et al., 1989). It has been suggested that the allosteric and competitive sites may lie within a single pocket of the receptor (Hulme et al., 1990). Slowing of the kinetics of the binding of competitive ligands might be explained in that case simply by steric hinderance. However, gallamine has been found to also accelerate the dissociation of [^{3}H]QNB (Ellis and Seidenberg, 1989; Ellis et al., 1991), which implies a greater degree of complexity in the interaction.

To evaluate the pharmacology of agonists at muscarinic receptors, their abilities to induce second messenger responses have been investigated in transfected cells. As is the case for endogenously expressed receptors, the properties of agonists are highly dependent on both receptor levels and the efficiency of coupling to functional responses (spare receptors). For example, the greater the number of spare receptors, the lower the ED_{50} of the agonist for inducing a given response, and the higher the apparent efficacy of partial agonists (Kenakin, 1986). When the agonist pharmacologies of the m1–m4 receptors are compared in A9 L cells (using conditions of low receptor spareness), a number of patterns emerge. First, most full agonists (e.g., oxotremorine M, oxotremorine, acetylcholine, muscarine) are more potent at m2 and m4 receptors (using cAMP decrease as an assay) than at m1 and m3 receptors (using cAMP elevations as an assay). Second, in the case of partial agonists (e.g., McN-A-343, RS 86, pilocarpine, arecoline), greater efficacy and potency is observed at m2 and m4 receptors than at the m1 and m3 receptors (Novotny and Brann, 1989). One explanation for these observations is that these differences are largely due to more efficient coupling of m2 and m4 receptors to their respective G-protein and effector enzymes than that observed for the m1 and m3 receptors (see Section 3), thus leading to a greater spareness for the resultant responses. It should be noted however, that m2 and m4 receptors have a higher affinity ($\sim$tenfold) for carbachol in binding assays, even in the absence of G-protein coupling (Brann et al., 1987, 1988a; Wess et al., 1989, 1990a). Overall, beyond these modest differences, little evidence of a marked subtype selectivity has been observed for any agonists. The one exception is the oxotremorine analog BM-5 which has greater efficacy at the m4 receptor than at the other subtypes (Novotny and Brann, 1989).

These data are generally comparable to data generated using second messenger responses in tissues. For example, McN-A-343, RS 86, arecoline, and pilocarpine are partial agonists for M1-stimulated PI metabolism in cerebral cortex, and have greater potency and efficacy at M2-inhibited adenylate cyclase in heart, brain, and various cell lines (Baumgold and

White, 1989; Baumgold and Drobnick, 1989; Freedman et al., 1988). On the other hand, these data conflict with many functional studies where physiological responses have been measured. For example, in various preparations, these drugs have been shown to have high efficacy and selectivity for M1 receptors (e.g., RS 86: Palacios et al., 1986). The reason for this apparent disagreement remains to be established. One possible explanation is that differences in receptor spareness have not been adequately controlled for in the physiological studies. Many M1-mediated responses have a very high degree of receptor spareness (Kenakin, 1986; Ringdahl et al., 1987). Another possible explanation is the ambiguity concerning the molecular identity of many M1 receptors (see Section 5). Finally, it is possible that, unlike antagonists, agonist pharmacology may depend on cell type. The similarity of second messenger data generated in brain and transfected cells is one argument against the latter possibility.

3. Coupling to Functional Responses

3.1. Biochemical Responses

Responses to muscarinic receptor stimulation include inhibition of adenylyl cyclase, stimulation of phosphatidylinositol (PI) and arachidonic acid metabolism, guanylate cyclase activation, and release of calcium from intracellular stores (Brann et al., 1990; Harden et al., 1986; Nathanson, 1987). Initially these effects were divided into two categories based on the sensitivity of each response to pertussis toxin (PTX). PTX selectively ADP-ribosylates certain signal transducing G-proteins. For example, in heart inhibition of adenylyl cyclase (M2) has been demonstrated to be mediated by a PTX-sensitive G-protein, while in cerebral cortex stimulation of PI metabolism (M1) is PTX-insensitive (Gil and Wolfe, 1985; Nathanson, 1987).

The linking of function with pharmacological receptor subtype has not always obeyed this straightforward correlation. In NG108–15 cells, muscarinic receptor stimulation inhibited adenylyl cyclase via a PTX-sensitive G-protein, but had an "M1-like" pharmacological profile. In N132N1 astrocytoma and SK-N-SH neuroblastoma cells, the PTX-insensitive stimulation of PI metabolism is mediated by a receptor with intermediate affinity for both PZ and AF-DX 116 (characteristic of M3 receptors). Overall, these data have suggested a greater complexity of pharmacological and functional diversity. Consideration of the cloned subtypes provides a potential molecular basis for these differences (see also anatomy and pharmacology sections). For example, m1 mRNA is the predominant subtype in cerebral cortex, m2 mRNA in heart, m3 in SK-N-SH and N132N1 cells, and m4 in NG108–15 cells. The functional differences between muscarinic receptors expressed by these tissues and cells could be due either to intrinsic differences between the receptors or to the cellular environment in which they are expressed.

When each of the receptors are expressed by the same cell type, they display functional differences which are mediated by distinct G-proteins. In transformed A9 L, CHO-K1, and HEK cells, m2 and m4 receptors reduce forskolin-induced cAMP levels in a PTX-sensitive manner (Brann et al., 1988b; Peralta et al., 1988; Novotny and Brann, 1989; Jones et al., 1991b). When expressed at higher levels, these receptors also weakly stimulate PI metabolism, again by a PTX-sensitive mechanism (Ashkenazi et al., 1989a; Peralta et al., 1988; Wess et al., 1990a).

Stimulation of m1, m3, and m5 receptors enhances PI metabolism (Brann et al., 1988b; Conklin et al., 1988; Peralta et al., 1988; Novotny and Brann, 1989; Jones et al., 1991b) and arachidonic acid release (Conklin et al., 1988, 1989), via PTX-*in*sensitive mechanisms. Stimulation of PI metabolism by these receptors results in elevated IP3 which releases calcium from intracellular stores (Neher et al., 1988; Jones et al., 1990). In A9 L cells, m1 and m3 receptors increase cAMP levels (Brann et al., 1988b; Novotny and Brann, 1989) by a mechanism which is believed to be mediated indirectly by the rise in intracellular calcium concentration (Felder et al., 1989). In transformed CHO-K1 cells, m1 and m3 receptors strongly increase cAMP levels by a PTX-insensitive mechanism, while m5 receptors only modestly increase cAMP levels. m4 receptors also strongly increase cAMP levels, but only when the cells have been treated with PTX. The large elevations in cAMP are only observed when high concentrations of agonist are used and do not correlate with PI metabolism (Jones et al., 1991b).

The m1, m3, and m5 receptors exert potent effects on mitogenesis and cellular transformation. In A9 L and CHO-K1 cells muscarinic receptors inhibit mitogenesis (Conklin et al., 1988; Conklin et al. 1989). In other CHO cells, the same receptors weakly stimulate mitogenesis (Ashkenazi et al., 1989b). The observed inconsistencies are most likely due to differences in assay conditions. Namely, the latter CHO experiments were performed with serum-starved cells, while the A9 L and CHO-K1 experiments were carried out in the presence of serum (the receptors may potentially interact with serum-derived growth factors). In NIH 3T3 cells, m1, m3, and m5 receptors potently stimulate mitogenesis and induce a cellular transformation characteristic of potent oncogenes. While the second messengers mediating cellular transformation and mitogenesis are unknown, these responses have the same dose-response characteristics as does stimulation of PI metabolism (Gutkind et al., 1991). These functional responses are summarized in Table 8.2.

3.2. Physiological Responses

Muscarinic receptors modulate a wide variety of ion channels, especially a large selection of potassium conductances (for reviews, see Brown, 1986,

TABLE 8.2. Functional responses of cloned muscarinic receptor subtypes[a]

Response	m1	m2	m3	m4	m5
Stimulate PI[b]	$+++''$	$+'$	$+++''$	$+'$	$+++''$
Stimulate arachidonic acid release[c]	$+++''$	0	$+++''$	0	$+++''$
Increase cAMP levels[d]	$+''/+++''$	0/nd	$+''/+++''$	$0/+++''$	$+''/+''$
Decrease cAMP levels [e]	0	$+++'$	0	$+++'$	0
Release intracellular Ca^{2+}[f]	$+++$	$+'/0$	$+++$	nd/0	$+++$
Inhibit mitogenesis[g]	$++$	0	$++$	0	$++$
Stimulate mitogenesis[h]	$+++/+$	0	nd/$+$	0	nd
Stimulate transformation[i]	$+++$	0	$+++$	0	$+++$
Inhibit m-current[j]	$+++$	0	$+++$	0	nd
Inhibit Ca^{2+} conductance[j]	0	$++$	0	$++$	nd
Activate Ca^{2+} dep. K,Cl[k]	$+++''$	$+'/0$	$+++''$	0	$+++''$
Activate cation conductance[l]	0	$++$	nd	nd	nd
Stimulate secretion[m]	$+++$	0	nd	0	nd
Stimulate inward rectifier[m]	nd	$++$	nd	nd	nd

[a] $''$, PTX insensitive; $'$, PTX sensitive; nd, not determined.
[b] Stimulation of PI metabolism by m2 and m4 receptors requires higher receptor levels and higher concentrations of agonist than do cAMP decreases mediated by these same receptors.
[c] Responses have been observed in both CHO-K1 and A9 L cells.
[d] Small elevations in cAMP (correlated with PI stimulation) in A9 L and HEK cells. Large elevations have been observed in CHO-K1 cells, which require higher agonist concentrations and higher receptor levels than do PI responses.
[e] This pattern of responses has been observed in many cell lines. One study has reported a PTX-sensitive decrease in cAMP for m1 receptors (Stein et al., 1988).
[f] Has now been observed in several cell lines.
[g] Observed in A9 L and CHO cells assayed in the presence of serum.
[h] Weak responses have been observed in serum-starved CHO cells; very strong responses in NIH 3T3 cells.
[i] Observed in NIH 3T3 cells.
[j] Observed in NG108-15 cells.
[k] Observed in oocytes, A9 L cells, CHO-K1 cells, and NG108-15 cells.
[l] Observed in oocytes.
[m] Observed in RBL cells.

1990; Christie and North, 1988; Nicoll, 1988; North, 1989). The mechanisms of action and the muscarinic receptor subtypes involved in each of these responses are largely unknown. As with biochemical responses, muscarinic antagonists have been used to determine receptor subtype, but the problems with drug specificities are even greater in the physiological experiments because the absolute affinities of drugs are more difficult to measure.

Of the effects of muscarinic receptors on potassium channels, muscarinic receptor-induced increases in an inwardly rectifying potassium conductance in heart have been the best characterized (Hartzell, 1988; Brown and Birnbaumer, 1990; Szabo and Otero, 1990). Modulation of this channel by muscarinic receptors has been shown to be transduced directly by a PTX-sensitive G-protein (Breitweiser and Szabo, 1985; Pfaffinger et al.,

1985; Kirsh et al., 1988). The pharmacology indicates that the response is mediated by an M2 receptor. In brain, a similar increase in an inward potassium conductance has been observed on application of muscarinic agonists, which also is thought to be mediated via an M2 receptor (Egan and North, 1986; McCormick and Prince, 1986; Christie and North, 1988). A similar response has been observed in m2-transfected RBL 2H3 cells. In these cells, muscarinic agonists stimulate increases in an inwardly rectifying potassium conductance which is both cesium and barium sensitive, and may be similar to that observed in heart (Jones, 1991). As the m4 muscarinic receptor couples with PTX-sensitive G-proteins in biochemical assays, it seems likely that the m4 receptor will also be able to couple with the G-protein mediating the increase in the inward potassium current. This is supported by the observation that carbachol stimulates an inwardly recti-fying potassium conductance in AtT20 cells (Dousmanis and Pennefather, 1989), which express only m4 mRNA (S. V. P. Jones, D. M. Weiner, and M. R. Brann).

The M-current, so named because it was discovered by observation of a novel effect of muscarine, is a voltage- and time-dependent potassium conductance that is inhibited by muscarinic agonists (see Brown, 1988a,b for review). The mechanism of action of muscarine on this conductance still remains unclear and appears to differ with tissue. The inhibition is PTX insensitive and can be mimicked by IP3 in hippocampal neurons but not in NG108-15 cells or sympathetic ganglia, where phorbol esters have been shown to inhibit the M-current, suggesting the involvement of diacyl-glycerol and protein kinase C (Brown, 1988a,b; Brown and Higashida, 1988; Dutar and Nicoll, 1988; Pfaffinger et al., 1988; Brown et al., 1989; Hille, 1989). m1 and m3, but not m2 and m4 receptors inhibit the M-current in transfected NG 108-15 cells (Fukuda et al., 1988). Unfortunately, the mechanism of action was not addressed in this study.

Muscarinic agonists also inhibit a resting or background potassium conductance (thought to be a leak conductance maintaining the resting potential) (Muller and Misgeld, 1986; Madison et al., 1987; Galligan et al., 1989). Pharmacologically, this response has been suggested to be mediated by the M1 receptor (North et al., 1985) and it can be mimicked both by phorbol esters and by increased cAMP (Malenka et al., 1986; Palmer et al., 1987).

A variety of calcium-dependent potassium, chloride, and cation conduc-tances are *activated* by muscarinic stimulation in glandular tissue, such as salivary or lacrimal glands and pancreas. These effects are mimicked by raising intracellular calcium or by intracellular application of IP3 (for review, see Marty, 1987; Petersen and Gallacher, 1988). A similar activation of calcium-dependent conductances has been shown to occur on stimulation of the m1, m3, and m5 receptors in transformed cell lines (Brann et al., 1987; Fukuda et al., 1988; Jones et al., 1988a,b, 1991b). Unlike the potassium conductance activated in heart, the mechanism of activation of the calcium-dependent potassium conductance has been shown to be indirect via a second messenger (Jones et al., 1990). This is to be expected, as all of these receptors stimulate phospholipase C and thus produce IP3 which in turn releases calcium from intracellular stores. This was confirmed by application of intracellular IP3, which mimicked the muscarinic receptor-activated conductances, and by demonstration of a rise in intracellular free calcium on muscarinic receptor stimulation, using the fluorescent calcium indicator dye, Fura-2 (Neher et al., 1988; Jones et al., 1990; 1991a). The predominant conductance activated by m1 and m3 in *Xenopus* oocytes (using mRNA injections) was the calcium-dependent chloride conductance. In transformed mammalian A9 L, CHO-K1, and NG 108-15 cells, the potassium conductance predominates, although increases in the calcium-dependent chloride conductance have been demonstrated in the A9 L cells (Jones et al., 1988a). These actions of m1, m3, and m5 were PTX-insensitive which is consistent with stimulation of PLC (Jones et al., 1990, 1991b). In smooth muscle (Benham et al., 1985) and in chromaffin cells (Inoue and Kuriyama, 1991) muscarinic receptors activate a nonspecific cation conductance. In the chromaffin study stimulation of the cation conductance was PTX sensitive and believed to be mediated by m4. In *Xenopus* oocytes this action has been demonstrated with the m2 muscarinic receptor (Fukuda et al., 1987).

Another type of calcium-dependent potassium conductance is *inhibited* by muscarinic receptor stimulation. Although this conductance is activated by the opening of voltage-dependent calcium channels upon action potential firing, muscarinic agonists have been shown to act primarily on the potassium conductance at fairly low concentrations, effects on calcium currents occurring at higher concentrations (Knopfel et al., 1990). The conductance is termed the slow after hyperpolarization (AHP) and has been shown to be mediated by M1 receptors (Galligan et al., 1989). As with the background potassium conductance, inhibition of the AHP by muscarine can be mimicked either by phorbol esters or by raised cAMP levels (Malenka et al., 1986; Madison et al., 1987). A voltage-dependent transient potassium conductance or A-current has also been shown to be inhibited by muscarinic agents (Nakajima et al., 1986; Cassell and McLachlan, 1987; Akins et al., 1990); the mechanism of action is unknown.

Calcium conductances are both increased and decreased by muscarinic receptor stimulation (see Dolphin, 1990; Schultz et al., 1990; Trautwein and Hescheler, 1990, for calcium channel reviews). The decrease has been well documented (Gahwiler and Brown, 1987; Wanke et al., 1987; Tse et al., 1990) and appears to be due to reduced cAMP or increased cGMP levels in heart (Fischmeister and Hartzell, 1986, 1987; Hartzell and Fischmeister, 1987). In the hippocampus, an N-type calcium conductance (high-voltage activated, HVA) was decreased by a PTX-sensitive G-protein (Toselli et al., 1989). In dorsal root ganglion neurons, both N- and L-type conductances were decreased (Wanke et al., 1989). Also in hippocampus a T-type calcium conductance (LVA) was shown to be increased via a PTX-insensitive G-protein (Toselli and Lux, 1989). In smooth muscle, the muscarinic receptor-stimulated *increase* in an HVA (N or L type) calcium current was mimicked by diacylglycerol (Clapp et al., 1987; Vivazudou et al., 1988), suggesting activation of the phospholipase C pathway. Thus, muscarinic receptor-induced modulation of calcium conductances appears to be as complex as their modulation of potassium conductances. In fact, the absolute identity of second messengers mediating calcium current inhibition seems to be as elusive as those mediating the M-current (Bernheim et al., 1991). Perhaps there are many pathways to the same result. Recently, however, the muscarinic-induced inhibition of calcium currents in GH3 cells was shown to be mediated via the G-protein Go (Kleuss et al., 1991). Perhaps this information will be of use in unraveling some of the complexity. To date the identity of the subtypes of muscarinic receptor involved has not been addressed in terms of calcium channel type in transformed cell lines, but Higashida et al. (1990) have demonstrated a PTX-sensitive decrease in calcium conductance with m2 and m4 in NG108-15 cells transfected with each receptor subtype. m1 and m3 were without effect on calcium conductances in these cells.

Muscarinic receptors are known to regulate the release of acetylcholine in the central (Pepeu, 1973; Szerb and Somogyi, 1973) and peripheral (Briggs and Cooper, 1982) nervous systems. In neocortex, hippocampus, and striatum, agonist binding to presynaptic M2 receptors inhibits the potassium evoked release of acetylcholine from synaptosomes (Mayer and Otero, 1985; James and Cubeddu, 1987; Quirion et al., 1989; Hoss et al., 1990; Raiteri et al., 1990b). Recent studies have also implicated the involvement of a heterogeneity of receptor subtypes (Pittel et al., 1990), potentially including M3 receptors (Marchi and Raiteri, 1989).

Muscarinic receptors also regulate the release of other neurotransmitters. Inhibition of evoked release of glutamate and aspartate in hippocampus is mediated by M2 receptors (Kilbinger et al., 1989; Marchi and Raiteri, 1989; Raiteri et al., 1990a). In striatum, GABA release is inhibited by M3 receptors (Raiteri et al., 1990c). This subtype resembles one of the inhibitory autoreceptors and is distinct from those regulating glutamate and aspartate release. Therefore, different subtypes may mediate inhibition of neurotransmitter release. As with regulation of acetylcholine release, activation of M1 muscarinic receptors also augments the release of other neurotransmitters, including norepinephrine in cortex (Diamont et al., 1990) and dopamine in striatum (Xu et al., 1990). In vas deferens, M1 receptors inhibit neurogenic contractions, a response which is believed to be mediated by inhibition of ATP release (Eltze et al., 1988). It has been suggested that subtypes linked to PI metabolism (Diamont et al., 1990) and activation of protein kinase C (Xu et al., 1990) enhance release. Based on the pharmacology and functional properties of the cloned receptors, the m1, m3, and m5 receptors might all enhance neurotransmitter release, and m2 and m4 might be inhibitory. Studies with transfected RBL 2H3 cell lines are consistent with this suggestion. m1 receptors stimulate release of secretory granules (Jones et al., 1991a). m2 and m4 do not activate secretion in these cells. Table 8.2 summarizes the physiological actions of the five cloned receptor subtypes.

4. Structure/Function

Largely by analogy with rhodopsin and the β-adrenergic receptor (see other chapters in this volume) where more detailed structural information is available, the following model of the muscarinic receptors has been proposed (Figure 8.1). The muscarinic receptor sequences start with a variable N-terminal sequence (21–65 amino acids) which faces the extracellular space. The receptor protein crosses the membrane seven times, forming seven transmembrane (TM) α-helices which are highly conserved among the subtypes. The α-helices are arranged in a bundle, progressing in order TM 1 through TM 7, and several of the helices are disrupted by the presence of prolines. This results in three loops (o1–o3) facing the extracellular space and three loops (i1–i3) and a C-terminal domain facing the cytoplasmic space. All but the i3 loop are small (13–22 amino acids) and conserved with respect to size among the subtypes. The i3 loop is large, variable in size, and is the region with the lowest degree of conservation among the subtypes (Bonner et al., 1987, 1988).

The receptor proteins are likely to be subject to several posttranslational modifications. The extracellular N-terminal region of each of the receptors have 2–5 consensus sequences predictive of N-linked glycosylation. In the case of the m2 muscarinic receptor, site-directed mutagenesis has indicated that even without glycosylation, the muscarinic receptors are able to bind radioligands and mediate functional responses. Also, these mutations have little effect on receptor levels and localization to the cytoplasmic membrane (Van Koppen and Nathanson, 1990). Muscarinic receptors have a conserved cysteine located in the C-terminal region of the receptors. In the cases of rhodopsin (Ovchinnikov et al., 1988) and the β-adrenergic receptor (O'Dowd et al., 1989), the analogous cysteine has been shown to be palmitoylated and is likely to serve as a site of membrane attachment, thus forming an "i4 loop." Replacement of this cysteine with a glycine in the C-terminal region of the m2 receptor has no effect on coupling to cAMP metabolism or localization to the membrane (Van Koppen and Nathanson, 1991). The sequence of the C-terminal region from TM 7 to this cysteine is well conserved for all of the subtypes; after the cysteine, those muscarinic receptors which couple efficiently to PI metabolism (m1, m3, m5) have a long nonconserved sequence, which is not present in the m2 and m4 receptors. Exchange of C-terminal regions between the m2 and m3 receptors does not influence receptor coupling to PI turnover (Wess et al.,

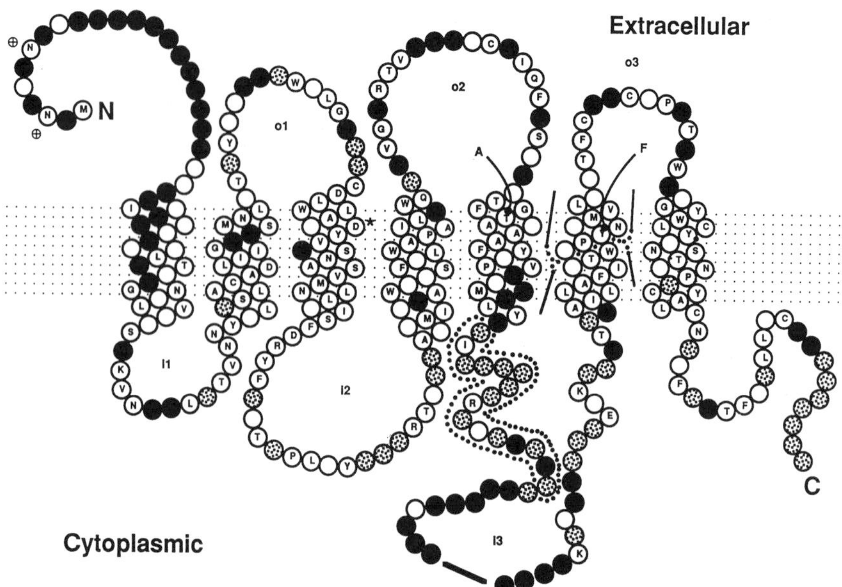

FIGURE 8.1. Model of muscarinic acetylcholine receptor structure. The receptors consist of an extracellular N-terminal region, and seven hydrophobic transmembrane domains that are linked by three outer (o1–o3) and three inner loops (i1–i3). Amino acids that are identical for all subtypes are indicated by letters and those with conserved substitutions by unshaded circles. Amino acids with nonconservative substitutions are indicated by filled circles, and those conserved for m1/m3/m5 vs. m2/m4 by shaded circles. The large variable region of the i3 loop is not shown. Dotted lines indicate regions of receptors that have been exchanged in chimeric receptors (see text). (+) indicates regions of N-linked glycosylation, and (*) marks the aspartic acid labeled by *N*-propylbenzilylcholine mustard. Hydroxyl-containing amino acids in TM 5 and TM 6 which are important in agonist binding are marked (arrows) by the substitutions which have been tested: A for T and F for Y, respectively.

1990b). Within the third cytoplasmic loop and the C-terminal region, several consensus sequences predictive of substrates for A and C kinases are present, and thus phosphorylation of the muscarinic receptors is likely to contribute to regulation of receptor function. As with rhodopsin and β-adrenergic receptors, kinases which are selective for agonist-occupied muscarinic receptors are also likely to exist (Benovic et al., 1989; Kwatra et al., 1989).

Ligands are likely to bind to a pocket formed by distal portions of the TM α-helices and adjacent extracellular epitopes. Several lines of evidence suggest that ligand binding involves many epitopes. The irreversible muscarinic receptor antagonist *N*-propylbenzilylcholine mustard covalently attaches to an aspartic acid located in the extracellular half of TM 3 (Curtis et al., 1989; marked in Figure 8.1). This, together with the characteristic presence of a cationic headgroup in most muscarinic ligands, suggests that ion-bond formation is important in ligand binding. Similarly, structure–function studies of muscarinic ligands have suggested that multiple hydrophobic interactions may contribute to ligand binding. A recent series of point mutations has indicated that hydroxyl groups present in a threonine in TM 5 and a tyrosine in TM 6 strongly contribute to agonist binding (Wess et al., 1991b).

Exploiting the pharmacological differences between the muscarinic receptor subtypes (see Section 2), chimeric receptors have been used to evaluate regions associated with subtype selectivity. Using m2/m3 chimeras the i3 loop has been shown not to contribute to antagonist binding, while the higher affinity of m2 than m3 for acetylcholine appears to be related to the different i3 loops (Wess et al., 1990a). On the other hand, C-terminal regions (distal TM 6, o3, TM 7, and C-terminal) of the m2 and m3 receptors appear to significantly contribute to the subtype selectivity of certain muscarinic antagonists (Wess et al., 1990b). In a more systematic study involving a large number of m2/m5 chimeras, multiple domains of the m2 receptor were shown to contribute to its higher affinity for himbacine and AQ-RA 741. On the other hand, the higher affinity of UH-AH 37 for the m5 receptor seems to be solely due to differences in the distal part of TM 6 and o3 (Wess et al., 1991a).

Regions which are involved in the coupling of the receptors to G-proteins have also been examined using chimeric receptors. Chimeric m1/m2 receptors have indicated that the i3 loop of the m1 receptor is able to induce efficient coupling of the m2 receptor to the chloride conductance (Kubo et al., 1988). We have recently shown using chimeric m2/m3 receptors that the i3 loops contain sequences that define selective coupling to G-proteins and second messengers. m2 receptors with the i3 loop of m3 couple to PI metabolism via a PTX-insensitive G-protein. Conversely, m3 receptors with the i3 loop of m2 couple to inhibition of adenylyl cyclase. Within the i3 loop, the region proximal to TM 5 seems to be the most important (Wess et al., 1989, 1990a). Within this region there is a series of amino acids which are conserved with respect to m1, m3, and m5 versus m2 and m4 (see Section 3). These data are consistent with results with m1 receptors, where the majority of the i3 loop can be deleted and still retain efficient stimulation of PI metabolism (Shapiro and Nathanson, 1989). The importance of the region proximal to TM 5 has recently been confirmed using electrophysiological (Lechleiter et al., 1991b) and calcium measurements (Lechleiter et al., 1991a) in *Xenopus* oocytes.

5. Anatomical Localization

As previously discussed, muscarinic receptors in different tissues can be distinguished pharmacologically. Three tissues which have been widely used to evaluate the pharmacology of the distinct receptor subtypes are the vas deferens, heart, and ileum: M1 receptors inhibit neurogenic contractions in vas deferens; M2 receptors slow the heart, and M3 receptors contract ileal smooth muscle. Using data from several of the most selective muscarinic compounds, we correlated the pharmacologies of these physiologically defined subtypes (M1–M3) with that of the genetically defined subtypes (m1–m5) (Table 8.3). Very good correlations are observed between the m2 and M2 receptors, and these receptors are the easiest to differentiate from the other subtypes. The pharmacology of the M3 receptor correlates the best with m3, but caution should be exercised as even the most discriminating compounds have a modest selectivity ($<$ tenfold). Unfortunately, even when this relatively large number of compounds are considered, it is impossible to unequivocally assign the M1 receptor of vas deferens to a genetically defined subtype. Clearly the pharmacological data alone is inadequate to evaluate the anatomical distribution of the muscarinic receptor subtypes. The pharmacological approach is particularly unreliable if the binding of a single ligand is considered, as in radioligand receptor autoradiography.

The molecular cloning of the receptor subtypes has provided new tools to evaluate their tissue-specific expression. For example, using cloned DNA or oligodeoxynucleotides it is possible to measure the distribution of the mRNA which encodes each of the receptor subtypes. Both approaches have been used to map the distribution of muscarinic receptor mRNAs (Brann et al., 1988a; Bonner et al., 1987, 1988; Buckley et al., 1988; Maeda et al., 1988; Peralta et al., 1987a; Weiner and Brann, 1989; Weiner et al., 1990). Similarly, the encoded receptor proteins can be measured by use of subtype selective antibodies. One approach is to prepare synthetic peptides based on the predicted sequences of the receptor proteins. These peptides have been conjugated to carrier protein and used as antigens (Luthin et al., 1988; Levey et al., 1989). Another approach is to express fragments of the cloned receptors as fusions with antigenic proteins in *E. coli,* and use these proteins as antigens (Levey et al., 1990, 1991).

TABLE 8.3. Correlation of antagonist affinities for pharmacological (M1–M3) and molecular muscarinic receptor subtypes (m1–m5)[a]

	M1	M2	M3	m1	m2	m3	m4	m5
M1	—	0.67	0.67	0.76	0.59	0.64	0.75	0.67
M2	0.67	—	0.55	0.43	0.90	0.51	0.75	0.43
M3	0.67	0.55	—	0.82	0.65	0.99	0.89	0.82

[a]Drugs and data listed in Table 8.1 were included in this comparison. Affinities for M1 (rabbit vas deferens), M2 (guinea pig atria), and M3 receptors (guinea pig ileum) were as reported (Lambrecht et al., 1988, Feifel et al., 1990: Dorje et al., 1990). Drugs and data from binding assays of cloned receptors expressed in CHO-K1 cells are illustrated in Table 8.1. Table indicates correlation coefficients.

Overall, the above approaches have indicated that the muscarinic receptors are differentially distributed in peripheral tissues and have provided evidence for the molecular identities of the pharmacologically defined receptor subtypes. In addition to inhibitory M1 receptors, vas deferens has M2 receptors which enhance neurogenic contractions (Eltze et al., 1988). Both m1 and m2 receptor proteins are present in vas deferens (Dorje et al., 1991a). Many studies have indicated a homogeneous population of M2 muscarinic receptors in heart (Hulme et al., 1990), and only m2 mRNA (Buckley et al., 1988; Maeda et al., 1988) and protein (Dorje et al., 1991a) have been detected in heart. Similarly, the majority of ileal receptors have an M2 pharmacology (Candell et al., 1990), and m2 mRNA (Maeda et al., 1988) and protein (Dorje et al., 1991a) are the predominant molecular species in ileum. As indicated above, ileum also has a functionally defined M3 subtype. Small amounts of m3 mRNA (Maeda et al., 1988) have been detected in ileum, but no m3 protein (Dorje et al., 1991a). M1 and M3 muscarinic receptors are present in submaxillary glands (Mei et al., 1990), and high levels of both m1 and m3 mRNAs (Maeda et al., 1988) and proteins (Dorje et al., 1991a) have also been observed. Sympathetic ganglia have both M1 and M2 receptor subtypes (Newberry and Priestly, 1987), and both m1 and m2 receptor proteins are present (Dorje et al., 1991a). Binding studies have indicated that muscarinic receptors in peripheral lung and NG108–15 cells have a unique "M1-like" pharmacology which has been termed the "M4" receptor (Lazareno et al., 1990). m4 mRNA is the predominant subtype in both lung (Lazareno et al., 1990) and NG108–15 cells (Peralta et al., 1987a). Both m2 and m4 proteins are present in peripheral lung (Dorje et al., 1991a).

All five of the receptor subtype mRNAs and proteins have been detected in the brain (Brann et al., 1988a; Buckley et al., 1988; Levey et al., 1991; Weiner and Brann, 1989; Weiner et al., 1990). m1 and m4 mRNA and protein are widely expressed in cerebral cortex, basal ganglia (including caudate-putamen), and hippocampus. Thus, these receptors are likely to play major roles as postsynaptic muscarinic receptors in various cognitive and motor functions and are likely to be major contributors to the M1 responses which have been measured in these brain regions. m4 mRNA and protein are the most abundant subtype in caudate-putamen, possibly explaining the anomalous M1-like "M4" pharmacology of binding sites in this brain region (Waelbroeck et al., 1990). m2 receptor and protein predominates in the brainstem and cholinergic cells of the basal forebrain and caudate-putamen. The distribution of m2 versus m4 receptors may account for differences in the pharmacology of cAMP inhibition by muscarinic receptors among brain regions (McKinney et al., 1989). These data are also consistent with receptor autoradiography of M2 binding sites (Cortes and Palacios, 1986; Mash and Potter, 1985; Mash et al., 1985). As discussed in the physiological section, pharmacological studies have indicated that M2 receptors inhibit acetylcholine release, and the finding of m2 mRNA in cholinergic cells (Weiner et al., 1990) and m2 protein in cholinergic neurons (Levey et al., 1991) establishes this subtype as a major presynaptic muscarinic receptor. It should also be noted that both m2 mRNA and protein are widely expressed by noncholinergic cells in various brain regions including the cerebral cortex (Levey et al., 1991; Weiner and Brann, 1989; Weiner et al., 1990). m3 mRNA is present in the cerebral cortex, hippocampus and thalamus, but not in basal ganglia (Brann et al., 1988a; Buckley et al., 1988; Weiner et al., 1990).

m1 mRNA is expressed by the majority of medium-sized neurons of the caudate-putamen and m4 is expressed by ~50% of these neurons. Within the caudate-putamen, the m4 receptor is co-expressed with dopamine receptors, implicating a direct interaction with dopaminergic neurotransmission and the control of dopamine-mediated psychomotor function. The m5 receptor is expressed by the dopaminergic neurons within the substantia nigra pars compacta, suggesting that this receptor may be the muscarinic receptor which mediates direct stimulation of dopamine release by acetylcholine (Weiner et al., 1990).

Overall, because of the complex expression patterns of muscarinic receptors within the brain and the paucity of cellular information concerning the behavioral function of the neuronal systems involved, it is difficult to unequivocally relate individual receptor subtypes with the individual behavioral effects of muscarinic drugs. For example, the anti-parkinsonian site of action of drugs such as trihexyphenidyl is likely to be within the basal ganglia. Since this drug has high affinity for both m1 and m4 receptors (Table 8.1) and both of these receptors are expressed by the basal ganglia, either or both of these receptors could be the relevant site of action.

The subtype selective targeting of cholinergic agonists for treatment of Alzheimer's disease is similarly problematic, because brain regions involved in cognitive function express all five of the muscarinic receptor subtypes. Studies in animal models using muscarinic antagonists have tended to rule out M2/m2 receptors, and considering their presynaptic location one would expect that M2/m2 receptor stimulation should be avoided. On the other hand, "M1" selective agonists have not proved to be more effective in clinical trials than acetylcholinesterase inhibitors. It should be noted that biochemical experiments suggest that these "M1" agonists are, in fact, weak partial agonists with selectivity for m2/m4 receptors. Thus the receptor subtype(s) which should be targeted for treatment of Alzheimer's disease remains to be established.

Because of their discrete patterns of expression, the m4 and m5 receptors represent compelling therapeutic targets. Within the periphery, expression of the m4 receptor is most prominent in the lung. Since "M1" selective drugs are useful in the treatment of asthma, an m4 selective antagonist may avoid m1-mediated side effects (e.g., in sympathetic ganglia). Similarly, m5 has a very limited distribution within the brain. If m5 receptors are those which enhance release of dopamine, then these receptors may be useful targets for therapeutic modulation of dopaminergic tone (e.g., Tourette's syndrome and schizophrenia).

6. Future Directions

1. Functional studies indicate that muscarinic receptors couple selectively with distinct G-proteins to mediate a diversity of responses. However, it is now clear that this selectivity is not absolute and is likely to involve complex equilibria among many receptors and G-proteins. Thus a quantitative evaluation of the interaction of the individual cloned receptors with each of the G-proteins is required. Potential approaches are coexpression of mammalian receptors and G-proteins in cultured cells which do not endogenously express either the receptor or G-protein (e.g., lower eukaryotes such as yeast; King et al., 1990) and/or reconstitution of pure recombinant receptors and G-proteins in phospholipid vesicles (Parker et al., 1991). Once these molecular interactions are quantified, the endogenous ratios of receptors to G-protein should be evaluated in tissues, thus allowing functional predictions.

2. Little is known concerning the mechanism by which muscarinic receptor expression is controlled, either in the context of cellular phenotype or upon adaptive changes in response to receptor and/or cellular activation. Analysis of the role of posttranslational modifications in receptor activity and protein turnover has been extensively studied in the case of β-adrenergic receptors (see other chapters in this volume), but similar study of muscarinic receptors has just begun. The complex relationship between receptor phosphorylation and receptor sensitivity is a particularly rich area for future study. Another largely unknown frontier is the mechanism by which transcription of each of the muscarinic receptors is so tightly controlled.

3. It is anticipated that a combination of physicochemical (X-ray crystallography, NMR), mutagenesis (chimeric receptors, point mutations), and drug structure–function approaches will eventually lead to an understanding of the molecular details of ligand binding to muscarinic receptors and their subsequent activation of G-proteins. This information will hopefully lead to rational design of muscarinic drugs with targeted selectivity and efficacy. Perhaps the most challenging application of this approach will be the design of the first potent agonists with real selectivity for muscarinic receptor subtypes.

4. The remarkable differences in the patterns of expression of muscarinic receptors within the brain suggests that subtype selective drugs are likely to

have distinct behavioral effects. Unfortunately, it is presently impossible to precisely predict behavioral effects from this data. It is hoped that future combinations of molecular neurobiological and psychopharmacological techniques will allow such predictions.

Acknowledgments

The authors wish to thank Dr. Neil Nathanson for reviewing the manuscript and for sharing unpublished information and the late Dr. Ernst Freese, whose support and direction of the Laboratory of Molecular Biology made this work possible.

References

Akiba I, Kubo T, Maeda A, Bujo H, Kakai J, Mishina M, Numa S (1988): Primary structure of porcine muscarinic acetylcholine receptor III and antagonist binding studies. *FEBS Lett* 235:257–261

Akins PT, Surmeier DJ, Kitai ST (1990): Muscarinic modulation of a transient K^+ conductance in rat neostriatal neurons. *Nature* 344:240–242

Ashkenazi A, Peralta EG, Winslow JW, Ramachandran J, Capon DJ (1989a): Functionally distinct G-proteins selectively couple different receptors to PI hydrolysis in the same cell. *Cell* 56:487–493

Ashkenazi A, Ramachandran J, Capon DJ (1989b): Acetylcholine analogue stimulates DNA synthesis in brain-derived cells via specific muscarinic receptor subtypes. *Nature* 340:146–150

Baumgold J, Drobnick A (1989): An agonist that is selective for adenylate cyclase-coupled muscarinic receptors. *Mol Pharmacol* 36:465–470

Baumgold J, White T (1989): Pharmacological differences between muscarinic receptors coupled to phosphoinositide turnover and those coupled to adenylate cyclase inhibition. *Biochem Pharmacol* 38:1605–1616

Benham CD, Bolton TB, Lang RJ (1985): Acetylcholine activates an inward current in single mammalian smooth muscle cells. *Nature* 316:345–347

Benovic JL, Deblasi A, Stone WC, Caron MG, Lefkowitz RJ (1989): β-Adrenergic receptor kinase: Primary structure delineates a multigene family. *Science* 246:235–246

Bernheim L, Beech DJ, Hille B (1991): A diffusible second messenger mediates one of the pathways coupling receptors to calcium channels in rat sympathetic neurons. *Neuron* 6:859–867

Birdsall NJM, Hulme EC, Kromer W, Stockton JM (1987): A second drug binding site on muscarinic receptors. *Fed Proc* 46:2525–2527

Bonner TI, Buckley NJ, Young AC, Brann MR (1987): Identification of a family of muscarinic acetylcholine receptor genes. *Science* 237:527–532

Bonner TI, Young A, Brann MR, Buckley NJ (1988): Cloning and expression of the human and rat m5 muscarinic receptor genes. *Neuron* 1:403–410

Brann MR (1989): Neuronal receptors, molecular biology approaches. In: *Neuroscience Year: The Yearbook of the Encyclopedia of Neuroscience*, Adelman G, ed. Boston: Birkhäuser, Suppl 1, pp 120–123

Brann MR, Buckley NJ, Bonner TI (1988a): The striatum and cerebral cortex express different muscarinic receptor mRNAs. *FEBS Lett* 230:90–94

Brann MR, Buckley NJ, Jones SVP, Bonner TI (1987): Expression of a cloned muscarinic receptor in A9 L cells. *Mol Pharmacol* 32:450–455

Brann MR, Conklin BR, Dean NM, Collins RM, Bonner TI, Buckley NJ (1988b): Cloned muscarinic receptors couple to different G-proteins and second messengers. *Soc Neurosci Abstr* 14:600

Brann MR, Wess J, Jones SVP (1990): Molecular genetics of signal transduction by muscarinic acetylcholine receptors. Proceedings of 43rd symposium on "G-Proteins and Signal Transduction." *J Gen Physiol* 45:106–115

Breitwieser GE, Szabo G (1985): Uncoupling of cardiac muscarinic and beta-adrenergic receptors from ion channels by a guanine nucleotide analogue. *Nature* 317:538–540

Briggs CA, Cooper JR (1982): Cholinergic modulation of the release of 3H-acetylcholine from synaptasomes of the myenteric plexus. *J Neurochem* 38:501–508

Brown AM, Birnbaumer L (1990): Ionic channels and their regulation by G protein subunits. *Annu Rev Physiol* 52:197–213

Brown DA (1986): ACh and brain cells. *Nature* 319:358–359

Brown DA (1988a): M currents. In: *Ion Channels* Narahashi T, ed. New York and London: Plenum Press, Vol 1, pp 55–94

Brown DA (1988b): M-currents: An update. *Trends Neurosci* 11:294–299

Brown DA (1990): G-proteins and potassium currents in neurons. *Annu Rev Physiol* 52:215–242

Brown DA, Higashida H (1988): Inositol 1,4,5-trisphosphate and diacylglycerol mimic bradykinin effects on mouse neuroblastoma × glioma hybrid cells. *J Physiol (Lond)* 397:185–207

Brown DA, Marrion NV, Smart TG (1989): On the transduction mechanism for muscarine-induced inhibition of M-current in cultured rat sympathetic neurons. *J Physiol (Lond)* 413:469–488

Buckley NJ, Bonner TI, Brann MR (1988): Localization of a family of muscarinic receptor mRNAs in rat brain. *J Neurosci* 8:4646–4652

Buckley NJ, Bonner TI, Buckley CM, Brann MR (1989): Antagonist binding properties of five-cloned muscarinic receptors expressed in CHO-K1 cells. *Mol Pharmacol* 35:469–476

Candell LM, Yun SH, Tran LL, Ehlert J (1990): Differential coupling of subtypes of the muscarinic receptor to adenylate cyclase and phosphoinositide hydrolysis in longitudinal muscle of the rat ileum. *Mol Pharmacol* 38:689–697

Cassell JF, McLachlan EM (1987): Muscarinic agonists block five different potassium conductances in guinea-pig sympathetic neurones. *Br J Pharmacol* 91:259–261

Christie MJ, North RA (1988): Control of ion conductances by muscarinic receptors. *Trends Pharmacol Sci [Suppl]* 3:30–34

Clapp LH, Vivaudou MB, Walsh JJ, Singer JJ (1987): Acetylcholine increases voltage-activated calcium currents in freshly dissociated smooth muscle cells. *Proc Natl Acad Sci USA* 84:2092–2096

Clark AL, Mitchelson F (1976): The inhibitory effect of gallamine on muscarinic receptors. *Br J Pharmacol* 58:323–331

Conklin BR, Brann MR, Buckley NJ, Bonner TI, Ma AL, Felder C, Axelrod J

(1989): Carbachol stimulation causes inhibition of mitogenesis and cell elongation in CHO cells transfected with muscarinic receptor genes. *Trends Pharmacol Sci [Suppl]* 4:Abstr 74

Conklin BR, Brann MR, Buckley NJ, Ma AL, Bonner TI, Axelrod J (1988): Stimulation of arachidonic acid release and inhibition of mitogenesis by cloned muscarinic receptor subtypes stably expressed in A9 L cells. *Proc Natl Acad Sci USA* 85:8698-8702

Cortes R, Palacios JM (1986): Muscarinic cholinergic receptor subtypes in the rat brain. I. Quantitative autoradiographic studies. *Brain Res* 362:227-238

Curtis CAM, Wheatley M, Basal S, Birdsall NJM, Eveleigh P, Pedder EK, Poyner D, Hulme EC (1989): Propylbenzilylcholine mustard labels an acidic residue in transmembrane helix 3 of the muscarinic receptor. *J Biol Chem* 264:489-495

Dale HH (1914): The occurrence in ergot and action of acetylcholine. *Proc Physiol Soc Lond* p iii

Dale HH, Ewin AJ (1914): Choline-esters and muscarine. *Proc Physiol Soc Lond* p xxiv

Diamont S, Schwartz S, Atlas D (1990): Potentiation of neurotransmitter release coincides with potentiation of phosphatidyl inositol turnover. A possible in vitro model for long term potentiation. *Neurosci Lett* 109:140-145

Dolphin AC (1990): G protein modulation of calcium currents in neurons. *Annu Rev Physiol* 52:243-255

Dorje F, Friebe T, Tacke R, Mutschler E, Lambrecht G (1990): Novel pharmacological profile of muscarinic receptors mediating contraction of guinea-pig uterus. *Naunyn Schmiedebergs Arch Pharmacol* 343:284-289

Dorje F, Levey A, Brann MR (1991a): Immunological detection of muscarinic receptor subtype proteins (m1-m5) in rabbit peripheral tissues. *Mol Pharmacol* 40:459-462

Dorje F, Wess J, Lambrecht G, Mutschler E, Brann MR (1991b): Antagonist binding studies at five cloned human muscarinic receptor subtypes. *J Pharmacol Exp Ther* 256:727-733

Dousmanis AG, Pennefather PS (1989): Characterization of the inwardly-rectifying conductances in AtT-20 cells. *Biophys J* 55:546

Dutar P, Nicoll, RA (1988): Stimulation of phosphatidylinositol (PI) turnover may mediate the muscarinic suppression of the m-current in hippocampal pyramidal cells. *Neurosci Lett* 85:89-94

Egan TM, North RA (1986): Acetylcholine hyperpolarizes central neurons by acting on an M2 muscarinic receptor. *Nature* 319:405-407

Ellis J, Huyler J, Brann MR (1991): Allosteric regulation of cloned m1-m5 muscarinic receptor subtypes. *Biochem Pharmacol* 42:1927-1932

Ellis J, Seidenberg M (1989): Gallamine exerts biphasic allosteric effects at muscarinic receptors. *Mol Pharmacol* 35:173-176

Eltze M, Gmelin G, Wess J, Strohmann C, Tacke R, Mutschler E, Lambrecht G (1988): Muscarinic M1 and M2 receptors mediating opposite effects on neuromuscular transmission in rabbit vas deferens. *Eur J Pharmacol* 151:205-221

Feifel R, Wagner-Roder M, Strohmann C, Tacke R, Waelbroeck M, Christophe J, Mutschler E, Lambrecht G (1990): Stereo selective inhibition of muscarinic receptor subtypes by the enantiomers of hexahydrodifenidol and acetylenic analogues. *Br J Pharmacol* 99:455-460

Felder CC, Kanterman RY, Ma AL, Axelrod J (1989): A transfected ml muscarinic acetylcholine receptor stimulates adenylate cyclase via phosphatidyl inositol hydrolysis. *J Biol Chem* 264:20356–20362

Fischmeister R, Hartzell HC (1986): Mechanism of action of acetylcholine on calcium current in single cells from frog ventricle. *J Physiol (Lond)* 376:183–202

Fischmeister R, Hartzell HC (1987): Cyclic guanosine 3',5'-monophosphate regulates the calcium current in single cells from frog ventricle. *J Physiol (Lond)* 387:453–472

Freedman SB, Harley EA, Iversen LL (1988): Biochemical measurement of muscarinic receptor efficacy and its role in receptor regulation. *Trends Pharmacol Sci [Suppl]* 3:54–60

Fukuda K, Higashida H, Kubo T, Maeda A, Akiba I, Bujo H, Mishina M, Numa S (1988): Selective coupling with K^+ currents of muscarinic acetylcholine receptor subtypes in NG108–15 cells. *Nature (Lond)* 335:355–358

Fukuda K, Kubo T, Akiba I, Maeda A, Mishina M, Numa S (1987): Molecular distinction between muscarinic acetylcholine receptor subtypes. *Nature* 327:623–625

Gahwiler BH, Brown DA (1987): Muscarine affects calcium currents in rat hippocampal pyramidal cells in vitro. *Neurosci Lett* 76:301–306

Galligan JJ, North RA, Tokimasa T (1989): Muscarinic agonists and potassium currents in guinea-pig myenteric neurones. *Br J Pharmacol* 96:193–203

Gil DW, Wolfe BB (1985): Pirenzepine distinguishes between muscarinic receptor-mediated phosphoinositide breakdown and inhibition of adenylate cyclase. *J Pharmacol Exp Ther* 232:608–616

Gutkind JS, Novotny EA, Brann MR, Robbins K (1991): Muscarinic acetylcholine receptor subtypes as agonist dependent oncogenes. *Proc Natl Acad Sci USA* 88:4703–4707

Haga K, Haga T (1983): Affinity chromatography of the muscarinic acetylcholine receptor. *J Biol Chem* 258:13575–13579

Hammer R, Berrie CP, Birdsall JM, Burgen ASV, Hulme EC (1980): Pirenzepine distinguishes between different subclasses of muscarinic receptors. *Nature* 283:90–92

Harden TK, Tanner LI, Martin MW, Nakahata K, Hugher AR, Kepler JR, Evans T, Masters SB, Brown JH (1986): Characteristics of two biochemical responses to stimulation of muscarinic cholinergic receptors. *Trends Pharmacol Sci [Suppl]* 7:14–18

Hartzell HC (1988): Regulation of cardiac ion channels by catecholamines, acetylcholine and second messenger systems. *Prog Biophys Mol Biol* 52:165–247

Hartzell HC, Fischmeister R (1987): Effect of forskolin and acetylcholine on calcium current in single isolated cardiac myocytes. *Mol Pharmacol* 32:639–645

Henis YI, Kloog Y, Sokolovsky M (1989): Allosteric interactions of muscarinic receptors and their regulation by other membrane proteins. In: *The Muscarinic Receptors,* Brown JH, ed. Clifton, New Jersey: Humana Press, pp 377–418

Higashida H, Hashii M, Fukuda K, Gaulfield MP, Numa S, Brown DA (1990): Selective coupling of different muscarinic acetylcholine receptors to neuronal calcium currents in DNA-transfected cells. *Proc R Soc Lond [Biol]* 242:68–74

Hille B (1989): Ionic channels: Evolutionary origins and modern roles. *Q J Exp Physiol* 74:785–804

Hoss W, Messer WS, Monsma FJ, Miller MD, Ellerbrock BR, Scranton T, Ghodsi-Hovsepian S, Price MA, Balan S, Mazloum Z, Bohnett M (1990): Biochemical and behavioral evidence for muscarinic autoreceptors in the CNS. *Brain Res* 517:195-201

Hulme EC, Birdsall NJM, Buckley NJ (1990): Muscarinic receptor subtypes. *Annu Rev Pharmacol Toxicol* 30:633-673

Inoue M, Kuriyama H (1991): Muscarinic receptor is coupled with a cation channel through a GTP-binding protein in guinea-pig chromaffin cells. *J Physiol (Lond)* 436:511-529

James MK, Cubeddu LX (1987): Pharmacological characterization and functional role of muscarinic autoreceptors in the rabbit striatum. *J Pharmacol Exp Ther* 240:203-215

Jones SVP (1991): Effects of muscarinic receptor subtypes on an inward potassium conductance and on exocytosis. *Neurosci Soc Abstr* 17:67

Jones SVP, Barker JL, Bonner TI, Buckley NJ, Brann MR (1988a): Electrophysiological characterization of the cloned m1 muscarinic receptor expressed in A9 L cells. *Proc Natl Acad Sci USA* 85:4056-4060

Jones SVP, Barker JL, Buckley NJ, Bonner TI, Collins R, Brann MR (1988b): Cloned muscarinic receptor subtypes expressed in A9 L cells differ in their coupling to electrical responses. *Mol Pharmacol* 34:421-426

Jones SVP, Barker J, Goodman M, Brann MR (1990): IP3 mediates muscarinic receptor-activated calcium-dependent conductances in m1- and m3-transfected A9 cells. *J Physiol (Lond)* 421:499-519

Jones SVP, Choi OH, Beaven MA (1991a): Carbachol induces secretion in a mast cell line (RBL-2H3) transfected with the m1 muscarinic receptor gene. *FEBS Lett* 289:47-50

Jones SVP, Heilman CJ, Brann MR (1991b): Functional responses of cloned muscarinic receptors expressed in CHO-K1 cells. *Mol Pharmacol* 40:242-247

Kenakin T (1986): Receptor reserve as a tissue misnomer. *Trends Pharmacol Sci* 5:323-346

Kenakin T, Boselli C (1989): Pharmacologic discrimination between receptor heterogeneity and allosteric interaction: Resultant analysis of gallamine and pirenzepine antagonism of muscarinic responses in rat trachea. *J Pharmacol Exp Ther* 250:944-952

Kilbinger H, Schworer H, Suss KD (1989): Muscarinic modulation of acetylcholine release: Receptor subtypes and possible mechanisms. *Experientia [Suppl]* 57:197-203

King K, Dohlman HG, Thorner J, Caron MG, Lefkowitz RJ (1990): Control of yeast mating signal transduction by a mammalian β2-adrenergic receptor and Gs α subunit. *Science* 250:121-123

Kirsh GE, Yatani A, Codina J, Birnbaumer L, Brown AM (1988): Alpha-subunit of GK activates atrial potassium channels of chick, rat and guinea pig. *Am J Physiol* 254:1200-1205

Kleuss C, Hescheler J, Ewel C, Rosenthal W, Schultz G, Wittig B (1991): Assignment of G-protein subtypes to specific receptors inducing inhibition of calcium currents. *Nature* 353:43-48

Knopfel T, Vranesic I, Gahwiler BH, Brown DA (1990): Muscarinic and β-adrenergic depression of the slow calcium-activated potassium conductance in hippocampal CA3 pyramidal cells is not mediated by a reduction of depolarization-induced cytosolic calcium transients. *Proc Natl Acad Sci USA* 87:4083–4087

Kubo T, Bujo H, Akiba I, Nakai J, Mishina M, Numa S (1988): Location of a region of the muscarinic acetylcholine receptor involved in selective effector coupling. *FEBS Lett* 241:119–125

Kubo T, Fukuda K, Mikami A, Maeda A, Takahashi H, Mishina T, Haga K, Haga A, Ichiyama A, Kangawa K, Kojima M, Matsuo H, Hirose T, Numa S (1986a): Cloning, sequencing and expression of complementary DNA encoding the muscarinic acetylcholine receptor. *Nature* 323:411–416

Kubo T, Maeda A, Sugimoto K, Akiba I, Mikami A, Takahasi T, Haga K, Haga A, Ichiyama A, Kanagawa K, Matsuo H, Hirose T, Numa S (1986b): Primary structure of porcine cardiac muscarinic acetylcholine receptor deduced from the cDNA sequence. *FEBS Lett* 209:367–372

Kwatra MM, Benovic JL, Caron MG, Lefkowitz RJ, Hosey MM (1989): Phosphorylation of chick heart muscarinic cholinergic receptors by the β-adrenergic receptor kinase. *Biochemistry* 28:4543–4547

Lambrecht G, Feifel R, Moser U, Aasen AJ, Waelbroeck M, Christophe J, Mutschler E (1988): Stereoselectivity of the enantiomers of trihexyphenidyl and its methiodide at muscarinic receptor subtypes. *Eur J Pharmacol* 155:167–170

Lazareno S, Buckley NJ, Roberts F (1990): Characterization of muscarinic M4 binding sites in rabbit lung, chicken heart, and NG108-15 cells. *Mol Pharmacol* 38:805–815

Lechleiter J, Girard S, Clapman D, Peralta E (1991a): Subcellular patterns of calcium release determined by G protein-specific residues of muscarinic receptors. *Nature* 350:505–508

Lechleiter J, Hellmiss R, Duerson K, Ennulat D, David N, Clapham D, Peralta E (1991b): Distinct sequence elements control the specificity of G protein activation by muscarinic acetylcholine receptor subtypes. *EMBO J* 9:4381–4390

Levey AI, Simonds W, Spiegel A, Brann MR (1989): Characterization of muscarinic receptor subtype specific antibodies. *Soc Neurosci Abstr* 15:64

Levey AI, Stormann TM, Brann MR (1990): Bacterial expression of human muscarinic receptor fusion proteins and generation of subtype-specific antisera. *FEBS Lett* 275:65–69

Levey AI, Kitt C, Simonds W, Price D, Brann MR (1991): Identification and localization of muscarinic receptor subtype proteins in rat brain. *J Neurosci* 11:3218–3226

Levine RR, Birdsall NJM, eds (1989): Subtypes of muscarinic receptors IV. *Trends Pharmacol Sci [Suppl]* 10:1–119

Liao C-F, Themmen APN, Joho R, Barberis C, Birnbaumer M, Birnbaumer L (1989): Molecular cloning and expression of a fifth muscarinic acetylcholine receptor. *J Biol Chem* 264:7328–7337

Luthin GR, Harkness J, Artymyshyn RP, Wolfe BB (1988): Antibodies to a synthetic peptide can be used to distinguish between muscarinic acetylcholine receptor binding sites in brain and heart. *Mol Pharmacol* 34:327–333

Madison DV, Lancaster B, Nicoll RA (1987): Voltage clamp analysis of cholinergic action in hippocampus. *J Neurosci* 7:733–741

Maeda A, Kubo T, Mishina M, Numa S (1988): Tissue distribution of mRNAs encoding mucarinic acetylcholine receptor subtypes. *FEBS Lett* 239:339–342

Malenka RC, Madison DV, Andrade R, Nicoll A (1986): Phorbol esters mimic some cholinergic actions in hippocampal pyramidal neurons. *J. Neurosci* 6:475–480

Marchi M, Raiteri M (1989): Interaction acetylcholine-glutamate in rat hippocampus: Involvement of two subtypes of M2 muscarinic receptors. *J Pharmacol Exp Ther* 248:1255–1260

Marty A (1987): Control of ionic currents and fluid secretion by muscarinic agonists in exocrine glands. *Trends Neurosci* 10:373–377

Mash DC, Flynn DD, Potter LT (1985): Loss of M2 muscarine receptors in the cerebral cortex in Alzheimer's disease and experimental cholinergic denervation. *Science* 228:1115–1117

Mash DC, Potter LT (1985): Autoradiographic localization of M1 and M2 muscarine receptors in the rat brain. *Neuroscience* 19:551–564

Mayer EM, Otero DH (1985): Pharmacological and ionic characterizations of the muscarinic receptors modulating [^{3}H]acetylcholine release from rat cortical synaptosomes. *J Neurosci* 5:1202–1207

McCormick DA, Prince DA (1986): Acetylcholine induces burst firing in thalamic reticular neurons by activating a potassium conductance. *Nature* 319:402–404

McKinney M, Anderson D, Forray C, El-Fakahany EE (1989): Characterization of the striatal M2 muscarinic receptor mediating inhibition of cyclic AMP using selective antagonists: A comparison with the brainstem M2 receptor. *J Pharmacol Exp Ther* 250:565–572

Mei L, Roeske WR, Izutsu KT, Yamamura HI (1990): Characterization of muscarinic acetylcholine receptors in human labial salivary glands. *Eur J Pharmacol* 176:367–370

Mitchelson F (1988): Muscarinic receptor differentiation. *Pharmacol Ther* 37:357–423

Muller W, Misgeld U (1986): Slow cholinergic excitation of guinea pig hippocampal neurones is mediated by two muscarinic receptor subtypes. *Neurosci Lett* 67:107–112

Nakajima Y, Nakajima S, Leonard RJ, Yamaguchi K (1986): Acetylcholine raises excitability by inhibiting the fast transient potassium current in cultured hippocampal neurons. *Proc Natl Acad Sci USA* 83:3022–3026

Nathanson NM (1987): Molecular properties of the muscarinic acetylcholine receptor. *Annu Rev Neurosci* 10:195–236

Neher E, Marty KA, Fukuda K, Kubo T, Numa S (1988): Intracellular calcium release mediated by two muscarinic receptor subtypes. *FEBS Lett* 240:88–94

Newberry NR, Priestly T (1987): Pharmacological differences between two muscarinic responses of the rat superior cervical ganglion in vitro. *Br J Pharmacol* 92:817–826

Nicoll RA (1988): The coupling of neurotransmitter receptors to ion channels in the brain. *Science* 241:545–551

North RA (1989): Muscarinic cholinergic regulation of ion channels. In: *Muscarinic Receptor Subtypes,* Brown JH, ed. Clifton, New Jersey: Humana Press, pp 341–375

North RA, Slack BE, Suprenant A (1985): Muscarinic M1 and M2 receptors mediate depolarization and presynaptic inhibition in guinea-pig enteric nervous system. *J Physiol (Lond)* 368:435–453

Novotny E, Brann MR (1989): Agonist pharmacology of cloned muscarinic receptors. *Trends Pharmacol Sci [Suppl]* 4:Abstr 66

O'Dowd BF, Hnatowich M, Caron MG, Lefkowitz RJ, Bouvier M (1989): Palmitoylation of the human β2-adrenergic receptor. *J Biol Chem* 264:7564–7569

Ovchinnikov YA, Abdulaev NG, Bogachuk AS (1988): Two adjacent cysteine residues in the C-terminal cytoplasmic fragment of bovine rhodopsin are palmitylated. *FEBS Lett* 230:1–5

Palacios JM, Bolliger G, Closse A, Enz A, Gmelin G, Malanowski J (1986): The pharmacological assessment of RS 86 (2-ethyl-8-methyl-2,8-diazaspiro-(4,5)-decan-1,3-dion hydrobromide) a potent, specific muscarinic acetylcholine receptor agonist. *Eur J Pharmacol* 125:45–62

Palmer JM, Wood JD, Zafirov DH (1987): Purinergic inhibition in the small intestinal myenteric plexus of the guinea-pig. *J Physiol (Lond)* 387:357–368

Parker EM, Kameyama K, Higashijima T, Ross M (1991): Reconstitutively active G protein-coupled receptors purified from baculovirus-infected insect cells. *J Biol Chem* 266:519–527

Pepeu G (1973): The release of acetylcholine from brain: An approach to the study of the central cholinergic mechanisms. *Prog Neurobiol* 2:257–288

Peralta EG, Ashkenazi A, Winslow JW, Ramachandran J, Capon DJ (1988): Differential regulation of PI hydrolysis and adenylyl cyclase by muscarinic receptor subtypes. *Nature* 334:434–437

Peralta EG, Ashkenazi A, Winslow JW, Smith DH, Ramachandran J, Capon DJ (1987a): Distinct primary structures, ligand-binding properties and tissue-specific expression of four human muscarinic acetylcholine receptors. *EMBO J* 6:3923–3929

Peralta EG, Winslow JW, Peterson GL, Smith DH, Ashkenazi A, Ramachandran J, Schimerlik MI, Capon DJ (1987b): Primary structure and biochemical properties of an M2 muscarinic receptor. *Science* 236:600–605

Petersen OH, Gallacher DV (1988): Electrophysiology of pancreatic and salivary acinar cells. *Annu Rev Physiol* 50:65–80

Peterson GL, Herron GS, Yamaki M, Fullerton DS, Schimerlik MJ (1984): Purification of the muscarinic acetylcholine receptor from porcine atria. *Proc Natl Acad Sci USA* 81:4993–4997

Pfaffinger PJ, Leibowitz MD, Subers EM, Nathanson NM, Almers W, Hille B (1988): Agonists that suppress M-current elicit phosphoinositide turnover and Ca^{2+} transients, but these events do not explain M-current suppression. *Neuron* 1:477–484

Pfaffinger PJ, Martin JM, Hunter DD, Nathanson NM, Hille B (1985): GTP-binding proteins couple cardiac muscarinic receptors to a potassium channel. *Nature* 317:536–538

Pittel Z, Heldman E, Rubinstein R, Cohen S (1990): Distinct muscarinic receptor subtypes differentially modulate acetylcholine release from corticocerebral synaptosomes. *J Neurochem* 55:665–672

Poyner DR, Birdsall NJ, Curtis C, Eveleigh P, Hulme EC, Pedder EK, Wheatley M (1989): Binding and hydrodynamic properties of muscarinic receptor subtypes solubilized in 3-(3-cholamidopropyl) dimethylammonio-2-hydroxy-1-propanesulfonate. *Mol Pharmacol* 36:420–429

Quirion R, Aubert I, Lapchak PA, Schaum RP, Teolis S, Gauthier S, Araujo DM (1989): Muscarinic receptor subtypes in human neurodegenerative disorders: Focus on Alzheimer's disease. *Trends Pharmacol Sci [Suppl]* 4:80–84

Raiteri M, Marchi M, Costi A, Volpe G (1990a): Endogenous aspartate release in the rat hippocampus is inhibited by M2 "cardiac" muscarinic receptors. *Eur J Pharmacol* 177:181–187

Raiteri M, Marchi M, Paudice P (1990b): Presynaptic muscarinic receptors in the central nervous system. *Ann NY Acad Sci* 604:113–129

Raiteri M, Marchi M, Paudice P, Pittaluga A (1990c): Muscarinic receptors mediating inhibition of γ-aminobutyric acid release in rat corpus striatum and their pharmacological characterization. *J Pharmacol Exp Ther* 254:496–501

Ringdahl B, Roch M, Jenden DJ (1987): Regional differences in receptor reserve for analogs of oxotremorine in vivo: Implications for development of selective muscarinic agonists. *J Pharmacol Exp Ther* 242:464–471

Schultz G, Rosenthal W, Hescheler J (1990): Role of G-proteins in calcium channel modulation. *Annu Rev Physiol* 52:275–292

Shapiro RA, Nathanson NM (1989): Deletion analysis of the mouse m1 muscarinic acetylcholine receptor: Effects on phosphoinositide metabolism and down-regulation. *Biochemistry* 28:8946–8950

Shapiro RA, Wakimoto BT, Subers EM, Nathanson NM (1989): Characterization and functional expression in mammalian cells of genomic and cDNA clones encoding a Drosophila muscarinic acetylcholine receptor. *Proc Natl Acad Sci USA* 86:9039–9043

Stein R, Pinkas-Kramarski R, Sokolovsky M (1988): Cloned M1 muscarinic receptors mediate both adenylate cyclase inhibition and phosphoinositide turnover. *EMBO J* 7:3031–3035

Stockton JM, Birdsall NJM, Burgen ASV, Hulme EC (1983): Modification of the binding properties of muscarinic receptors by gallamine. *Mol Pharmacol* 23:551–557

Sunderland T, Tariot PN, Newhouse PA (1988): Differential responsivity of mood, behavior, and cognition to cholinergic agents in elderly neuropsychiatric populations. *Brain Res Rev* 13:371–389

Szabo G, Otero AS (1990): G-protein mediated regulation of potassium channels in heart. *Annu Rev Physiol* 52:293–305

Szerb JC, Somogyi GT (1973): Depression of acetylcholine release from cortical slices by cholinesterase inhibition and oxotremorine. *Nature* 241:121–123

Tietje KM, Goldman PS, Nathanson NM (1990): Cloning and functional analysis of a gene encoding a novel muscarinic acetylcholine receptor expressed in chick heart and brain. *J Biol Chem* 265:2828–2834

Toselli M, Lang J, Costa T, Lux HD (1989): Direct modulation of voltage-dependent calcium channels by muscarinic activation of a pertussis toxin sensitive G-protein in hippocampal neurons. *Pfluegers Arch* 415:255–261

Toselli M, Lux HD (1989): Opposing effects of acetylcholine on two classes of voltage-dependent calcium channels in hippocampal neurons. *Experientia [Suppl]* 57:97–103

Trautwein W, Hescheler J (1990): Regulation of cardiac L-type calcium current by phosphorylation and G-proteins. *Annu Rev Physiol* 52:257–274

Tse A, Clark RB, Giles WR (1990): Muscarinic modulation of calcium current in neurones from the interatrial septum of bull-frog heart. *J Physiol (Lond)* 427:127–149

Van Koppen CJ, Nathanson NM (1990): Site-directed mutagenesis of the m2 muscarinic acetylcholine receptor. *J Biol Chem* 265:20887–20892

Van Koppen CJ, Nathanson NM (1991): The cysteine residue in the carboxyl terminal domain of the m2 muscarinic acetylcholine receptor is not required for formation of adenylate cyclase. *J Neurochem* In press.

Vivaudou MB, Clapp LH, Walsh JV, Singer JJ (1988): Regulation of one type of calcium current in smooth muscle cells by diacylglycerol and acetylcholine. *FASEB J* 2:2497–2504

Waelbroeck M, Tastenoy M, Camus J, Christophe J (1990): Binding of selective antagonists to four muscarinic receptors (M1–M4) in rat forebrain. *Mol Pharmacol* 38:267–273

Wanke E, Ferroni A, Malgaroli A, Ambrosini A, Pozzan T, Meldolesi J (1987): Activation of a muscarinic receptor selectively inhibits a rapidly inactivating calcium current in rat sympathetic neurons. *Proc Natl Acad Sci USA* 84:4313–4317

Wanke E, Sardini A, Ferroni A (1989): L and N Ca^{2+} channels coupled to muscarinic receptors in rat sensory neurons. *Ann NY Acad Sci* 560:398–400

Weiner DM, Brann MR (1989): Distribution of m1–m5 muscarinic acetylcholine receptor mRNAs in rat brain. *Trends Pharmacol Sci [Suppl]* 4:115

Weiner DM, Levey A, Brann MR (1990): Expression of muscarinic acetylcholine and dopamine receptor mRNAs within the basal ganglia. *Proc Natl Acad Sci USA* 87:7050–7054

Wess J, Bonner TI, Dorje F, Brann MR (1990a): Delineation of muscarinic receptor domains conferring selectivity of coupling to G proteins and second messengers. *Mol Pharmacol* 38:517–523

Wess J, Bonner TI, Brann MR (1990b): Chimeric m2/m3 muscarinic receptors: Role of carboxyl terminal receptor domains in selectivity of ligand binding and coupling to phosphoinositide hydrolysis. *Mol Pharmacol* 38:872–877

Wess J, Brann MR, Bonner TI (1989): Identification of a small intracellular region of the muscarinic m3 receptor as a determinant of selective coupling to PI turnover. *FEBS Lett* 258:133–136

Wess J, Gdula D, Brann MR (1991a): Chimeric m2/m5 muscarinic receptors: Identification of receptor domains confering antagonist binding selectivity. *Mol Pharmacol* 41:369–374

Wess J, Gdula D, Brann MR (1991b): Site-directed mutagenesis of the m3 muscarinic receptor: Identification of a series of threonine and tyrosine residues involved in agonist but not antagonist binding. *EMBO J* 10:3729–3734

Wess J, Lambrecht G, Mutschler E, Brann MR, Dorje F (1991c): Selectivity profile of the novel muscarinic antagonist UH-AH 37 determined by the use of cloned receptors and isolated tissue preparations. *Br J Pharmacol* 102:246–250

Whitehouse PJ, Price DL, Struble RG, Clark AW, Coyle JT, DeLong MR (1982): Alzheimer's disease and senile dementia: loss of neurons in the basal forebrain. *Science* 215:1237–1239

Wolfe BB (1989): Subtypes of muscarinic cholinergic receptors: Ligand binding, functional studies, and cloning. In: *The Muscarinic Receptors,* Brown JH, ed. Clifton, New Jersey: Humana Press, pp 125–150

Xu M, Yamamoto T, Kato T (1990): In vivo striatal dopamine release by M1 muscarinic receptors is induced by activation of protein kinase C. *J Neurochem* 54:1917–1919

9

Molecular Biology of Peptide and Glycoprotein Hormone Receptors

DAVID R. POYNER AND MICHAEL R. HANLEY

1. General Introduction

Although over sixty mammalian peptides have been identified, their receptor actions are, with the notable exception of the atrial peptide family, mediated exclusively by G-protein-linked mechanisms (Hanley, 1989). The full spectrum of nonphotoreceptor G-protein mechanisms have been described for peptide signaling, including positive and negative regulation of adenylyl cyclase, stimulation of phospholipase C, and positive and negative regulation of ion channels (Table 9.1).

With the cloning of mammalian opsins and the β-adrenergic receptor (see this volume), it became apparent that G-protein-coupled receptors had a conserved domain structure, with seven predicted transmembrane α-helices (Figure 9.1). On the basis of shared second messenger mechanisms, peptide receptors were also anticipated to belong to this class. This was confirmed with the cloning of the substance K receptor (Masu et al., 1987). Earlier, however, two eukaryotic peptide receptors, those for mating factors in the yeast *Saccharomyces cerevisiae,* had been cloned and were found to have seven predicted transmembrane domains (Nakayama et al., 1985). Eight of the cloned and sequenced peptide receptors belonging to the G-protein coupled class are discussed in this chapter (Table 9.2). The number of cloned receptors is increasing rapidly, but concepts of receptor structure have not dramatically altered.

Peptides have attracted attention in the past for a number of reasons linked to their common properties. Their construction as linear polymers with the associated flexibility suggested to many that their mode of binding could not be explained by the classic lock-and-key model (Goldstein et al., 1974) where the ligand has to adopt a single fixed conformation in solution

TABLE 9.1. Peptide receptor mechanisms[a]

Peptide receptor[b]	G-Protein class	Effector	Reference
*Calcitonin	Gs	Adenylyl cyclase;	Lin et al. (1991)
CGRP		channel modulation	
Glucagon		(voltage-sensitive	
*LH		Ca^{2+} channels)	Loosefelt et al. (1989)
*PTH			Juppner et al. (1991)
*Secretin			Ishihara et al. (1991)
*TSH			Straub et al. (1991)
*Vasopressin (V$_2$)			Lolait et al. (1991)
*VIP			Steedharan et al. (1991)
Galanin	Gi	Adenylyl cyclase;	
*NPY (Y1)		channel modulation	Herzog et al. (1992)
Opioids (μ, δ)		(voltage sensitive	
*Somatostatin		Ca^{2+} channels,	Yamada et al. (1992)
(SST1, SST2)		K^+ channels	
*Angiotensin			Sasaki et al. (1991)
(AIIa/AIIb)			
*Angiotensin			Jackson et al. (1988)
(AIII/mas)			
*Bombesin/GRP			Spindel et al. (1991);
			Wada et al. (1991)
*Bradykinin (B2)			McEachern et al. (1991)
CCK			
*Endothelin			Arai et al. (1991);
(ET$_A$, ET$_B$)			Sakurai et al. (1991)
*F-Met-Leu-Phe			Boulay et al. (1990)
LH–RH			
*Neurotensin			Tanaka et al. (1990)
*NPY (Y2)	Gq (Gi$_2$?, Go ?)	PtdInsP$_2$	Herzog et al. (1992)
*Oxytocin		breakdown (Ca^{2+}	Kimura et al. (1992)
*TRH		mobilization/protein	Straub et al. (1990)
*Thrombin		kinase C activation	Vu et al. (1991)
* Vasopressin (VI)			Morel et al. (1992)

[a]CGRP, Calcitonin-gene-related protein; LH, luteinizing hormone; PTH, parathyroid hormone; VIP, vasoactive intestinal peptide; NPY, neuropeptide Y; CCK, choleocystokinin; LH–RH, luteinizing hormone-releasing hormone; TRH, thyrotropin-releasing hormone. All other abbreviations as defined in text.
[b]Asterisk (*) indicates that a cloned structure is available.

before binding to the receptor. This led to the development of the zipper model, whereby the association of a single residue of the peptide ligand with its complementary acceptor site on the receptor would be sufficient to promote a sequence of interactions leading to the correct receptor-bound activating conformation (Burgen et al., 1975). The advent of techniques capable of studying the solution structure of molecules has shown that relatively small peptides do have well-defined regions of stable tertiary

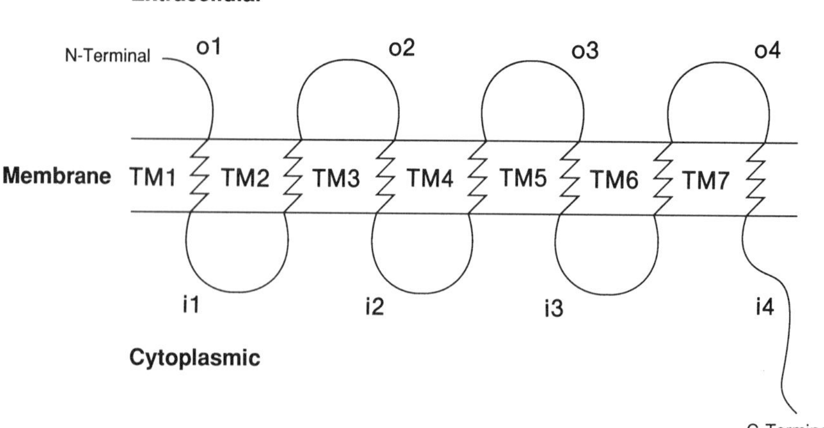

FIGURE 9.1. Structure of a generalized G-protein-coupled receptor indicating the nomenclature used throughout this chapter. o, outer (extracellular) loops; i, inner (cytoplasmic) loops, TM, membrane-spanning helices.

structure, placing constraints on their true conformational flexibility. Nonetheless, the zipper model remains an influential concept and it is likely that it is an accurate approximation to the mechanism of at least some peptide–receptor interactions.

A second theoretical concept that has arisen to a large extent from the study of peptide–receptor interactions is the "address-message" theory (Schwyzer, 1987). This divides the residues in a peptide into two discrete and nonoverlapping functional categories: those which serve to direct and bind the peptide to the appropriate receptor (the "address") and those which contain the information necessary to activate the receptor (the "message"). There has been speculation that, because of their size, peptides are in an unique position to carry "more" information than a small molecule messenger. In its more extreme presentations, the address-message concept is misleading. There is no theoretical basis for why information should be coded linearly along a peptide, given that most ligand binding sites in proteins are made up of residues brought together from very different parts of the primary structure. Equally, there is nothing particularly compelling about the amount of "information" contained in a peptide as compared to the small molecule (e.g., acetylcholine), on the one hand, and large proteins (e.g., growth factors), on the other. However, the basic idea behind the address-message concept, namely, that different parts of the ligand have different functions, is still useful. It is entirely possible that certain functional groups on an agonist are primarily concerned with docking the ligand into the receptor, whereas others produce productive conformational changes within the receptor. Certainly it is possible to identify groups on the

TABLE 9.2. Cloned peptide receptors

Receptor	Species	Cloning strategy	Residue number	Predicted or known effector[a]	Reference
Substance P	Rat	Cross-hybridization	407	PI	Yokota et al. (1989)
Substance K	Bovine	Expression	384	PI	Masu et al. (1987)
	Rat	Cross-hybridization	390	PI	Sasai and Nakanishi (1989)
Neuromedin K	Rat	Cross-hybridization	452	PI	Shigemoto et al. (1990)
LH–CG	Rat	Protein purification	674	cAMP	Macfarland et al. (1989)
	Porcine	Antibody screening	669	cAMP	Loosfelt et al. (1989)
TSH	Dog	Cross-hybridization (PCR generated probe)	744	cAMP	Parmentier et al. (1989)
	Human	Cross-hybridization	744	cAMP	Libert et al. (1989a)
mas/angiotensin	Human	Expression	325	PI	Jackson et al. (1988)
	Rat	Cross-hybridization	324	PI	Young et al. (1988)
STE2	S. cerevisiae	Complementation	430	$\beta\gamma$-subunits of G-protein	Nakayama et al. (1985)
	S. kluyveri	Cross-hybridization	426	$\beta\gamma$-subunits of G-protein	Marsh and Herskowitz (1988)
STE3	S. cerevisiae	Complementation	470	$\beta\gamma$-subunits of G-protein	Nakayama et al. (1985)

[a]PI, breakdown of $PtdInsP_2$; cAMP, stimulation of adenylyl cyclase.

receptors which seem to correspond to these functions (Strader et al., 1988; Chung et al., 1988).

However, the temptation to regard peptides as conforming to a linear "code" reached its zenith with the proposition that the complementary, antisense strand to a peptide coding sequence would define the binding site on its corresponding receptor (Mulchahey et al., 1986). Indeed, antibodies prepared to predicted "complementary peptides" were claimed to interact with several receptors, including an angiotensin binding protein (Elton et al., 1988). In Table 9.3, we show two examples where this hypothesis has been applied to cloned receptors, of known sequence, using predictions derived from the peptide precursor nucleotide sequence within the same species. For both the bovine substance K receptor, and the human mas/angiotensin receptor, the resulting peptides from either 5' to 3' or 3' to 5' translation of complementary codons (Bost and Blalock, 1989) do not identify any portion of the receptor sequence. Thus, this eccentric notion cannot assist in defining peptide recognition sites which, in any event, are unlikely to be simply colinear with the natural agonist peptide, but are rather likely to have contributions from many parts of the sequence.

The zipper model and the address-message concept have, of necessity, remained rather theoretical concepts. The elucidation of the structures of several peptide receptors is an important step forward in allowing their detailed experimental evaluation. Moreover, cloning of the peptide receptors allows questions to be raised beyond the immediate world of peptide hormone biology. The other G-protein-coupled receptors which have been sequenced recognize primarily amines and other small positively charged ligands. The peptides are clearly very different to these ligands, and their receptor sequences should help the understanding of what parts of the conserved intracellular domain structure are essential for the recognition of specific signal transduction G-proteins and whether there are any common themes to extracellular stimulant recognition, which are independent of the chemical nature of the ligand itself. In this regard, the cloning of the receptors for the glycoprotein hormones luteinizing hormone-chorio-gonadotropic hormone (LH–CG) and thyroid-stimulating hormone (TSH) (Macfarland et al., 1989; Loosfelt et al., 1989; Parmentier et al., 1989; Libert et al., 1989a) is particularly valuable, as the cognate ligands are highly distinctive, with molecular weights in excess of 30 kDa. In this chapter, we will emphasize the protein sequences predicted from molecular cloning and will not address issues of gene organization, transcriptional regulation, or translational control of the peptide receptors.

2. The Cloned Receptors

2.1. Introduction

The receptors for the mammalian tachykinin peptides, substance P, substance K, and neuromedin K, collectively define an emerging family.

TABLE 9.3. Failure to find predicted binding site sequences in cloned peptide receptors

Human angiotensinogen and its predicted complementary peptides[a]

Angiotensinogen	N-terminal-Ala-Ala-Gly-Asp-Arg-Val-Tyr-Ile-His-Pro-Phe-His- · · ·			
			Angiotensin II	
(+) mRNA sequence		5′	GAC CGG GUG UAC AUA CAC CCC UUC CAC	3′
(−) mRNA sequence		3′	CUG GCC CAC AUG UAU GUG GGG AAG GUG	5′
Complementary peptide (read 3′ to 5′)	N-		Leu-Ala-His-Met-Tyr-Val-Gln-Lys	C-terminal
Complementary peptide (read 3′ to 5′)			Glu-Gly-Val-Tyr-Val His-Pro-Val	

Bovine substance K receptor and its predicted complementary peptides [b]

Preprotachykinin	N-terminal- · · ·His-Lys-Thr-Asp-Ser-Phe-Val-Gly-Leu-Met- · · ·			
			Substance K	
(+) mRNA sequence	N-	5′	CAU AAA ACA GAU UCC UUU GUU GGA CUA AUG	3′ C-terminal
(−) mRNA sequence		3′	GUA UUU UGU CUA AGG AAA CAA CCU GAU UAC	5′
Complementary peptide (read 3′ to 5′)			Val-Phe-Cys-Leu-Arg-Lys-Gly-Pro-Asp-Tyr	
Complementary peptide (read 5′ to 3′)			His-stop-Ser-Asn-Lys-Gly-Ile-Cys-Phe-Met	

[a]Neither sequence shown is found in human mas.
[b]Neither sequence shown is found in bovine substance K receptor.

Similarly the receptors for the glycoprotein hormones LH–CG and TSH, appear to define a novel convergent family. mas, first identified as an oncogene using genomic DNA in a tumorigenicity assay (Young et al., 1986), has now been shown to be an angiotensin receptor with sequence and functional similarities to the tachykinin receptors (Jackson et al., 1988). Last, STE2 and STE3 are peptide mating factor receptors found in budding yeast.

In addition to indicating explicitly the specific receptors cloned, Table 9.2 demonstrates two other general points. First, the only receptor to be cloned exclusively by initial receptor purification, using traditional methods of protein chemistry, was the rat LH–CG receptor. All other receptor sequences were obtained by expression cloning, cross-hybridization, or genetic analysis. This is a key point which reflects the difficulty of working with peptide receptors. Many of the problems of initial characterization stem from the fact that almost all of the ligands available for defining a peptide receptor are the natural peptide agonists and their analogues. Peptides are physically and enzymatically difficult to handle, exhibiting problems of instability and nonspecific binding (Hanley, 1985). Furthermore, agonists are in general less desirable than antagonists as starting points for affinity ligands, as they only bind with high affinity to receptor–G-protein complexes. Moreover, most neuropeptide receptors are expressed at low densities (Hanley, 1985), giving, at most, 1–10 pmol/mg protein in even the most highly enriched tissues. Not surprisingly, the purification of peptide receptors has proved an extremely difficult task. A second point is that to date only two major second messenger systems, adenylyl cyclase and phosphoinositide hydrolysis, are overrepresented among the cloned receptors, but so far, only a limited number of peptide receptors have been cloned which act through Gi to inhibit adenylyl cyclase or regulate ion channels. The yeast receptors may fit into this picture by coupling to gene expression by as yet undefined transduction pathways, but initiating responses by yeast homologues of vertebrate G-proteins (Herskowitz and Marsh, 1987). In this regard it is worth noting that activation of STE2 or STE3 leads to growth arrest, but a number of vertebrate peptides such as angiotensin (Jackson et al., 1988) or the tachykinins (Payan, 1985) are positive regulators of cell proliferation. The growth control aspect of eukaryotic peptide signaling is poorly understood and may call attention to novel biochemical pathways of action.

2.2. The Tachykinin Receptor Family

2.2.1. Substance K Receptor (NK-2 Tachykinin Receptor)

The substance K receptor was the first vertebrate peptide receptor to be cloned (Masu et al., 1987). The sequence was obtained by mRNA expression cloning in *Xenopus* oocytes, a strategy particularly useful for single subunit receptors. A fetal bovine stomach cDNA library was produced and used to generate a library of synthetic mRNAs, which were injected as a mixture into *Xenopus* oocytes. Injected oocytes were screened for their ability to produce an electrophysiological response to substance K, which had previously been shown with poly A^+ mRNA to be a characteristic activation of an inward chloride current (Harada et al., 1987). Successive subtractive fractionation steps led to the isolation of a single clone, conferring responsiveness to substance K, which was then sequenced. The rat homologue (Sasai and Nakanishi, 1989) was subsequently obtained by cross-hybridization with a bovine cDNA probe. Substance K receptors have been characterized pharmacologically for their rank order of tachykinin preference by electrophysiological techniques after transient expression in *Xenopus* oocytes, but direct data on actions on second messengers, or on radioligand binding, is a conspicuous need.

Some of the basic structural features of the substance K receptor are shown in Tables 9.4 and 9.5 and Figure 9.2. It has relatively short interhelix loops, but a long C-terminus. It has two potential N-glycosylation sites at the N-terminus. The transmembrane (TM) helices include the conserved aspartate in TM 2 in a position shown in the β-receptor to play an important role in agonist binding (Chung et al., 1988), and proline and aromatic amino acids which are highly conserved in nonpeptide receptors. A potential disulfide bridge between TM 2 and the o3 (extracellular) loop [present in muscarinic (Curtis et al., 1989)] and β-receptors [Dixon et al., 1987] is found in the substance K receptor, as is a cysteine in the C-terminus, palmitoylation of which plays an important role in receptor–G-protein coupling in the β-receptor (O'Dowd et al., 1989). Although there are no identified kinase phosphorylation consensus sites, the C-terminus is rich in serine and threonine residues, which could act as phosphate acceptors. At the start of the i2 (cytoplasmic) loop there is the highly conserved DRY

TABLE 9.4. Loop sizes in receptor structures

Receptor	Species	o1	i1	o2	i2	o3	i3	o4	i4
					Number of amino acids in:[a]				
Substance P	Rat	29	10	9	19	21	29	9	98
Substance K	Bovine	30	11	9	19	19	33	9	79
Neuromedin K	Rat	69	11	9	19	17	28	8	106
mas	Human	26	10	10	20	11	20	5	42
LH	Rat	364	11	16	23	19	18	11	66
TSH	Dog	398	10	17	22	18	24	10	80
STE2	*S. cerevisiae*	48	7	23	13	20	20	9	131
STE3	*S. cerevisiae*	0	6	18	21	12	26	30	187

[a]i1–i4, inner (cytoplasmic) loops; o1–o4, outer (extracellular) loops.

TABLE 9.5. Conserved features of receptor structures

Receptor	Glycosylation sites	Acidic residues		DRY motif	Predicted extracellular S–S bonds	Palmitoylation of G-terminus	Prolines		
		TM 2	TM 3				TM 5	TM 6	TM 7
Substance P	2	+	−	+	+	+	+	+	+
Substance K	2	+	−	+	+	+	+	+	+
Neuromedin K	4	+	−	+[a]	+	+	+	+	+
LH–CG	5	+	+	+[b]	+	+	−	+	+
TSH	5	+	+	+[b]	+	+	−	+	+
mas/angiotensin	3	+	−	−[c]	−[d]	−	+	+	+
STE2	3	−	+	−	−	−	−	+	+
STE3	−	−	−	−	−	−	−	+	−

[a]Present as ERY.
[b]Present as E/D RW.
[c]Sequence is ERC.
[d]Cysteines in helix TM 4 and loop i3.

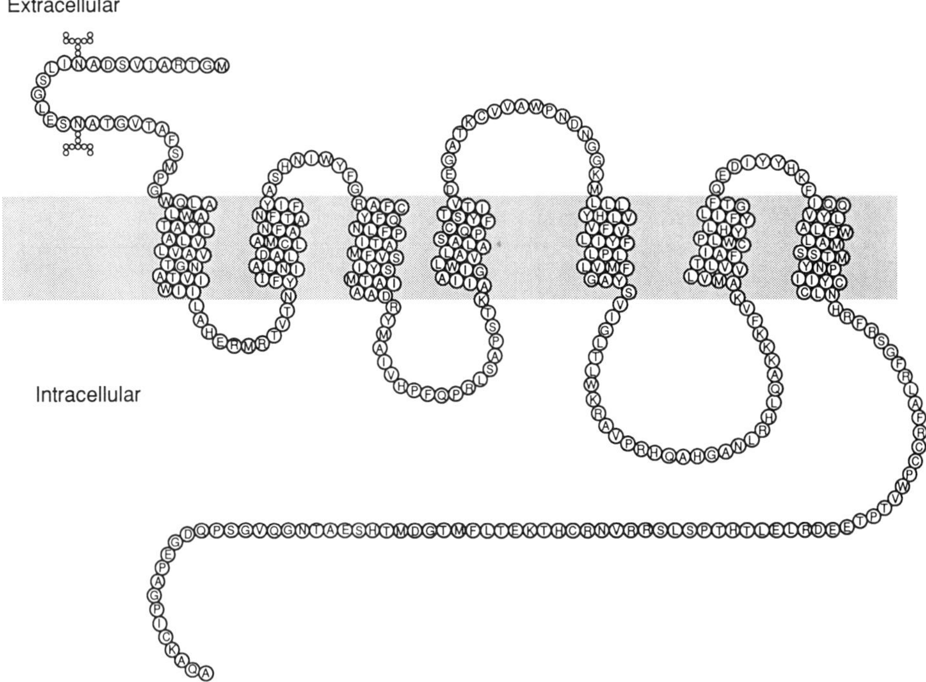

FIGURE 9.2. Structure of substance K receptor, showing putative glycosylation sites.

sequence, which may be involved in G-protein coupling and activation (Fraser et al., 1988). The bovine and rat sequences show strong conservation, having 83% amino acid sequence identity. The rat receptor has one predicted N-glycosylation site, unlike the bovine receptor. The bulk of the sequence differences cluster to the o1 and i4 (N- and C-terminals) regions. A probe based on the rat substance K receptor recognized two transcripts in poly A^+ RNA from the rat gastrointestinal tract (2.7 kb, 3.1 kb), but, interestingly, it recognized no transcripts in poly A^+ RNA preparations from rat brain. This implies that there are no substance K receptors in the rat brain, a result in accordance with some binding studies (Saffroy et al., 1988). The complication is that a presumptive endogenous ligand for substance K receptors is present in brain, although it may not be identical to authentic substance K, and may correspond to an N-terminally extended form, γ-preprotachykinin–(72–92)–peptide amide (Dam et al., 1990). This may imply that there are novel tachykinin receptors, selective for these extended forms of substance K, which may be expressed in brain.

2.2.2. Substance P (NK-1 Tachykinin Receptor) and Neuromedin K (NK-3 Tachykinin Receptor)

The substance P receptor was cloned by first identifying a poly A$^+$ RNA fraction which gave responses to substance P when injected into *Xenopus* oocytes, and then screening the corresponding cDNA library with a probe derived from the bovine substance K receptor. This allowed the isolation of a single clone which was sequenced to give the structure of the rat substance P receptor (Yokota et al., 1989). The rat neuromedin K receptor was identified by cross-hybridization with the same bovine substance K receptor probe, with final confirmation of its identity by expression of the selected clone in *Xenopus* oocytes (Shigemoto et al., 1990). Both substance P and neuromedin K receptors were characterized electrophysiologically in the oocyte expression system, but the former was subject to a distinctive rapid desensitization which limited functional analysis in oocytes. Consequently, both clones were expressed in mammalian COS-7 cells and characterized by radioligand binding. These results showed that the cloned receptors had ligand binding properties similar to the receptor subtypes found in rat brain membranes. The sequence of the rat substance P receptor was independently obtained by using degenerate polymerase chain reaction (PCR) primers to isolate cDNA clones from brain and submandibular gland libraries (Hershey and Krause, 1990). The sequence results are identical to those of Yokota et al. (1989) except at residue 214, which has an alanine instead of a glycine residue. The second messenger pathways to which each cloned receptor is linked have not been investigated, but in each case, the electrical responses in oocyte expression are the trademark characteristics of a phosphoinositide hydrolysis-coupled receptor.

As can be seen from Tables 9.4 and 9.5 and Figures 9.3 and 9.4, the substance P and neuromedin K receptors resemble the substance K receptor very closely, including the distribution of acidic residues, cysteines, prolines, and other conserved residues or sequences. Both of these tachykinin receptors are larger than the substance K receptor (substance P: 407 amino acids, mw 43,364; neuromedin K: 452 amino acids, mw 51,104) In the case of the substance P receptor, this arises from a longer C-terminus, but the neuromedin K receptor has not only (by far) the longest C-terminus of the three but also a greatly extended N-terminus. Both receptors retain the possible C-terminal tail palmitoylation site. Histidine residues found in TM 5 and 6 appear to be characteristic of the tachykinin receptors (Shigemoto et al., 1990).

2.2.3. Sequence Comparisons between the Tachykinin Receptors

As is the case with all G-protein-coupled receptors, the greatest similarities are found in the hydrophobic transmembrane core regions. Shigemoto et al. (1990) has shown sequence identities of 66.3% between neuromedin K and substance P receptors, 54.9% between neuromedin K and substance K

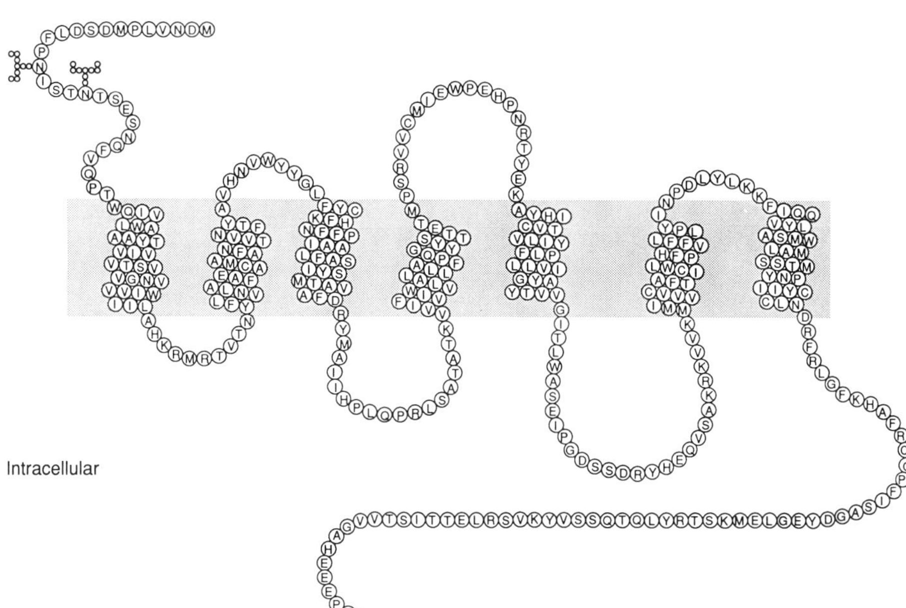

FIGURE 9.3. Structure of substance P receptor.

receptor, and 53.7% between substance P and substance K receptors. These figures are similar to those found between different members of the muscarinic or the β-adrenengic receptor families. With the amine receptors, it is this core domain which is widely held to contain the agonist binding site, and the acidic residues in TM 2 and TM 3 are potentially important in acting as counterions to the positive charge of the physiological amines. The only charged conserved residues in the tachykinin receptors are an acidic residue in TM 2 and a basic residue in TM 3. In view of its role in agonist binding in the β-receptor, the acidic residue in TM 2 is likely to play a similar role in the tachykinin receptors. However, the absence of any conserved acidic residue in TM 3 is worth noting, as this plays a pivotal role in all ligand binding to the β- and muscarinic receptors. Clearly the model of binding worked out for these small molecules does not apply to peptides. In this model, the whole ligand binding site is in a pocket formed by the helices. Studies on α_2, β_2 chimeric receptors have shown that TM 6 and TM 7 play important roles in discriminating between different ligands. In the light of this it is interesting to note that the degree of similarity varies between the different helices. There is almost complete homology in TM 7, with over 50% in helices TM 2, 3, and 6. This leaves most variety in TM 1 and 5. How this relates to receptor function is unclear. Shigemoto et al.

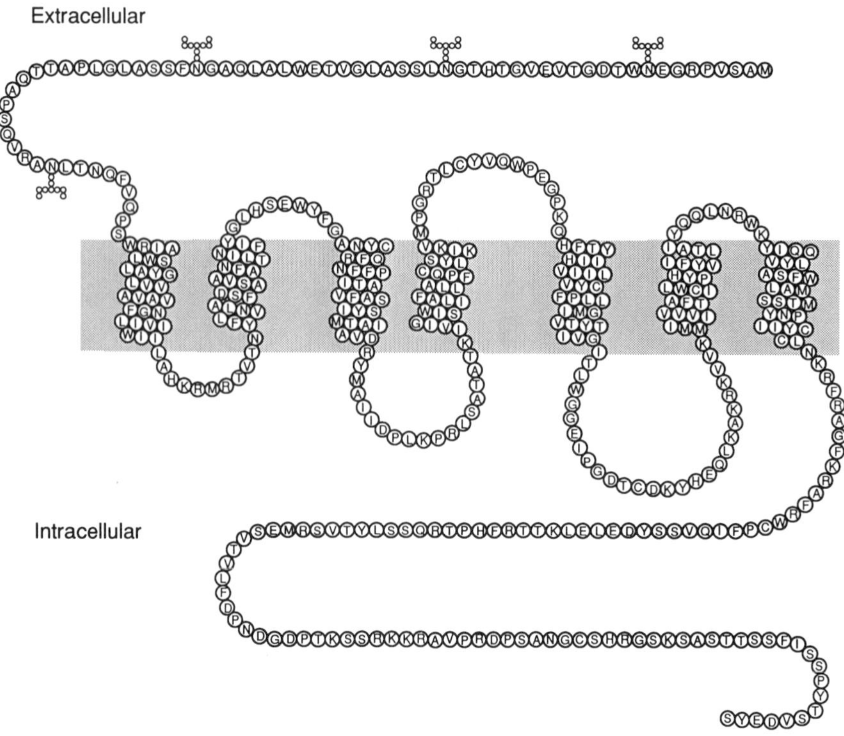

FIGURE 9.4. Structure of neuromedin K receptor.

(1990) have suggested that the conserved C-terminus of the tachykinin peptides may interact with TM 7. The size of peptides means that they cannot be entirely accommodated within the interhelical pocket, and so the extracellular portions of the tachykinin receptor deserve attention. The interhelical loops are short and each receptor has a unique pattern of charged acidic and basic residues, especially loops o3 and o4. It may be worth directing future effort at elucidating what role, if any, these might play.

The evidence from endogenous receptors indicates that all three couple to inositol lipid hydrolysis by phospholipase C, via a pertussis toxin-insensitive G-protein, leading to production of inositol 1,4,5-trisphosphate and diacylglycerol. This shared mechanism fits with the considerable sequence resemblance on the cytoplasmic segments of the receptors. Loop i1 has almost total conservation, loop i2 has about 75% sequence identity, and there are considerable sequence similarities in the proximal and distal ends of the i3 loop near the transmembrane segments. It has been suggested that the proximal and distal ends of this loop are particularly important in confering G-protein specificity (O'Dowd et al., 1988). Another region similarly implicated is the proximal (membrane) end of the C-terminus, and again this shows a high degree of conservation among the tachykinin receptor subtypes. The main differences between the tachykinin receptor subtypes in functional terms is their rate of desensitization: substance P > neuromedin K > substance K (Shigemoto et al., 1990). The C-terminus in the β-receptor has been shown to be of importance in desensitization (Cheung et al., 1989), since it is the site at which the kinase specific for agonist-occupied receptor, β-adrenergic receptor kinase (β-ARK), appears to inactivate normal receptor function (Bouvier et al., 1988). There is clear divergence in the sequence of the C-terminal tails of the tachykinin receptors. Shigemoto et al. (1990) have drawn attention to the different number and distribution of serine and threonine residues in the C-terminus among the three subtypes (26 in the substance P receptor, 28 in the substance K receptor, and 14 in the neuromedin K receptor). In addition, there are different patterns of serine/threonine residues in the i3 loop. As kinases of several types are now held to be the final common pathway of desensitization (Sibley et al., 1988), these residues may be points at which regulatory kinases act. In these two regions, the substance P and neuromedin K receptors have much greater similarity compared to each other than compared to substance K receptor. The size of the C-terminal tails, especially for the neuromedin K receptor, raises questions as to whether the tails may have additional functions apart from G-protein coupling and control of desensitization, perhaps in the recognition of other proteins associated with signaling.

2.3. The Pituitary Glycoprotein Hormones

2.3.1. Introduction

The anterior pituitary secretes a number of glycoprotein hormones, which regulate endocrine target glands. These hormones are α,β dimers with molecular weights in the region of 30,000 Da, and include LH–CG, TSH, and follicle-stimulating hormone (FSH). Those proteins all have a common α subunit but differ in their β subunits. They all act through Gs-coupled receptors to stimulate adenylyl cyclase. The LH–CG and TSH receptors have been cloned, and define a new subfamily of G-protein coupled receptors. Their major unique feature is an enormous extracellular N-terminus, probably necessary to bind the large glycoprotein ligands. The transmembrane and C-terminal portions of the receptors are more like those for other receptors in the superfamily, although they do show differences in detail when compared with other peptide receptors.

2.3.2. LH–CG Receptors

The cloning of the porcine and rat LH–CG receptors was achieved by two separate groups (Loosfelt et al., 1989; Macfarland et al., 1989). The rat hormone receptor was cloned using information derived from classic protein chemistry. The receptor was purified by lectin and choriogonado-tropic hormone affinity columns and sequenced directly to obtain the NH_2-terminus, as well as after proteolytic or chemical cleavage to obtain smaller peptides. These were used to generate oligonucleotide primers for the PCR which produced a large probe. This was then used to isolate a clone from a cDNA library. The porcine receptor was first purified to a stage sufficient to allow the production of a monoclonal antibody. A cDNA library was prepared from porcine testis, and this was screened by the monoclonal antibody to identify a clone which contained the complete sequence. Both clones have been expressed in cell lines (rat in human kidney 293 cells; porcine in COS-7 cells) and bind LH–CG with an affinity similar to that of the native receptor. In addition, the rat receptor was shown to couple to stimulation of adenylyl cyclase (Macfarland et al., 1989). Several tissues were examined for expression of the receptor mRNA and, for receptors from both species, expression was restricted to gonadal sites — ovaries and testes.

The cloning of the porcine receptor revealed three clones (B, C, and D) coding for truncated forms of the receptor. These all had the large extracellular domain intact up to residue 315, but then had a deletion removing all of the transmembrane domains, and most of the C-terminus. Forms B and C also had frameshifts, resulting in different C-terminal sequences. Form D retained the original reading frame and so resumed the normal C-terminal sequence. The authors suggested that these variations arise by alternative splicing. They estimated that the most common mRNA (4.4–4.7 kb) was for the full-length clone (60%) followed by D (20%), C (15%), and B (5%). They suggested that some of the minor bands on Northern blots could represent truncated receptor, and also claimed that their monoclonal antibody recognized additional proteins besides the 85 kDa receptor. As the variants lack the transmembrane domain, they would be expected to be soluble. No comparable clones were isolated related to the rat receptor, despite the fact that the probe used to screen the cDNA library was upstream of the divergence point. However, this cDNA library was luteal in origin, whereas the porcine library came from the testis, leaving the possibility open that there is tissue-specific processing.

Both receptors are large (rat:674 amino acids + 26 residue signal peptide, mw 75,000; porcine: 669 amino acids + 27 residue signal peptide, mw 72,025). The most striking feature is the enormous extracellular N-terminal, containing over 300 residues (Fig. 9.5). Macfarland et al. (1989) point out that the extracellular region shows most homology to PG40, a proteoglycan found in the extracellular matrix of connective tissue. Both PG40 and the N-terminus of the LH–CG receptor are leucine-rich glycoproteins (LRG). They contain leucine-rich repeats, which in the LH–CG receptor are about 25 residues long and repeated 14 times. The repeats can form amphipathic helices able to interact with hydrophilic and hydrophobic surfaces. The

Extracellular

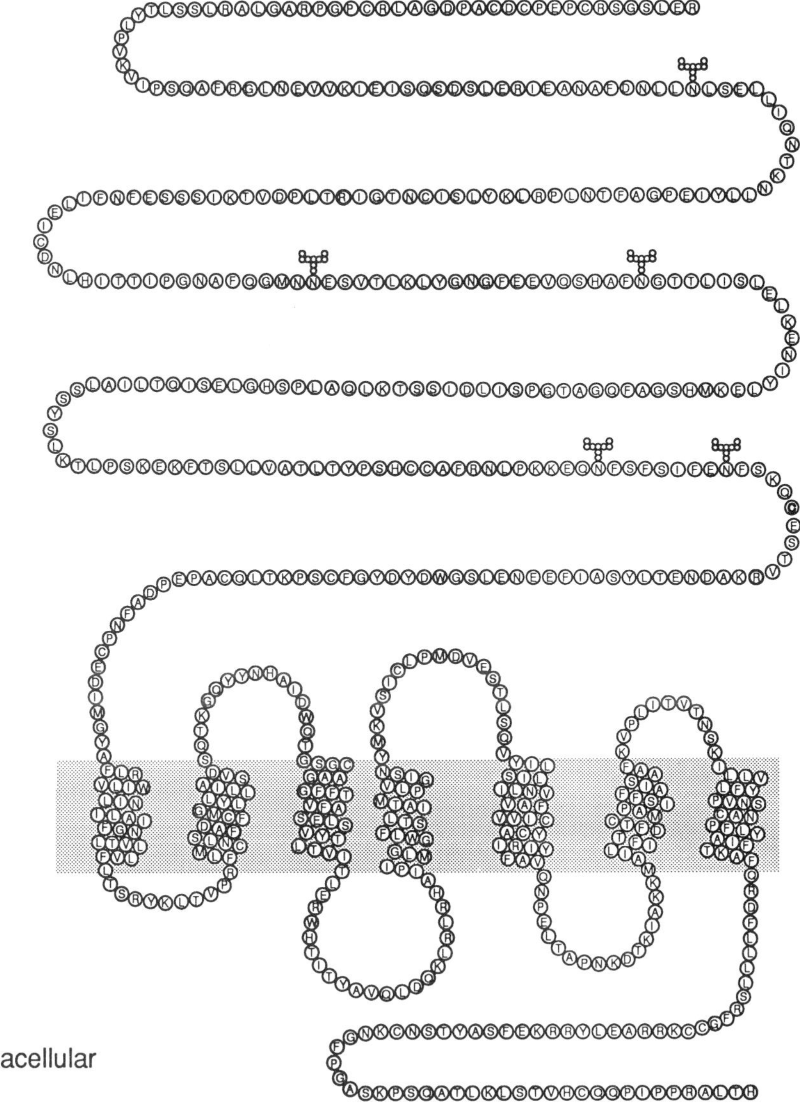

Intracellular

FIGURE 9.5. Structure of LH–CG receptor.

extracellular domain of the LH–CG receptor is responsible for ligand binding (Keinanen and Rajaniemi, 1986; Kim et al., 1987) and these amphipathic repeats may then interact with the hydrophobic core of the receptor. Macfarland et al. (1989) also note that the N-terminal sequence contains a ten-residue run which is identical to a sequence in soya bean lectin. This may be significant because deglycosylated LH and CG bind to the LH receptor but do not activate it. Hence, this region may be important in agonist recognition and signal transduction.

The hydrophobic domain of the receptor shows limited homology to other receptors including peptide receptors such as mas and the tachykinin receptors. The greatest similarity is with the β_2-adrenergic receptor, with 26% identity. The structural features are highlighted in Table 9.5. There are conserved acidic residues (glutamate) in TM 2 and 3, with an additional aspartate in TM 6, and an arginine in TM 5; TM 7 is notable for containing two prolines, distorting the predicted α-helixal structure. The DRY sequence at the cytoplasmic side of TM 3 is replaced by ERW. Although the receptor couples to adenylyl cyclase through Gs, there are no particularly strong homologies in the cytoplasmic loops to other Gs-coupled receptors. The C-terminus contains nine serine or threonine residues, and two of these lie within consensus phosphorylation sites for protein kinase C. There is also the possibility that the tail may be palmitoylated at a cysteine residue. Two possible proteolytic cleavage regions in the tail can be recognized (residues 623–625 and 630–632) which have clusters of basic amino acids (Macfarland et al., 1989).

2.3.3. THYROID-STIMULATING HORMONE RECEPTORS

The receptor for dog TSH/thyrotropin was first described by Parmentier et al. (1989). The sequence was obtained by cross-hybridization with a probe generated from degenerate PCR primers to consensus sequences from other G-protein-coupled receptors. The first step was to construct a probe from human genomic DNA PCR to amplify a sequence defined by two degenerate primers targeted to the most convergent regions of TM 2 and TM 7. This produced a clone with a previously undescribed structure. As this did not identify any thyroid mRNA on a Northern blot under conditions of high stringency, it was obviously not the TSH receptor itself. However, under conditions of reduced stringency, it hybridized to a cDNA clone from the thyroid, suggesting it might be able to identify the true TSH receptor. This clone was isolated, sequenced, and expressed in adrenocortical Y1 cells and *Xenopus* oocytes. It conferred TSH sensitivity on both these systems, in accordance with it encoding a TSH receptor, and recognized a 4.9-kb transcript which was present only in poly A^+ RNA made from the thyroid gland or thymocytes. It failed to recognize any transcripts in a variety of other tumors such as heart, liver, or stomach. The initial probe which cross-reacted with this clone is of interest in its own right, as it only recognized a transcript from the ovary (and testis, to a lesser extent) on Northern blot done at high stringency, and is probably a partial sequence of the human LH receptor or, perhaps more likely, the FSH receptor (Parmentier et al., 1989).

The human TSH receptor amino acid sequence has now been reported by several groups (Libert et al., 1989a; Misrahi et al., 1990), and the nucleotide sequence has been published by Nagayama et al. (1989). Its identity has been confirmed by expression in CHO and COS-7 cells. Unlike the situation found with the dog, there are two mRNA species in the human (4.4 and 4.6 kb). It has been suggested that this could be due to alternative splicing (Libert et al., 1989a), as is found with the LH–CG receptor.

The TSH receptors resemble the LH-CG receptors in both size and sequence (Figure 9.6 and Tables 9.4 and 9.5). From both species, the receptors have 744 amino acids, with calculated molecular weights of about 86,000 Da. Like the LH–CG receptor, there is a huge N-terminus of 398 residues before the first of the transmembrane helices. The N-terminal region has a 52-amino acid insert when compared to the LH–CG receptor, located close to the TM 1 helix, but even so, the entire extracellular domain still shows about 40% homology to the LH–CG receptor. The extracellular domain (o1) of the TSH receptor has been analyzed by Misrahi et al. (1990). They have divided it into five sections, placing the insert in a 94-amino acid stretch they call E3. The corresponding portion of the LH–CG receptor has only 41 residues, and shows a limited 7% homology to the TSH sequence. The other extracellular domains show homologies which vary from 33 to 85%. Since TSH and LH–CG have common α subunits but different β subunits, it is suggested that the domains with only 7% homology may mediate the interactions with the β subunits, while the α subunits recognize other parts of the molecule. The remainder of the molecule has 70% homology to the LH–CG receptor. The structural similarities here can be seen by examining Tables 9.4 and 9.5. It has similarly sized intracellular and extracellular loops, the common DRY sequence at the beginning of i3 is replaced by ERW (DRW in the LH–CG receptor), the disulfide predicted between TM 3 and loop o3 is present, and there are similar distribution patterns of prolines and charged residues. Indeed, including conservative replacements, the transmembrane helices and their connecting loops have 85% homology to the corresponding structures in the LH–CG receptor. This is greater similarity than is found among members of the tachykinin family. As LH–CG and TSH both stimulate adenylyl cyclase, this degree of conservation is not unexpected. The intracellular loops are involved in coupling to the G-protein—in the case of these two receptors, this G-protein will be Gs. The first loop, i1 has almost total conservation; i2 has 65% identity, and there is also strong conservation at the proximal and distal ends of i3. This pattern is very similar to that seen within the tachykinin family. The main differences between the two receptors are seen in the C-terminus (i4), where after a short stretch of homology, the polypeptide chains leave the membrane and the sequences diverge. By analogy with other systems, this may predict differences in rates of desensitization between the two receptors. The human and dog TSH receptors have 90% homology, with the bulk of the substituents being in the N-terminus (o1) and C-terminal tail (i4).

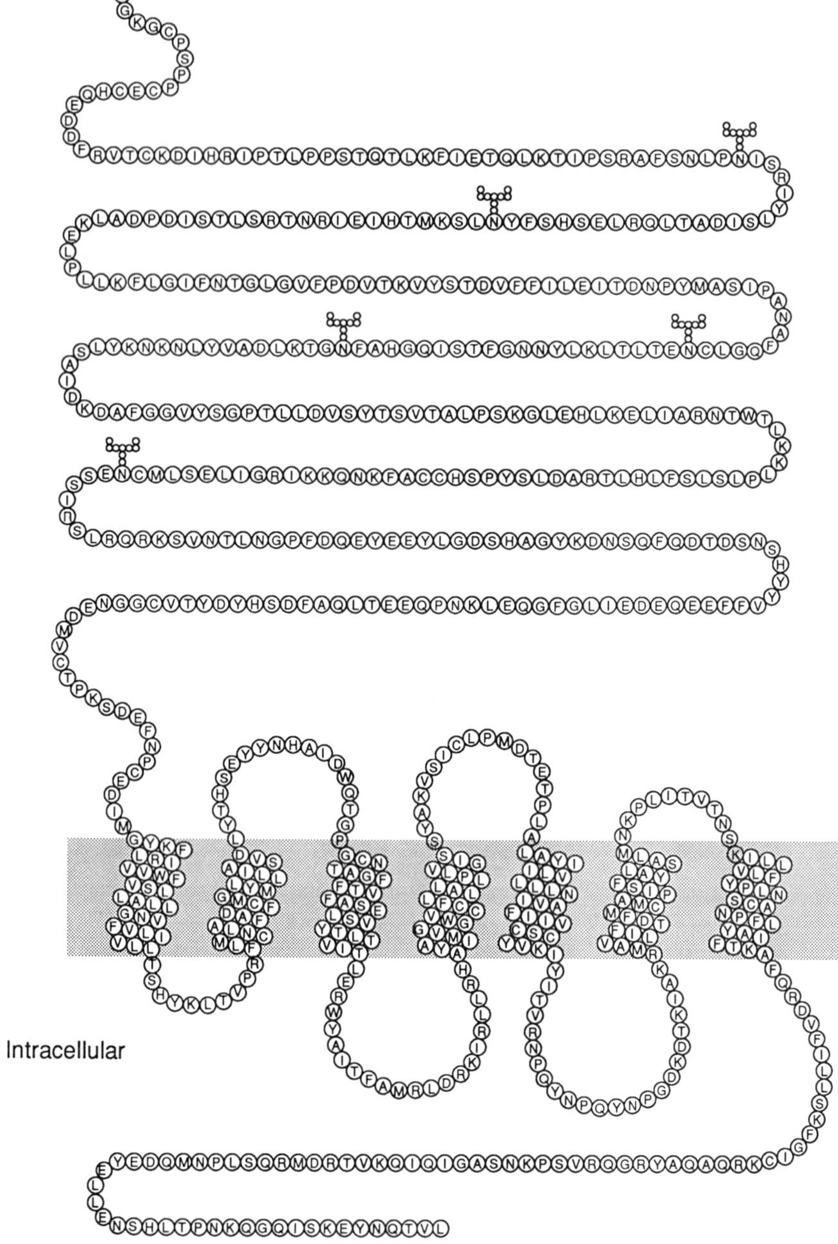

FIGURE 9.6. Structure of TSH receptor.

An unusual feature unique to the TSH receptor is sequence homology to nonreceptor type protein kinases (Misrahi et al., 1990). In the amino terminal part of the i3 loop is the sequence YITVRNPQYNPGDKD (residues 605–619). In the proteins c-fgr, c-slk, c-syn, c-yes, and c-src there are twelve residues at the C-terminus with the consensus sequence YFTS/A A/T EPQYQPGD/E Q/N T/L which has 75% identity to the TSH sequence. Furthermore, in c-src, phosphorylation of a tyrosine in this motif regulates its kinase activity. It would clearly be of great interest to learn whether this region allows the TSH receptor to be modulated by tyrosine kinases.

2.4. mas/Angiotensin

The mas/angiotensin receptor has an unusual history among receptor genes as its structure was known before its ligand specificity or mechanism could be assigned. With rapid improvements in PCR-based cloning, this may be a situation more frequently encountered in the future. mas has added interest because it was originally identified as an "oncogene" by Young et al. (1986). They showed that when human mas was transfected into NIH 3T3 cells, these cells became tumorigenic in nude mice. The cDNA responsible for this activity encoded a 325-amino acid protein, which had no explicit homology to any known oncogene. However, hydrophobicity plots demonstrated that it resembled rhodopsin in having a seven-hydrophobic domain pattern and it was likely to be a membrane-associated receptor. Later it was proposed that mas was a neural peptide receptor since it showed strongest sequence similarity to the substance K receptor (Hanley and Jackson, 1987). When synthetic mas mRNA was generated from a plasmid containing the mas coding sequence and injected into *Xenopus* oocytes, expression of a novel sensitivity to angiotensin peptides was detailed, which manifested itself as a stimulation of an inward current with the characteristics of the endogenous calcium-activated chloride current (Jackson et al., 1988). The results on the assignment of mas as an angiotensin receptor were extended by stable expression of the human mas oncogene in a mammalian neural cell line, NG 115-40IL (Jackson et al., 1988). Transfected cells responded to angiotensin with a breakdown of phosphatidylinositol 4, 5-bisphosphate (PtdInsP$_2$) and an increase in cytosolic calcium levels and DNA synthesis (Jackson et al., 1988; Jackson and Hanley, 1989; Poyner et al., 1990).

The stable expression of mas in a mammalian cell line has allowed a detailed examination of its pharmacology and regulation. The potency rank order for naturally occurring mammalian angiotensins was AIII > AII ≫ AI. The response was blocked by a broad-spectrum peptide antagonist [D-Arg^1, D-Pro^2, D-Trp^{7,9}, Leu^{11}]–substance P, but not by angiotensin analogues thought to act as competitive antagonists. Indeed, two peptide-based antagonists [Sar^1, Val^5, Ala^8]–AII and [Sar^1, Val^5]–AII were weak partial agonists. This data suggests that the receptor represents a distinct subtype among angiotensin receptors. The pharmacological profile resembles that seen in mammalian brain and adrenal cortex but is unlike that observed in cardiovascular sites or in liver hepatocytes (Peach, 1977; Hanley, 1990). The molecular cloning of the "AT1-angiotensin" (vascular type) AII receptor confirms that it is a structurally distinct receptor (Sasaki et al., 1991). Surprisingly, there is very limited sequence similarity between the AT1 receptor and the mas gene product, suggesting they may not be evolutionarily related. Consistent with the mas gene product pharmacology, recent work has shown that mas transcript is selectively expressed in neurons and is found at highest levels in the frontal cortex and hippocampus, as well as in a limited number of other of brain nuclei (Bunnemann et al., 1990; Young et al., 1988).

Stable expression in a neuronal cell line has allowed the biochemistry of early responses to be examined in detail, and to be compared with an endogenous receptor working through the same phosphoinositide pathway, the bradykinin receptor. The calcium response is of similar magnitude in both cases, but is slightly slower when angiotensin is applied. The distinctive mitogenic signal generated by angiotensin stimulation is unclear, but it is unlikely to be activation of type 1 phosphatidylinositol kinase, as has been suggested for the platelet-derived growth factor (PDGF) receptor (Coughlin et al., 1989) because detailed analysis has now shown no regulation of the product of this enzyme in transfected cells (Poyner et al., 1990). The angiotensin response can be abolished by acute pretreatment of the cells with protein kinase C activators (e.g., phorbol esters) — a phenomenon not seen with the bradykinin response (Jackson and Hanley, 1989).

The mas gene is intronless and located on human chromosome 6, in a region associated with neuroblastoma and other tumors (Rabin et al., 1987). The proto-oncogene is activated by rearrangement of noncoding 5′-flanking region, approximately 3.5 kb upstream from the start of translation. This rearrangement might cause changed levels or sites of expression of the gene product. The fact that the protein coding region is unaltered in this oncogenic activation means that induced tumor formation would be likely to be dependent on an endogenous activating signal, that is, angiotensin. Thus, it is tempting to speculate on a role for mas in tissue hyperplasia and vascular hypertrophy in essential hypertension and other syndromes associated with angiotensin dysfunctions.

Structural information on mas is summarized in Tables 9.4 and 9.5, and Figure 9.7. It is a structure that does not exhibit several of the "consensus" features of mammalian G-protein receptors. For example the DRY sequence at the C-terminal end of TM 3 is replaced by the sequence ERC, containing a nonconservative replacement, as is also the case in LH and TSH receptors. There can be no disulfide bond between TM 3 and the o3

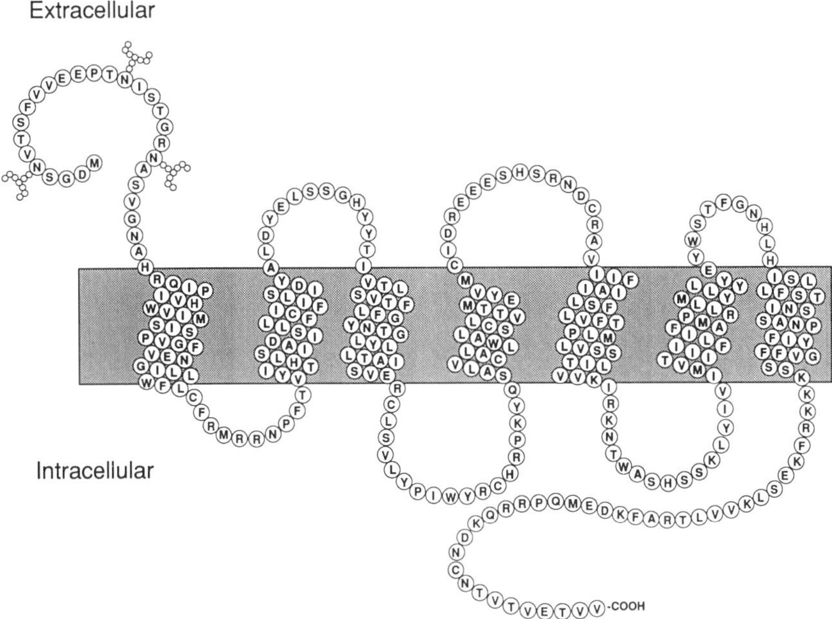

FIGURE 9.7. Structure of mas/angiotensin receptor.

loop, as the former lacks a cysteine residue. TM 4 contains what might be
a suitable cysteine, but if that were to take part in a bond with the o3 loop,
then inspection of the primary structure suggests that it would distort TM
5. The only cysteine present in the C-terminus is not in a palmitoylation
consensus sequence. The membrane-spanning regions contain three buried
acidic residues, one in TM 2 corresponding to that found in nonpeptide
receptors and the others in TM 4 and TM 1. There is also a striking stretch
of three consecutive glutamic acid residues in the proximal, juxtamembrane
portion of the o3 loop. This site is an appealing candidate for docking the
basic N-terminal of angiotensin III. However, any model for binding of the
ligand to its receptor must take account of the fact that an additional amino
acid at the N-terminus, such as in AII, does not greatly reduce the potency
of the resulting peptide. mas contains three conserved proline residues in the
TM 5, 6, and 7 and unusually, two in TM 1. It also shows reasonable
sequence similarity to the tachykinin receptors (e.g., based on conserved
residues: 44% homology with the bovine substance K receptor). A similar
level of homology is seen with the LH–GC receptor. The human and rat
mas sequences are over 90% identical, with the bulk of the changes
concentrated at the N- and C-terminals. As mas is coupled to breakdown of
$PtdInsP_2$, it might be expected that the i3 loop would show greatest
similarity to receptors also coupled to that response. Up to a point this is

true: taking account of inserted residues, there is over 60% conservation between mas and the bovine substance K receptor, compared to less than 40% between mas and the LH–CG receptor, and 20% conservation between mas and human β_1-adrenergic receptor. However, there is only about 10% conservation between mas and human M1 muscarinic receptor, which is also coupled to PtdInsP$_2$ breakdown. Clearly there is no predictive primary structural feature correlated with receptor coupling to phospholipase C through the relevant G-protein.

Unlike many of the peptide receptors, the actual protein product of the mas gene has been identified using a monospecific antibody to the C-terminal tail. On immunoblots, the protein appears as a single band of 45,000 to 47,000 Da—larger than the predicted size of 37,500 Da. This may be evidence for glycosylation, which would contribute the added mass (Hanley et al., 1990).

2.5. STE Gene Family

2.5.1. INTRODUCTION

Yeast contains a signaling system analogous to the G-protein coupled pathways found in mammalian cells (Whiteway et al., 1989). The yeast conjugation response is also described as "mating," to form single diploid cells from two haplotypes. In the life cycle of the budding yeast, *Sacchararomyces cerevisiae,* haploid *a* and α cells are produced. For conjugation to occur, *a* cells need to respond to a peptide pheromone, α factor, secreted by α cells. Likewise, α cells respond to a similar pheromone *a* factor, produced by *a* cells. These pheromones cause arrest of the cell cycle in G$_1$ phase, and cause the cells to distort and assume a pear shape. These fuse with each other and complete the process of conjugation. The α and *a*-factor receptors are encoded by the STE2 and STE3 genes, respectively (Jenness et al., 1983; Hagen et al., 1986). These receptors have seven predicted transmembrane helices, and thus are members of the G-protein-coupled receptor superfamily. This relationship to the vertebrate genes has been strengthened with the discovery that at least three other gene products are involved in the mating hormone response: SCG1 or GPA1, which is similar to G-protein α subunits (Nakafuku et al., 1987); STE4, analogous to G-protein β subunits, and STE18, analogous to G-protein γ subunits (Whiteway et al., 1989). However, there is clear genetic evidence that in *S. cerevisiae* the equivalent of the β,γ complex initiates the mating response, not the α subunit. Evidence for a comparable effector action of G-protein β,γ subunits in mammalian systems is controversial. The effector protein which is activated by β,γ subunits in *S. cerevisiae* is unknown.

Extracellular

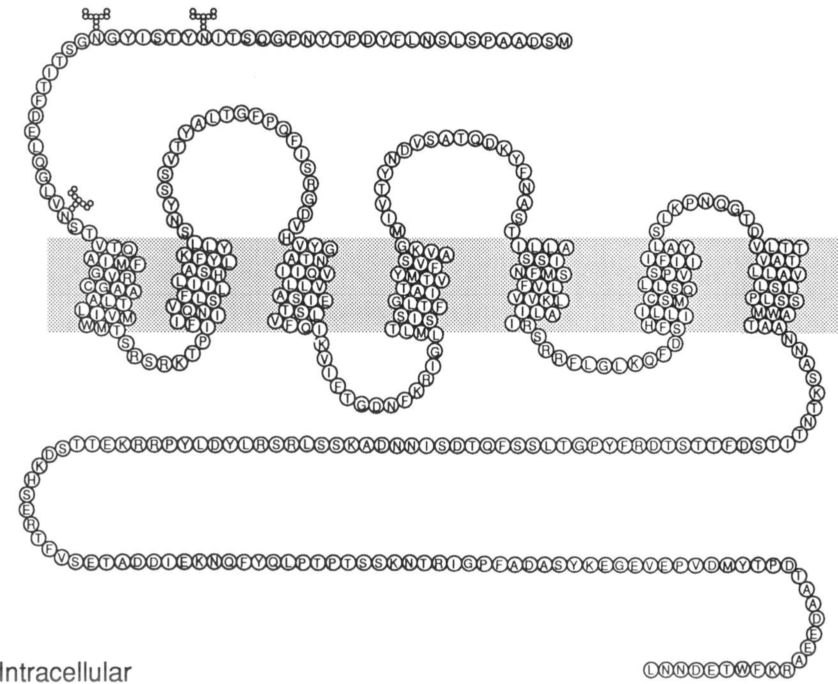

Intracellular

FIGURE 9.8. Structure of STE2 receptor.

2.5.2. STE2

STE2 encodes a 431-amino acid protein (Figure 9.8 and Tables 9.4 and 9.5). It was initially cloned by a complementation assay involving correcting the mating deficiency of the STE2 mutant (Nakayama et al., 1985). Hydrophobicity analysis suggests that the protein contains seven transmembrane helices. STE2 has a moderate sized N-terminus, lacking a signal peptide and with three possible N-linked glycosylation sites. The interhelical linking loops are short, especially those on the cytoplasmic side of the membrane. There is a single acidic residue, a glutamic acid in TM 3 buried within the membranes, and this may represent the equivalent aspartate found in a similar position in many mammalian G-protein-coupled receptors. There are a number of buried basic residues—an arginine in TM 1 and lysines in TM 2, 3, and 4. There is no conservation of cysteine residues or retention of the DRY sequence in the i2 loop. The most striking feature of the STE2 gene product is a long C-terminus, which is rich in serine, threonine, and charged amino acids. There is little sequence homology in this domain with any of the mammalian receptors.

The genomic structure of STE2 has revealed two possible promotor sequences, 49 and 114 bp before the initiation codon. The former of these, with the sequence TATAAA, has been judged to be the most likely promotor in vivo based on analysis of STE2-coded mRNA. An appropriate termination sequence was found at the 3'-end of the gene. Expression studies revealed that the major transcript from the STE2 gene was a 1.6- to 1.7-kb mRNA, present only in *a* cells, as predicted. The expression of STE2 is thought to be controlled by a regulated repressor, the α_2 protein, and a binding site for this protein was identified on the STE2 gene, 200 bp upstream from the start codon (Nakayama et al., 1985).

A second STE2 gene has recently been cloned from another species of yeast *S. kluyveri* (Marsh and Herskowitz, 1988), which produces similar pheromones to *S. cerevisiae,* (they can cross-react with each other). The *S. kluyveri* STE2 sequence was obtained by cross-hybridization of a *S. cerevisiae* STE2 probe to *S. kluyveri* DNA. The gene product contains 426 amino acids, and has 50% identity with that of *S. cerevisiae.* This rises to 67% if only the transmembrane helices are considered. This sequence homology is similar to that found between different subtypes of, for example, the muscarinic receptor or the β-adrenergic receptor. The major differences occur in the C-terminus, which, however, retains the general features of being rich in charged and hydroxy-containing amino acids. The best region of identity is the i3 loop which might imply that the two receptors couple to the same G-protein, which is suggested by genetic complementation evidence (Marsh and Herskowitz, 1988).

A limited amount of work has been carried out on the STE2 protein, and on deletion mutants. The studies on the protein have used an antibody directed against the C-terminus; the major species identified in cells appeared to have a molecular weight about 6 kDa larger than the expected value and this was attributed to covalently added carbohydrate, consistent with three predicted glycosylation sites (Blumer et al., 1988). It was also shown that a mutant which lacked the C-terminus (residues 297 onward) was able to bind α factor and mate, albeit less efficiently. This shows that the C-terminus is not essential for the basic conjugation response. More detailed studies were carried out on a series of C-terminal mutations, which all had in common a deletion of the tail after residue 360. Although these could bind α factor and mate normally, they had increased sensitivity to the pheromone, increased receptor number, and did not undergo characteristic morphological changes in response to the hormone (Konopka et al., 1988). The basic lesion seemed to be a failure of the receptor to desensitize, coupled with increased stability. The mechanism behind mating-factor-induced morphological changes may involve some additional function of the C-terminal tail, as mutants which fail to adapt (i.e., desensitize) also fail to show shape alterations.

STE2 has been expressed in *Xenopus* oocytes (Yu et al., 1989), in the first experiment in which a surface membrane protein from a unicellular eukaryote has been expressed in a vertebrate cell. Neither STE2 on its own, nor STE2 in the presence of mRNA coding for the yeast G-protein α subunit (GPA1), could couple to any conductance change in the *Xenopus* membrane. Given that, in yeast mating, activation proceeds by β,γ effector subunits, this demonstration of the inability of its signaling mechanism to function in *Xenopus* may not be surprising, but this oocyte system comprises a novel expression strategy for further analysis of the responses to STE2. It may be that relatively modest modifications to the STE2 gene product will permit it to couple to a vertebrate transduction pathway.

2.5.3. STE3

STE3 was cloned by a complementation assay, similar to that used for STE2 (Nakayama et al., 1985). The protein contains 470 amino acids (Figure 9.9). Among the family of G-protein-coupled receptors it is unique in having virtually no N-terminal sequence before TM 1. There are few charged residues, although an aspartate residue is located superficially in TM 2 and

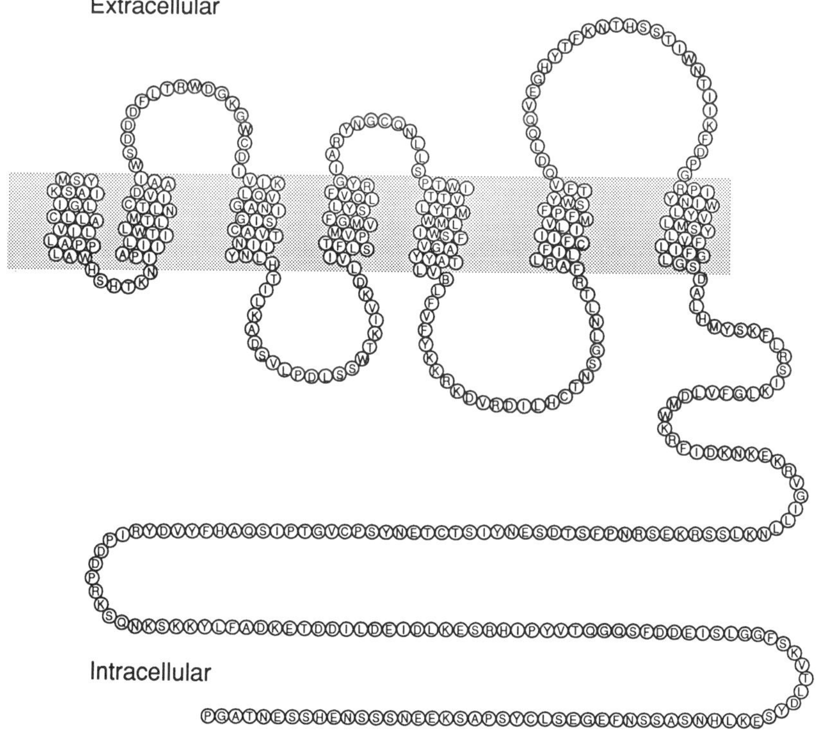

FIGURE 9.9. Structure of STE3 receptor.

three basic residues may be within membrane segments. Like STE2, there are none of the conserved cysteines or the DRY sequence of vertebrate counterparts, and there is virtually no homology with any mammalian receptor, or, surprisingly, STE2. TM 1 must be distorted as there are two adjacent prolines. The C-terminus is the longest of any G-protein-coupled receptor so far identified (187 residues).

Expression of STE3 is restricted to α cells, where it is present as a 1.6- to 1.7-kb transcript. The promotor may be contained in a sequence $(AT)_{10}GTA$ followed by TATAGA, found 145–130 bp above the start codon (Nakayama et al., 1985). A deletion mutation was isolated lacking the last 106 amino acids of the C-terminus—this was able to complement the STE3 mutant for the mating response. Thus an analogous situation exists to that found for STE2, where large parts of the C-terminus are not essential for mating but must play some other role in regulating the response, perhaps in the control of desensitization, as has been suggested for the β-receptor

3. Conclusions

The cloned peptide receptors exhibit sequence and domain similarities within their diverse structures which may relate to the ligands they bind and the effectors they activate. This poses the question as to how well they can be integrated into the schemes worked out for the monoamine-activated receptors. Briefly, the results with amine receptors identify the conserved acidic residue in TM 3 as essential for binding, and, further, that acidic residues in conserved positions in TM 2 and 3 play important roles in the binding and mode of action of agonists. The ligand binding site is predicted to be formed within a pocket of all seven helices. The specificity of agonist binding may be a function of several of these helices, but discrimination between antagonists is largely determined by TM 6 and 7. A disulfide bond linking helix TM 3 with the o3 loop may be essential for normal ligand recognition. Coupling to G-proteins is mediated by the cytoplasmic loops of the receptor. The i1 and i2 loops are probably important for specifying a G-protein interaction, whereas the proximal and distal ends of the i3 loop determines the exact molecular species of the G-protein to which coupling will take place. The C-terminal tail has been identified with an important role in regulation of desensitization in amine receptors (Lefkowitz et al., 1989) and appears to retain this function in several, perhaps all, of the peptide receptors.

Certain features of this modular scheme of domains stand up better than others. The general model of ligand binding, whereby the drug makes direct contact with residues located exclusively in the hydrophobic core of the molecule, does not appear to apply to all peptide receptors. At the most extreme end of the spectrum, the glycoprotein hormones LH–CG and TSH interact predominantly with the large N-terminus of their receptors, as can be inferred by the retention of binding activity in the isolated N-terminal region. How this leads to activation of the receptor is not clear; perhaps either part of the ligand or the N-terminus is exposed and able to insert itself into the hydrophobic core of the receptor to mimic activation by smaller ligands acting on their target receptors. All of the animal peptide receptors have an acidic residue corresponding to Asp^{130} of the β-receptor, mutation of which abolishes receptor–G-protein coupling in that system (Venter et al., 1989). This is also always followed by a basic residue, and then usually an aromatic amino acid. Perhaps the paired acidic–basic residues represent a common element in G-protein activation. Following from this, it is interesting that this pair of residues is absent in the yeast receptors, which may be related to the different recognition of the G-proteins which operates in these unicellular organisms.

Further inspection of the receptor structures gives more clues as to ligand binding. LH and TSH are at one extreme, but it is difficult to envisage conventional neuropeptides (seven to ten amino acids) accommodating to binding entirely within the interhelical pocket. Equally, the all important acidic residue in TM 3 is often absent. Studies on adrenergic β_2–α_2 chimeric receptors placed particular emphasis of the role of TM 7 in conferring ligand specificity. This is unlikely to be the case for tachykinin receptors, where TM 7 is almost completely conserved among the different subtypes. Thus, while it was argued above that the mechanism of receptor activation may have been conserved, the mode of ligand binding is clearly different. The role of charged and noncharged residues in the helices is probably very important in binding, but attention should be directed to the extracellular "loops." The largest of these is the N-terminus — clearly of importance for LH and TSH, and probably important for smaller ligands as well. The other extracellular regions of the receptors are much smaller, but they have unique patterns of charged residues, making them attractive candidates for ligand interactions. Indeed, in the case of STE3 the N-terminus is completely absent, but there is, on the other hand, an enlarged o3 loop. This may help to explaining the discrimination between α and a factors by the STE2 and STE3 receptors. A final point worth considering is whether or not peptide receptors can be modulated by ligands acting at an allosteric site. Gallamine acts on the M2 muscarinic receptor subtype in an allosteric fashion (Stockton et al., 1983), but little consideration has been given as to whether the finding might be more generally applicable. Gallamine is thought to exert its effects by binding to residues on the extracellular portions of the muscarinic receptor. If peptides recognize extracellular regions of their receptors, this might suggest that worthwhile antagonists could be developed which block binding at these regions. Equally, at receptors with large N-termini, such as LH-CG or neuromedin K, other allosteric interactions might be possible, allowing the possible convergence of multiple physiological extracellular stimuli on a single receptor target, such as both diffusible ligands and immobilized determinants from the extracellular matrix.

The role of disulfide bonds has been emphasized in many G-protein-coupled receptors, but it is not universal since an extracellular disulfide cannot occur in STE2; thus it is not essential to the stability or functioning of the entire class of receptors. Equally, the widely found TM 3–loop o3 disulfide bond is not found in mas, although there is very likely to be an equivalent surface disulfide between TM 4 and loop o3. The usual disulfide bond is predicted by extracellular cysteines in other receptors, and so a disulfide linkage tethering one of the middle helices to an extracellular loop seems a recurrent feature of peptide receptors—it is in other G-protein-coupled receptors.

The topography of the receptors suggests that cytoplasmic loops must interact with appropriate G-proteins. The hypothetical role of a pair of acidic–basic amino acids at the cytoplasmic junction of TM 3 has already been discussed. Marsh and Herskowitz (1988) have pointed out that in the i1 loop the sequence KRXYKRTP is conserved among *S. cerevisiae* STE2, *S. kluyveri* STE2, and bovine, human, and *Drosophila* rhodopsins. This sequence is not found in other receptors, however, and thus cannot constitute a consensus recognition signal. This short i1 loop often contains at least two closely spaced basic amino acids at the top. In the nonpeptide receptors much interest has focused on the i3 loop. This shows great variability in size, but the proximal and distal ends seem to be of considerable importance in determining the specificity of the G-protein interaction. However, there is no recognized convergence in primary structure to identify a candidate G-protein recognition site. The peptide receptors seem to reinforce this point. There is no extensive similarity between the i3 loop of peptide receptors coupling to $PtdInsP_2$ hydrolysis, when compared with that loop in nonpeptide receptors coupling to the same response, and a similar conclusion holds for adenylyl cyclase-coupled receptors. However the extent of heterogeneity among G-proteins is an unknown factor, and this might hinder identification of simple primary recognition sequences. There is no discernible correlation between loop size, the size of the C-terminus, and the effector system. However, a general feature of all peptide G-protein-coupled receptors so far cloned is that the intracellular loops are very short. Thus i1 generally has 10–15 residues, i2 has about 20 residues, and i3 usually has 20–30 residues. In nonpeptide receptors, the i3 loop is normally significantly longer than this.

The C-termini of peptide receptors may further contribute to multipoint binding in coupling to G-proteins, and also in controlling rates of desensitization. The largest tails belong to the STE family. With these receptors, large sections of the C-terminal tails have been deleted without destroying the basic function of the receptor in the control of mating. However, STE2 mutants having truncated tails did not undergo desensitization and also failed to show characteristic morphological changes. This strongly supports the notion that the C-termini are involved in modulation of the response to receptor activation and may also be essential in obtaining a complete biological response to agonist. By analogy, similar functions may be expected for the C-terminal tails of other peptide hormone receptors and they may interact with distinct regulatory proteins. The prototype for such an interaction may be the binding of arrestin to the C-terminal tail of rhodopsin.

4. Summary of Consensus Features of Peptide Receptors

1. The mode of ligand binding to peptide to receptors is predicted to be different from that of nonpeptide receptors. Interactions with the extracellular portions of the receptor are likely to be important.
2. Predicted surface disulfide bridge formation may not be essential for the stability or function of all eukaryotic peptide receptors, but there does seem to be an evolutionary tendency toward retention of a potential disulfide bond linking an extracellular loop to a transmembrane helix, particularly in mammalian peptide receptors.
3. A paired acidic–basic residue is conserved at the base of TM 3 in the mammalian receptors and may play an important role in receptor–G-protein activation.
4. There are no readily identifiable G-protein recognition sequences on the cytoplasmic loops of receptors of shared mechanisms.
5. The C-terminal tails of the receptors may be involved in controlling desensitization and modulation of biological responses, perhaps by specifying additional intracellular protein interactions.
6. Peptide receptors do have characteristic sequence elements, although none are unique to this class. To date, most examples share an i3 loop size range which has fewer than 30 amino acids, while the C-terminus (i4) can be either long or short. Using this criteria, a PCR consensus probe-derived clone, RDC1 from dog thyroid gland (Libert et al., 1989b) may be predicted to be a peripheral peptide receptor. Indeed, the human vasoactive intestinal polypeptide (VIP) receptor is 95% identical to the dog RDC1 clone, suggesting that RDC1 may encode the VIP receptor (Steedham et al., 1991). A large N-terminus (larger than 200 residues?) might predict a glycoprotein hormone-like ligand. Thus the *Drosophila Frizzled* gene product (Vinson et al., 1989), another invertebrate example of a seven transmembrane domain protein, may encode a cell surface receptor, coupled to a specific G-protein which is related to the TSH/LH receptor subfamily. This would carry the additional prediction that its activating stimulus could be an immobilized or diffusible glycoprotein hormone homologue. On the other hand, a novel precedent was established by the metabotropic glutamate receptor, which has an extended N-terminus, even though the ligand is an amino acid. However, it has been noted that the extended N-terminus has significant sequence

similarity to the extracellular binding domain of the sea urchin guanylate cyclase peptide receptor (Houamed et al., 1991). It may be that the metabotropic glutamate receptor in fact responds to more than one extracellular ligand, and that a candidate for a second type of ligand would be a peptide related to the sea urchin RESACT/SPERACT class.

Acknowledgments

We would like to thank Simon Pledger, Patrick Sadler, and all other members of the Illustration Department, LMB, for preparing the figures. We would also like to thank Claire Searle and Lucy Howes for typing the manuscript.

References

Arai H, Hori S, Aramori I, Ohkubo H, Nakanishi S (1990): Cloning and expression of a cDNA encoding an endothelin receptor. *Nature* 348:730–732

Blumer KG, Rencke JE, Thorner J (1988): The STE 2 gene product is the ligand-binding component of the α-factor receptor of *Saccharomyces cerevisiae*. *J Biol Chem* 263:10836–10842

Bost KL, Blalock JE (1989): Preparation and use of complementary peptides. *Methods Enzymol* 168:16–28

Bouvier M, Hausdorff WP, DeBlasi A, O'Dowd BF, Kobilka BK, Caron MG, Lefkowitz FJ (1988); Removal of phosphorylation sites from the β_2-adrenergic receptor delays onset of agonist-promoted desensitization. *Nature* 333:370–373

Bunnemann B, Fuxe K, Metzger R, Mullins J, Jackson TR, Hanley MR, Ganten D (1990): Autoradiographic localization of MAS proto-oncogene mRNA in adult rat brain using *in situ* hybridization. *Neurosci Lett* 114:147–153

Burgen ASV, Roberts GCK, and Feeney J (1975). Binding of flexible ligands to macromolecules. *Nature* 253:753–755

Cheung GH, Sigal IS, Dixon RGJ, Strader CD (1989): Agonist-promoted sequestration of the β_2-adrenergic receptor requires regions involved in functional coupling with Gs. *Mol Pharmacol* 34:132–138

Chung FC, Wang CD, Potter PC, Venter JC, Fraser CM (1988): Site-directed mutagenesis and continuous expression of human β-adrenergic receptors: Identification of a conserved aspartate residue involved in agonist binding and receptor activation. *J Biol Chem* 263:4052–4055

Coughlin SR, Escobedo JA, Williams LT (1989): Role of phosphatidylinositol kinase in PDGF receptor signal transduction *Science* 243:1191–1193

Curtis CAM, Wheatley M, Bansal S, Birdsall NGM, Eveleigh P, Peddar EK, Poyner DR, Hulme EC (1989): Propylbenzilylcholine mustard labels an acidic residue in transmembrane helix 3 of the muscarinic receptor. *J Biol Chem* 264:489–495

Dam TV, Takeda Y, Krause JE, Escher E, Quirion R (1990): γ-Preprotachykinin-(72–92)-peptide amide: An endogenous preprotachykinin I gene-derived peptide that preferentially binds to neurokinin-2 receptors. *Proc Natl Acad Sci USA* 87:246–256

Dixon RGJ, Sigal ID, Candelore MR, Register RB, Scatterwood W, Rands E, Strader CD (1987): Structural features required for ligand binding to the β-adrenergic receptor. *EMBO J* 6:3269–3275

Elton TS, Dion LD, Bost KL, Oparil S, Blalock JE (1988): Purification of an angiotensin II binding protein using antibodies to a peptide encoded by angiotensin II complementary RNA. *Proc Natl Acad Sci USA* 85:2518-2552

Fraser CM, Chung FZ, Wang CD, Venter JD (1988): Site-directed mutagenesis of human β-adrenergic receptors: Substitution of aspartic acid-130 by asparagine produces a receptor with high-affinity agonist binding that is uncoupled from adenylate cyclase. *Proc Natl Acad Sci USA* 85:5478-5482

Goldstein A, Aronow L, Kalman S (1974): *Principles of Drug Action,* 2nd ed. New York: Wiley

Hagen DC, McCaffrey G, Sprague GF (1986): Evidence that the yeast STE3 gene encodes a receptor for the peptide pheromone *a* factor: Gene structure and implications for the structure of the presumed receptor. *Proc Natl Acad Sci USA* 83:1418-1422

Hanley MR (1985): Peptide binding assays. In: *Neurotransmitter Receptor Binding* Yamamura HI, Enna SJ, Kuhar MJ, eds. New York: Raven Press, 2nd ed, pp 91-102

Hanley MR (1989): Neuropeptide receptors: Structure and transduction mechanisms. In: *Hormones and Cell Regulation,* Nuñez J, Dumont JE, Denton R, eds. London: John Libbey, Vol 13, pp 3-9

Hanley MR (1990): Molecular and cellular characterisation of the MAS oncogene as a neural peptide receptor. In: *Neuropeptides and Their Receptors,* Schwartz TW, Hilsted LM, Rehfeld JF, eds. Alfred Benzon Symp 29. Copenhagen: Munksgaard, pp 325-329

Hanley MR, Jackson T (1987): Substance K receptor. *Nature* 329:766-767

Hanley MR, Cheung WT, Hawkins P, Poyner D, Benton HP, Blair L, Jackson TR, Goedert M (1990): The MAS oncogene as a neural peptide receptor: Expression, regulation and mechanism of action. *Ciba Found Symp* 150:23-46

Harada Y, Takahashi T, Kuno M, Nakayama K, Masu Y, Nakanishi S (1987): Expression of two different tachykinin receptors in *Xenopus* oocytes by exogenous mRNAs. *J Neurosci* 7:3265-3273

Hershey AD, Krause JE (1990): Molecular characterization of a functional cDNA encoding the rat substance P receptor. *Science* 247:958-962

Herskowitz I, Marsh L (1987): Conservation of a receptor/signal transduction system. *Cell* 50:995-996

Herzog H, Hort YJ, Ball HJ, Hayes G, Shine J, Selbie LA (1992): A cloned human neuropeptide Y1 receptor couples to two different second messenger systems. *Proc Natl Acad Sci USA* In press

Houamed KM, Kuijper JL, Gilbert TL, Haldeman BA, O'Hara PJ, Mulvihill ER, Almers W, Hagen FS (1991): Cloning, expression and gene expression of a G-protein coupled glutamate receptor from rat brain. *Science* 252:1318-1321

Ishihara T, Nakamura S, Yoshito K, Takahashi T, Takahashi K, Nagata S (1991): Molecular cloning and expression of a cDNA encoding the secretin receptor. *EMBO J* 10:1635-1641

Jackson TR, Blair LAC, Marshall J, Goedert M, Hanley MR (1988): The mas oncogene encodes an angiotensin receptor. *Nature* 335:437-440

Jackson TR, Hanley MR (1989): Tumour promotor 12-O-tetradecanoylphorbol 13-acetate inhibits mas/angiotensin receptor-stimulated inositol phosphate production and intracellular calcium elevation in the 401L-C3 neuronal cell line. *FEBS Lett* 251:27-30

Jenness DD, Burkholder AC, Hartwell LN (1983): Binding of the α factor receptor. *Cell* 35:521–529

Juppner H, Abou-Samra A-B, Freeman M, Kong XF, Schipani E, Richards J, Kolakowski LF, Hock J, Potts JT, Kronenberg HM, Serge GV (1991): A G-protein linked receptor for parathyroid hormone and parathyroid hormone-related peptide. *Science* 254:1024–1026

Keinanen KP, Rajaniemi HJ (1986): Rat ovarian lutropin receptor is a transmembrane protein. *Biochemistry* 239:83–87

Kim IC, Ascoli M, Segaloff DL (1987): Immunoprecipitation of the lutropin/choriogonadotrophic receptor from biosynthetically labelled Leydig tumour cells. *J Biol Chem* 262:470–477

Konopka JB, Jenness DD, Hartwell LH (1988): The C-terminus of *S. cerevisiae* α-pheromone receptor mediates an adaptive responsive to pheromone. *Cell* 54:609–620

Lefkowitz J, Kobilka BK, Caron MG (1989): The new biology of drug receptors. *Biochem Pharmacol* 38:2941–2948

Libert F, Lefort A, Gerard C, Parmentier M, Perret J, Ludgate M, Dumont J, Vassart G (1989a): Cloning, sequence and expression of the human thyrotropin (TSH) receptor: Evidence for the binding of autoantibodies. *Biochem Biophys Res Commun* 165:150–1255

Libert F, Parmentier M, Lefort A, Dinsart C, Van Sande J, Maenhart C, Simons M-J, Dumont JE, Vassart G (1989b): Selective amplification and cloning of four new members of the G-protein-coupled receptor family. *Science* 244:569–572

Lin HY, Harris TL, Flannery MS, Aruffo A, Kaji EH, Gorn A, Kolakowski LF, Lodish HF, Goldring SR (1991): Expression cloning of an adenylate cyclase coupled calcitonin receptor. *Science* 254:1022–1024

Lolait SJ, O'Carroll AM, McBride OW, Konig M, Morel A, Brownstein MJ (1992): Cloning and characterisation of a vasopressin V2 receptor: Chromosomal localization of gene suggests link to hereditary diabetes insipidus. *Nature* In press.

Loosfelt H, Misrahi M, Atger M, Salesse R, Thi M, Jolivet A, Guiochon-Mantel A, Sar S, Jallal B, Garnier J, Milgrom E (1989): Cloning and sequencing of porcine LH-hCG receptor cDNA: Variants lacking transmembrane domain. *Science* 245:525–528

Macfarland KC, Sprengel R, Phillips HD, Kohler M, Rosemblit N, Nikolics K, Segaloff DL, Seeburg PH (1989): Lutropin-choriogonadotropin receptor: An unusual member of the G-protein-coupled receptor family. *Science* 245:494–499

Marsh L, Herskowitz I. (1988): STE2 protein of *Saccharomyces kluyveri* is a member of the rhodopsin/β-adrenergic receptor family and is responsible for recognition of the peptide ligand α factor. *Proc Natl Acad Sci USA* 85:3855–3859

Masu Y, Nakayama K, Yamaki H, Harada Y, Kuno M, Nakanishi S (1987): cDNA cloning of bovine substance-K receptor through oocyte expression system. *Nature* 329:836–838

McEachern AE, Shelton ER, Shakta S, Obernolte R, Bach C, Zuppan P, Fujisaki J, Aldrich RW, Jarnagin K (1991): Expression cloning of a rat B2 receptor. *Proc Natl Acad Sci USA* 88:7724–7728

Misrahi M, Loosefelt H, Atger M, Sar S, Guiochon-Mantel A, Milgrom E (1990): Cloning sequencing and expression of human TSH receptor. *Biochem Biophys Res Commun* 166:394–403

Morel A, O'Carroll AM, Brownstein MI, Lolait SJ, (1992): Molecular cloning and expression of a rat V1a arginine vasopressin receptor. *Nature* In press

Mulchahey JJ, Neil JD, Dion LD, Bost KL, Blalock JE (1986): Antibodies to the binding site of the receptor for lutenizing hormone releasing hormone (LHRH): Generation with a synthetic decapeptide encoded by an mRNA complementary to LHRH mRNA. *Proc Natl Acad Sci USA* 83:9714–9718

Nagayama Y, Kaufman KD, Seto P, Rapoport B (1989): Molecular cloning, sequence and functional expression of the cDNA for the human thyrotropin receptor. *Biochem Biophys Res Commun* 165:1184–1190

Nakafuku M, Itoh H, Nakamura S, Kaziro Y (1987): Occurrence in *Saccharomyces cerevisiae* of a gene homologous to the cDNA coding for the α-subunit of mammalian G-proteins. *Proc Natl Acad Sci USA* 84:2140–2144

Nakayama N, Miyajima A, Arai K (1985): Nucleotide sequences of STE 2 and STE 3, cell type-specific sterile genes from *Saccharomyces cerevisiae*. *EMBO J* 4:2643–2648

O'Dowd BJ, Hnatowich M, Regan JW, Leader WM, Caron MG, Lefkowitz RJ (1988): Site-directed mutagenesis of the cytoplasmic domains of the human β_2-adrenergic receptor: Localization of regions involved in G-protein receptor coupling. *J Biol Chem* 263:15985–15992

O'Dowd BJ, Hnatowich M, Caron MG, Lefkowitz RJ, Bouvier M (1989): Palmitoylation of the human β_2 adrenergic receptor. *J Biol Chem* 264:7564–7569

Parmentier M, Libert F, Maenhaut C, Lefort A, Gerard C, Perret J, Van Sande J, Dumont J, Vassart G (1989): Molecular cloning of the thyrotropin receptor. *Science* 246:1620–1622

Payan DG (1985): Receptor-mediated mitogenic effects of substance P on cultured muscle cells. *Biochem Biophys Res Commun* 130:104–109

Peach M (1977): Renin-angiotensin system: Biochemistry and mechanisms of action. *Physiol Rev* 57:313–370

Poyner DR, Hawkins PT, Benton H, Hanley MR (1990): Changes in inositol lipids and phosphates after stimulation of the MAS-transfected NG115-401L-C3 cell line by mitogenic and non-mitogenic stimuli. *Biochem J* 271:605–611

Rabin M, Birnbaum D, Young D, Birchmeier C, Wigler M, Ruddle FH (1987): Human Res1 and MAS1 oncogene located in regions of chromosome 6 associated with tumour-specific rearrangements. *Oncogene Res* 1:169–178

Saffroy M, Beaujounan JC, Torrens Y, Besseyre J, Bergström L, Glowinski J (1988): Localisation of tachykinin binding sites (NK$_1$, NK$_2$, NK$_3$ ligands) in the rat brain. *Peptides* 9:227–241

Sasai Y, Nakanishi S (1989): Molecular characterisation of rat substance K receptor and its mRNA. *Biochem Biophys Res Commun* 165:695–702

Sasaki K, Yamono Y, Bardhan S, Iwai N, Murray JJ, Hasegawa M, Matsuda Y, Inagami T (1991): Cloning and expression of a complementary DNA encoding a bovine adrenal angiotensin II type-1 receptor. *Nature* 351:230–233

Sakurai T, Yanagisawa M, Takuma Y, Miyazaki H, Kimura S, Goto K, Masaki T (1990): Cloning of a cDNA encoding a non-isopeptide-selective subtype of endothelin receptor. *Nature* 348:732–735

Schwyzer R (1987): Membrane-assisted molecular mechanism of neurokinin receptor subtype selection. *EMBO J* 6:2255–2259

Shigemoto R, Yokota Y, Nakanishi S (1990): Cloning and expression of a rat neuromedin K receptor cDNA. *J Biol Chem* 265:623–628

Sibley DR, Strasser RH, Benovic JL, Kiefer D, Lefkowitz RJ (1988): Phosphorylation/ dephosphorylation of the β-adrenergic receptor regulates its coupling to adenylate cyclase and subcellular distribution. *Proc Natl Acad Sci USA* 83:9408–9412

Spindel ER, Giladi E, Brehm P, Goodman RH, Segerson TP (1990): Cloning and functional characterization of a complementary DNA encoding the murine fibroblast bombesin/gastrin releasing peptide hormone receptor. *Mol Endocrinol* 4:1956–1963

Steedharan SP, Robichon A, Paterson KE, Goetel EJ (1991): Cloning and expression of the human vasoactive intestinal polypeptide receptor. *Proc Natl Acad Sci USA* 88:4986–4990

Stockton JM, Birdsall NJM, Burgen ASV, Hulme EC (1983): Modification of the binding properties of muscarinic receptors by gallamine. *Mol Pharmacol* 23:551–557

Strader CD, Sigal IS, Candelor MR, Rands E, Hill WS, Dixon RGF (1988): Conserved aspartic acid residues 79 and 113 of the β-adrenergic receptors have different roles in receptor function. *J Biol Chem* 263:10267–10271

Straub RE, Frech GC, Joho RH, Gerschenghorn MC (1990): Expression cloning of a cDNA encoding the mouse pituitary thyrotropin-releasing hormone receptor. *Proc Natl Acad Sci USA* 87:9514–9518

Venter JC, Fraser CM, Kerlavage AR, Buck MA (1989): Molecular biology of adrenergic and muscarinic cholinergic receptors. *Biochem Pharmacol* 38: 1197–1208

Vinson CR, Conover S, Adler PN (1989): A *Drosophila* tissue polarity locus encodes a protein containing seven potential transmembrane domains. *Nature* 338: 263–264

Vu T-H, Hung DT, Wheaton VI, Coughlin SR (1991): Molecular cloning of a functional thrombin receptor reveals a novel proteolytic mechanism of receptor activation. *Cell* 64:1057–1068

Wada E, Way J, Shapira H, Kusano K, Lebacq VAM, Coy D, Jensen R, Battery J (1991): cDNA cloning, characterization, and brain specific expression of a neuromedin-B-preferring bombesin receptor. *Neuron* 6:421–430

Whiteway M, Hougan L, Dignard D, Thomas DY, Bell L, Saari GC, Grant FJ, O'Hara P, and Mackay VL (1989): The STE4 and STE18 genes of yeast encode potential β and γ subunits of the mating factor receptor coupled G-protein. *Cell* 56:467–477

Yamada Y, Post SR, Wang K, Tager HS, Bell GI, Seino S (1992): Cloning and functional characterization of a family of human and mouse somatostatin receptors expressed in brain, gastrointestinal tract and kidney. *Proc Natl Acad Sci USA* 89:251–255

Yokota Y, Sasi Y, Tanaka K, Fujiwara J, Tsuchida K, Shigemoto R, Kakizuka A, Ohkubo H, Nakanishi S (1989): Molecular characterisation of a functional cDNA for rat substance P receptor. *J Biol Chem* 264:17649–17652

Young D, Waitches G, Birchmeier C, Fasano G, Wigler M (1986): Isolation and characterization of a new cellular oncogene encoding a protein with multiple potential transmembrane domains. *Cell* 45:711–719

Young D, O'Neill K, Jessell TM, Wigler M (1988): Characterization of the rat *mas* oncogene and its high-level expression in the hippocampus and cerebral cortex of rat brain. *Proc Natl Acad Sci USA* 85:5339–5342

Yu L, Blumer KJ, Davidson N, Lester HA, Thorner J (1989): Functional expression of the yeast α-factor receptor in *Xenopus* oocytes. *J Biol Chem* 264:20847–20850

10

Signal Transducing G-Proteins: α Subunits

Yoshito Kaziro

Signal transducing GTP-binding proteins are classified largely into two groups, i.e., heterotrimeric GTP-binding proteins which are referred to as G-proteins, and low-molecular-weight monomeric GTP-binding proteins (LMG) including Ras, Rap, Rho, Ral, ARF, YPT, and Rab proteins. The basic mechanism of the reaction catalyzed by these proteins appears to be analogous to that proposed for translational elongation factors (Kaziro, 1978). The GTP bound form is an active conformation which turns on the transmission of signals, and the hydrolysis of bound GTP to GDP is required to shift the conformation to an inactive form, i.e., to shut off the signal transduction.

G-proteins are involved in a variety of transmembrane signaling systems as transducers (for reviews, see Gilman, 1987). Two G-proteins, Gs and Gi, are involved in hormonal stimulation and inhibition, respectively, of adenylyl cyclase, whereas Go (o = other G-protein), which is present predominantly in brain tissues, may be involved in regulation of Ca^{2+} channels. Two transducins, Gt1 and Gt2, which are present in retinal rods and cones, respectively, regulate cGMP phosphodiesterase activity and mediate visual signal transduction. There is evidence indicating the presence of additional G-proteins, which are involved in the activation of phospholipase C and phospholipase A_2, as well as the gating of K^+ and Ca^{2+} channels.

This chapter briefly reviews the structure of cDNAs for various G-protein α subunits (Gα's) from mammalian cells, the organization of human genes for Gα's, and the occurrence of other Gα genes in lower eukaryotes.

1. Isolation of cDNA Clones for G-Protein α Subunits from Mammalian Cells

Recently, much effort has been focused on the cloning of cDNAs coding for various G-protein α subunits. Table 10.1 lists the cDNA as well as genomic clones isolated from mammalian cells [for developments after 1989, see Kaziro et al. (1991)]. These studies have revealed that there are at least three subtypes of Giα, designated as Gi1α, Gi2α, and Gi3α, the structures of which are closely related but distinct (reviewed in Kaziro et al., 1988).

The presence of multiple Giα subtypes had been suggested from the molecular heterogeneity, immunological distinction, and functional differences of the pertussis toxin substrates in mammalian cells (reviewed in Itoh et al., 1988a,b). By comparing the deduced amino acid sequences of each Giα subtype with the partial amino acid sequences of purified proteins, it was found that the 41-kDa pertussis toxin substrate, mainly expressed in brain tissues, corresponds to Gi1α, whereas 40-kDa and 41-kDa species expressed in most tissues correspond to Gi2α and Gi3α, respectively.

Transducins (Gt) also have two subtypes, Gt1α and Gt2α, which are expressed in rods and cones, respectively (Lerea et al., 1986). On the other hand, four different Gsα cDNAs (Bray et al., 1986) are generated from a Gsα (Gs1α) gene by alternative splicing (Kozasa et al., 1988) (see below). More recently, Jones and Reed (1989) reported the occurrence of another subtype of Gsα coded by a distinct gene (Golfα or Gs2α) which is expressed specifically in olfactory cells.

We have recently isolated a new Gα clone (designated Gxα) which is apparently insensitive to pertussis toxin (Matsuoka et al., 1988). The Cys residue at the fourth position from the C-terminus, which is common to all pertussis toxin-sensitive Gα's, is replaced by Ile in Gxα. Human Gzα cDNA, isolated independently by Fong et al. (1988) from retina, may be the counterpart of Gxα. The molecular weights, number of amino acid residues, and other properties of Gα's are listed in Table 10.2, and the amino acid sequences of rat Gs1α, Gi2α, Gi3α, Gi1α, Goα, and Gxα deduced from the nucleotide sequences are shown in Figure 10.1 together with bovine Gt1α and Gt2α.

Figure 10.2 shows a schematic representation of the structure of *E. coli* EF-Tu, G-protein α subunits (Gα), yeast RAS2 protein, and mammalian Ha-ras p21 protein. A remarkable homology was found in three regions, designated as P, G′, and G sites of all GTP-binding proteins. Earlier biochemical studies indicated that the region around Cys[137] of EF-Tu (G site) is responsible for interaction with guanine nucleotide (Kaziro, 1978). It was later found, using x-ray analysis, that the four residues Asn-Lys-Cys-Asp are situated close to the guanine ring (Jurnak, 1985). On the other hand, it has been shown that the mutation of amino acid residue 12 of p21 from Gly to Val decreases GTPase activity and increases transforming activity. The sequence homologous to this region (P site) was found in all GTP-binding proteins. The consensus sequence, Gly-(Xaa)$_4$-Gly-Lys (GXXXXGK), was located close to phosphoryl residues of bound guanine nucleotide (Jurnak, 1985; LaCour et al., 1985; De Vos et al., 1988). G' site is highly conserved in all G-proteins but less remarkable in other GTP-binding proteins except for ARF proteins (Sewell and Kahn, 1988). Only the sequence Asp-Xaa-Xaa-Gly-Gln (DXXGQ) is conserved in most of GTP-binding proteins. The DXXGQ sequence is found in p21 at residues 57 to 61, the mutation sites which induce malignant transformations, and in *E. coli* EF-Tu at residues 80 to 84 (in this case DCPGH). This region of EF-Tu is essential for interaction with aminoacyl-tRNAs and ribosomes (Kaziro, 1978).

The deduced amino acid sequences of Gα's in the P, G', and G sites are shown in Figure 10.3A, B, and C, respectively. It must be noted that the predicted amino acid sequence of new Gxα was different from other Gα proteins at the P site. As shown in Figure 10.3A, three amino acid residues in the consensus GTP hydrolysis site of Gxα (Thr-Ser-Asn, at position 41–43) are different from the corresponding residues in all other known Gα proteins (Ala-Gly-Glu). It remains to be seen whether the kinetics of the Gxα-mediated signal transduction may be different from other systems due to the replacement of Gly (which corresponds to Gly^{12} of p21) by Ser.

2. Isolation of Human Gα Genes

We have screened human genomic libraries with the above rat cDNA clones and isolated human genes coding for Gsα, Gi1α, Gi2α, Gi3α, Goα and Gzα. So far, we have determined the gene organization and nucleotide sequences of total exons of Gsα, Gi2α, and Gi3α, and obtained the partial sequences (exons 1, 2, and 3) of Gi1α (Kozasa et al., 1988; Itoh et al., 1988b) (see Figure 10.4). The structure of human Gzα gene has been reported by Matsuoka et al. (1990). Recently, we have clarified the structure of human Goα gene (Tsukamoto, et al., 1991). The human Goα gene is a huge gene spanning at least 90 kb; however, the organization of their exons is completely identical to those of Giα subfamily.

3. Structure of the Human Gsα Gene and Generation of Four Gsα cDNAs by Alternative Splicing

The human Gsα gene isolated by Kozasa et al. (1988) contained 13 exons and 12 introns and span about 20 kb of genomic DNA (Figure 10.4A). It has been known that there are two species of Gsα protein with different molecular masses (45 and 52 kDa) (Northup et al., 1980). Recently, Bray et

TABLE 10.1. Molecular cloning of Gs, Gi, Go, Gx, and Gt α subunit genes and cDNAs from mammalian cells

DNA library	Classification and nomenclature									References
	Gs1α	Gs2α	Gi1α	Gi2α	Gi3α	Goα	Gzα	Gt1α	Gt2α	
Genomic library										
Human	Gsα									Kozasa et al. (1988)
Human			Gi1α	Gi2α	Gi3α					Itoh et al. (1988b)
Human				Gi2α						Weinstein et al. (1988)
Human						Goα				Lavu et al. (1988)
Human						Goα				Tsukamoto et al. (1991)
Human							Gxα			Matsuoka et al. (1988)
Mouse								Trα		Raport et al. (1989)
cDNA library										
Human brain	Gsα	αi1								Bray et al. (1986, 1987)
Human brain						Goα				Lavu et al. (1988)
Human T cells				αi2	αi3					Beals et al. (1987)
Human T cells					αi3					Kim et al. (1988)
Human monocytes (U-937)				Giα						Didsbury et al. (1987)
Human granulocytes (HL-60)					Gxα					Didsbury and Snyderman (1987)
Human liver	Gsα									Mattera et al. (1986)
Human liver					αi3					Suki et al. (1987)
Human liver					αi3					Codina et al. (1988)
Human retina							Gzα			Fong et al. (1988)

Bovine brain	Gsα									Harris et al. (1985)
Bovine adrenal gland	Gsα									Robishaw et al. (1986b)
Bovine cerebral cortex	Gsα		Giα							Nukada et al. (1986b)
Bovine pituitary gland			αi	αh						Michel et al. (1986)
Bovine retina						Goα				Van Meurs et al. (1987)
Bovine cerebellum						G39				Ovchinnikov et al. (1987)
Bovine retina						Goα				Price et al. (1990)
Bovine retina								Tα		Tanabe et al. (1985)
Bovine retina								Tα		Medynski et al. (1985)
Bovine retina								Gtα		Yatsunami and Khorana (1985)
Bovine retina									Tα	Lochrie et al. (1985)
Hamster insulinoma						αo1, αo2				Hsu et al. (1990)
Hamster lung fibroblasts (CCL39)	Gsα									Mercken et al. (1990)
Rat brain							Gxα			Matsuoka et al. (1988)
Rat glioma cells (C6)	Gsα			Gi2α	Gi3α	Goα-1				Itoh et al. (1986, 1988b)
Rat PC12 cells						Goα-2				Tsukamoto et al. (1991)
Rat olfactory epithelium	Gαs	Golf	Gαi1	Gαi2	Gαi3	Gαo				Jones and Reed (1987, 1989)
Mouse macrophages (PU-5)	αs			αi						Sullivan et al. (1986)
Mouse lymphoma cells (S49)	Gsα									Sullivan et al. (1987)
Mouse lymphoma cells (S49)	Gsα									Rall and Harris (1987)
Mouse brain						GoAα GoBα				Strathmann et al. (1990)

TABLE 10.2. Properties of G-protein α subunits[a,b]

Species	No. of amino acids	M_r	Size (kDa)[c]	Expression	Substrate for:	Function
Mammalian						
hGsα1	394	45,664	52			
hGsα2	395	45,769	52	All tissues	CTX	ACase (+)
hGsα3	379	44,189	45			Ca^{2+} channel (+)
hGsα4	380	44,294	45			
rGs2α	381	44,322	45	Olfactory neurons	CTX	ACase (+)
hGi1α	354	40,345	41	Brain	PTX	ACase (−)
hGi2α	355	40,479	40	All tissues	PTX	PLC (+)
hGi3α	354	40,522	41	All tissues	PTX	K$^+$ channel (+)
hGoα1	354	40,053	39	Brain	PTX	Ca^{2+} channel (−)
hGoα2	354	40,075	39			
bGt1α	350	39,971	39	Retina (rod)	CTX, PTX	Cyclic GMP PDEase (+)
bGt2α	354	40,143	40	Retina (cone)	CTX, PTX	
hGxα (Gzα)	355	40,920	41	Brain	−	
Drosophila (D. melanogaster)						
DGsαL	385	45,003	51	Brain	CTX	ACase (+)?
DGsαS	382	44,704	48		CTX	
DGiα (DGα1)	355	40,612	41	Embryo, pupae	(−)	?

DGoα1	354	40,430	40	Brain	PTX	?
DGoα2	354	40,491	40		PTX	?
Yeast						
S. cerevisiae						
GP1α *(GPA1)*	472	54,075	54	Haploid cells	?	? (Mating)
GP2α *(GPA2)*	449	50,516	50	also diploid cells	?	? (cAMP pathway)
S. pombe						
GP1α *(gpa1)*	407	46,254	46	Inducible by nitrogen starvation	?	? (Mating)
Dictyostelium *(D. discoideum)*						
Gα1	356	40,621	41	Vegetative cells	?	?
Gα2	357	41,323	41	Aggregated cells. Inducible by cAMP pulse	?	PLC (+)?
Plant *(A. thaliana)*						
GPα1	383	44,582	45	Leaves and roots	?	?

[a]Modified from Kaziro et al. (1991).

[b]Abbreviations: h, human; r, rat; b, bovine; CTX, cholera toxin; PTX; pertussis toxin; ACase, adenylyl cyclase; PLC, phospholipase C; PDEase, phosphodiesterase; (+), activation; (−), inhibition.

[c]Size refers to the position of migration on the SDS polyacrylamide gel.

```
hGs1a:   1 -  48:  MGCLGNS-KT-EDQRNEEKAQREANKKIEKQLQKDKQVYRATHRLLLLGA
rGs2a:   1 -  50:  MGCLGNSSKTAEDQGVDEKERREANKKIEKQLQKERLAYKATHRLLLLGA
hGi1a:   1 -  41:  MGCTLSA---------EDKAAVERSKMIDRNLREDGEKAAREVKLLLLGA
hGi2a:   1 -  41:  MGCTVSA---------EDKAAAERSKMIDKNLREDGEKAAREVKLLLLGA
hGi3a:   1 -  41:  MGCTLSA---------EDKAAVERSKMIDRNLREDGEKAAKEVKLLLLGA
hGoa1:   1 -  41:  MGCTLSA---------EERAALERSKAIEKNLKEDGISAAKDVKLLLLGA
hGxa:    1 -  41:  MGCRQSS---------EEKEAARRSREIDRHLRSESQRQRREIKLLLLGA
bGt1a:   1 -  37:  MGAGASA---------EEK----HSRELEKKLKEDAEKDARTVKLLLLGA
bGt2a:   1 -  41:  MGSGASA---------EDKELAKRSKELEKKLQEDADKEAKTVKLLLLGA

hGs1a:  49 -  98:  GESGKSTIVKQMRILHVNGFNGEGGEEDPQAARSNSDGEKATKVQDIKNN
rGs2a:  51 -  85:  GESGKSTIVKQMRILHVNGFNPE---------------EKKQKILDIRKN
hGi1a:  42 -  76:  GESGKSTIVKQMKIIHEAGYSEE---------------ECKQYKAVVYSN
hGi2a:  42 -  76:  GESGKSTIVKQMKIIHEDGYSEE---------------ECRQYRAVVYSN
hGi3a:  42 -  76:  GESGKSTIVKQMKIIHEDGYSED---------------ECKQYKVVVYSN
hGoa1:  42 -  76:  GESGKSTIVKQMKIIHEDGFSGE---------------DVKQYKPVVYSN
hGxa:   42 -  76:  SNSGKSTIVKQMKIIHSGGFNLE---------------ECKEYKPLIIYN
bGt1a:  38 -  72:  GESGKSTIVKQMKIIHQDGYSLE---------------ECLEFIAIIYGN
bGt2a:  42 -  76:  GESGKSTIVKQMKIIHQDGYSPE---------------ECLEYKAIIYGN

hGs1a:  99 - 148:  LKEAIETIVAAMSNLVPPVELANPENQFRVDYILSVMNVPDFDFPPEFYE
rGs2a:  86 - 135:  VKDAIVTIISRAMSTIIPPVPLANPENQFRSDYIKSIAPITDFEYSQEFFD
hGi1a:  77 - 125:  TIQSIIAIIRAMGRLKIDFGDSARADDARQLFVLAGAAEE-GFMTAELAG
hGi2a:  77 - 128:  TIQSIMAIIVKAMGNLQIDFADPSRADDARQLFALSCTAEEQGVLPDDLSG
hGi3a:  77 - 125:  TIQSIIAIIRAMGRLKIDFGEAARADDARQLFVLAGSAEE-GVMTPELAG
hGoa1:  77 - 128:  TIQSLAAIVRAMDTLGIEYGDKERKADAKMVCDVVSRMEDTEPFSAELLSG
hGxa:   77 - 128:  AIDSLTRIIRALAALRIDFHNPDRAYDAVQLFALTGPAESKGEITPELLG
bGt1a:  73 - 121:  TLQSILAIIRAMTTLNIQYGDSARQDDARKLMHMADTIEE-GTMPKEMSD
bGt2a:  77 - 125:  VLQSILAIIRAMPTLGIDYAEVSCVDNGRQLNNLADSIEE-GTMPPPELVE
```

```
hGs1α: 149 - 198: H A K A L W E D E G V R A C Y E R S N E Y Q L I D C A Q Y F L D K I D V I K Q A D Y V P S D Q D L L
rGs2α: 136 - 185: H V K K L W D D E G V K A C F E R S N E Y Q L I D C A Q Y F L E R I D R I S Q P N Y T P T D Q D L L
hGi1α: 126 - 175: V I K R L W K D S G V Q A C F N R S R E Y Q L N D S A A Y Y L N D L D R I A Q P N Y I P T Q Q D V L
hGi2α: 127 - 176: V I R R L W A D H G V Q A C F G R S R E Y Q L N D S A A Y Y L N D L E R I A Q S D Y I P T Q Q D V L
hGi3α: 126 - 175: V I K R L W R D G G V Q A C F S R S R E Y Q L N D S A A Y Y L N D L E R I S Q S N Y I P T Q Q D V L
hGoα1: 127 - 176: A M M R L W G D S G I Q E C F N R S R E Y Q L N D S A K Y Y L D S L D R I G A A D Y Q P T E Q D I L
 hGxα: 127 - 176: V M R R L W A D P G A Q A C F S R S S E Y H L E D N A A Y Y L N D L E R I A A A D Y I P T V E D I L
bGt1α: 122 - 171: I I Q R L W K D S G I Q A C F D R A S E Y Q L N D S A G Y Y L S D L E R L V T P G Y V P T E Q D V L
bGt2α: 126 - 175: V I R K L W K D G G V Q A C F D R A A E Y Q L N D S A S Y Y L N Q L D R I T A P D Y L P N E Q D V L

hGs1α: 199 - 248: R C R V L T S G I F E T K F Q V D K V N F H M F D V G G Q R D E R R K W I Q C F N D V T A I I F V V
rGs2α: 186 - 235: R C R V L T S G I F E T R F Q V D K V N F H M F D V G G Q R D E R R K W I Q C F N D V T A I I Y V A
hGi1α: 176 - 225: R T R V K T T G I V E T H F T F K D L H F K M F D V G G Q R S E R K K W I H C F E G V T A I I F C V
hGi2α: 177 - 226: R T R V K T T G I V E T H F T F K D L H F K M F D V G G Q R S E R K K W I H C F E G V T A I I F C V
hGi3α: 176 - 225: R T R V K T T G I V E T H F T F K D L Y F K M F D V G G Q R S E R K K W I H C F E G V T A I I F C V
hGoα1: 177 - 226: R T R V K T T G I V E T H F T F K N L H F R L F D V G G Q R S E R K K W I H C F E D V T A I I F C V
 hGxα: 177 - 226: R S R D M T T G I V E N K F T F K E L T F K M V D V G G Q R S E R K K W I H C F E G V T A I I F C V
bGt1α: 172 - 221: R S R V K T T G I I E T Q F S F K D L N F R M F D V G G Q R S E R K K W I H C F E G V T C I I F I A
bGt2α: 176 - 225: R S R V K T T G I I E T K F S V K D L N F R M F D V G G Q R S E R K K W I H C F E G V T C I I F C A

hGs1α: 249 - 298: A S S Y N M V I R E D N Q T N R L Q E A L N L F K S I W N N R W L R T I S V I L F L N K Q D L L A
rGs2α: 236 - 285: A C S Y N M V I R E D N N T N R L R E A L D L F K S I W N N R W L R T I S I I L F L N K Q D M L A
hGi1α: 226 - 275: A L S D Y D L V L A E D E E M N R M H E S M K L F D S I C N N K W F T D T S I I L F L N K K D L F
hGi2α: 227 - 276: A L S A Y D L V L A E D E E M N R M H E S M K L F D S I C N N K W F T D T S I I L F L N K K D L F
hGi3α: 226 - 275: A L S D Y D L V L A E D E E M N R M H E S M K L F D S I C N N K W F T E T S I I L F L N K K D L F
hGoα1: 227 - 276: A L S G Y D Q V L H E D E T T N R M H E S L M L F D S I C N N K F F I D T S I I L F L N K K D L F
hGoα2: 242 - 276:                 N R M H E S L K L F D S I C N N K W F T D T S I I L F L N K K D I F
 hGxα: 227 - 276: E L S A Y D L K L Y E D N Q T S R M A E S L R L F D S I C N N N W F I N T S L I L F L N K K D L A
bGt1α: 222 - 271: A L S A Y D M V L V E D D E V N R M H E S L H L F N S I C N H R Y F A T T S I V L F L N K K D V F
bGt2α: 226 - 275: A L S A Y D M V L V E D D E V N R M H E S L H L F N S I C N H K F F A A T S I V L F L N K K D L F
```

FIGURE 10.1. (pp. 240–242). See p. 242 for legend.

```
hGs1α: 299 - 348:  EKVLAGKSKIEDYFPEFARYTTPEDATPEPGEDPRVTRAKYFIRDEFLRI
rGs2α: 286 - 335:  EKVLAGKSKIEDYFPEYANYTVPEDATPDAGEDPKVTRAKFFIRDLFLRI
hGi1α: 276 - 310:  EKIK--KSPLTICYPEYAGSNTYEE-------------AAAYIQCQFEDL
hGi2α: 277 - 311:  EKIT--HSPLTICFPEYTGANKYDE-------------AASYIQSKFEDL
hGi3α: 276 - 310:  EKIK--RSPLTICYPEYTGSNTYEE-------------AAAYIQCQFEDL
hGoα1: 277 - 310:  EKIK--KSPLTICFPEYTGPNTYED-------------AAAYIQAQF-ES
hGoα2: 277 - 310:  EKIK--KSPLTICFPEYTGPSAFTE-------------AVAYIQAQY-ES
 hGxα: 277 - 311:  EKIR--RIPLTICFPEYKGQNTYEE-------------AAVYIQRQFEDL
bGt1α: 272 - 306:  EKIK--KAHLSICFPDYNGPNTYED-------------AGNYIKVQFLEL
bGt2α: 276 - 310:  EKIK--KVHLSICFPEYDGNNSYED-------------AGNYIKSQFLDL

hGs1α: 349 - 394:  STASGDGRHYCYPHFTCAVDTENIRRVFNDCRDIIQRMHLRQYELL
rGs2α: 336 - 381:  STATGDGKHYCYPHFTCAVDTENIRRVFNDCRDIIQRMHLKQYELL
hGi1α: 311 - 354:  NKRKDTKEIY--THFTCATDTKNVQFVFDAVTDVIIKNNLKDCGLF
hGi2α: 312 - 355:  NKRKDTKEIY--THFTCATDTKNVQFVFDAVTDVIIKNNLKDCGLF
hGi3α: 311 - 354:  NRRKDTKEIY--THFTCATDTKNVQFVFDAVTDVIIKNNLKECGLY
hGoα1: 311 - 354:  KNRSPNKEIY--CHMTCATDTNNIQVVFDAVTDIIIANNLRGCGLY
hGoα2: 311 - 354:  KNKSAHKEIY--SHVTCATDTNNIQFVFDAVTDVIIAKNLRGCGLY
 hGxα: 312 - 355:  NRNKETKEIY--SHFTCATDTSNIQFVFDAVTDVIIQNNLKYIGLC
bGt1α: 307 - 350:  NMRRDVKEIY--SHMTCATDTQNVKFVFDAVTDIIIKENLKDCGLF
bGt2α: 311 - 354:  NMRKDVKEIY--SHMTCATDTQNVKFVFDAVTDIIIKENLKDCGLF
```

FIGURE 10.1. Deduced amino acid sequences of human Gs1α, Gi1α, Gi2α, Gi3α, Goα1, Goα2, and Gxα, rat Gs2α, and bovine Gt1α and Gt2α. Sources for sequences are as follows: hGsiα, Kozasa et al. (1988); rGs2α, Jones and Reed (1989); hGi1α, Bray et al. (1987) and Itoh et al. (1988b); hGi2α, Beals et al. (1987), Didsbury et al. (1987), and Itoh et al. (1988b); hGi3α, Beals et al. (1987); Itoh et al. (1988b), and Kim et al. (1988); hGoα2, Tsukamoto et al. (1991); hGzα, Fong et al. (1988) and Matsuoka et al. (1990); hGt1α, Tanabe et al. (1985), Medynski et al. (1985), and Yatsunami and Khorana (1985); bGt2α, Lochrie et al. (1985).

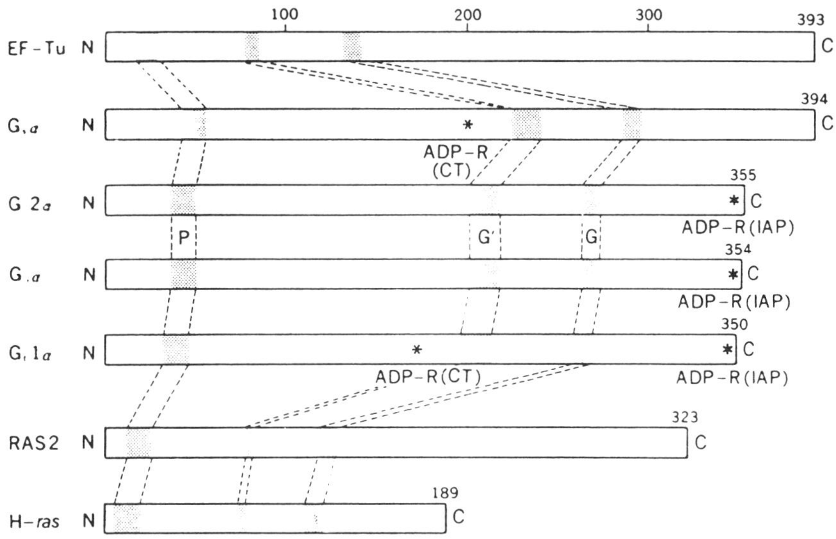

FIGURE 10.2. Schematic representation of structures of EF-Tu, Gsα, Gi2α, Goα, Gt1α, RAS2, and Ha-ras p21.

al. (1986) isolated four different Gsα cDNAs (Gsα1 to -4) from human brain and characterized the partial structure. Gsα1 and Gsα3 are identical except that Gsα3 lacks a single stretch of 45 nucleotides. Gsα2 and Gsα4 have three additional nucleotides (CAG) to Gsα1 and Gsα3 at 3' to the above 45 nucleotides. Robishaw et al. (1986a,b) also isolated two Gsα cDNAs from bovine adrenal that correspond to Gsα1 and Gsα4, and showed that these two cDNAs generated a 52- and a 45-kDa protein when expressed in COS-m6 cells. Mattera et al. (1986) also isolated two Gsα cDNAs from human liver that correspond to Gsα1 and Gsα4.

Comparison of the four types of human Gsα cDNAs reported by Bray et al. (1986) with the sequence of the human Gsα gene of Kozasa et al. (1988) suggests that four types of Gsα mRNAs may be generated from a single Gsα gene by alternative splicing, as shown in Figure 10.5. Gsα1 has a sequence identical to exons 2, 3, and 4, whereas Gsα3 lacks exon 3. Gsα2 and Gsα4 have three additional nucleotides (CAG) to Gsα1 and Gsα3, respectively, at the 5'-end of exon 4. This CAG sequence is found in the genomic sequence of the 3'-splice site of intron 3. Although the 5' adjacent nucleotides to the CAG are TG and do not match with the 3'-splice consensus sequence AG, this 3'-splice site may be used for the production of Gsα2 and Gsα4. One additional serine residue which is inserted upstream of exon 4 in Gsα2 and Gsα4 may be the potential site for phosphorylation, and the alternative use of these splice sites may confer Gsα proteins with differential regulatory properties.

(A)

```
 Gsα:    42-57:   R L L L L G A G E S G K S T I V
Gi1α:    35-50:   K L L L L G A G E S G K S T I V
Gi2α:    35-50:   K L L L L G A G E S G K S T I V
Gi3α:    35-50:   K L L L L G A G E S G K S T I V
 Goα:    35-50:   K L L L L G A G E S G K S T I V
 Gxα:    35-50:   K L L L L G T S N S G K S T I V
GP1α:    43-58:   K L L L L G A G E S G K S T V L
GP2α:  125-140:   K V L L L G A G E S G K S T V L
```

(B)

```
 Gsα:  223-240:   D V G G Q R D E R R K W I Q C F N D
Gi1α:  200-217:   D V G G Q R S E R K K W I H C F E G
Gi2α:  201-218:   D V G G Q R S E R K K W I H C F E G
Gi3α:  200-217:   D V G G Q R S E R K K W I H C F E G
 Goα:  201-218:   D V G G Q R S E R K K W I H C F E D
 Gxα:  201-218:   D V G G Q R S E R K K W I H C F E G
GP1α:  319-336:   D A G G Q R S E R K K W I H C F E G
GP2α:  296-313:   D V G G Q R S E R K K W I H C F D N
```

(C)

```
 Gsα:  285-296:   I S V I L F L N K Q D L
Gi1α:  262-273:   T S I I L F L N K K D L
Gi2α:  263-274:   T S I I L F L N K K D L
Gi3α:  262-273:   T S I I L F L N K K D L
 Goα:  263-274:   T S I I L F L N K K D L
 Gxα:  263-274:   T S L I L F L N K K D L
GP1α:  381-392:   T P F I L F L N K I D L
GP2α:  358-369:   T S V V L F L N K I D L
```

FIGURE 10.3. Conserved sequences of G-proteins. Sequences of P site (A), G' site (B) and G site (C) are shown.

4. Human Genes for Giα Subtypes

The coding region of the human Gi2α and Gi3α genes splits into 8 exons and 7 introns (Itoh et al., 1988b) (Figure 10.4B). There is an additional exon (exon 9) in the 3'-noncoding region of Gi2α and Gi3α, but this is not included in the figure. At present, for human Gi1α, we only have the sequences of exons 1 to 3. Remarkably, the positions of the splice junctions on the sequence of cDNA for Gi2α and Gi3α were completely identical (Figure 10.6), although the length of introns are different (Itoh et al., 1988b). The same splice sites are also conserved in the partial sequence (exons 1, 2, and 3) of the human Gi1α gene, as well as in the sequence of human Goα gene (Tsukamoto et al., 1991). From the Southern blot analysis, it appears that each of the three Giα genes occurs as a single copy per haploid human genome.

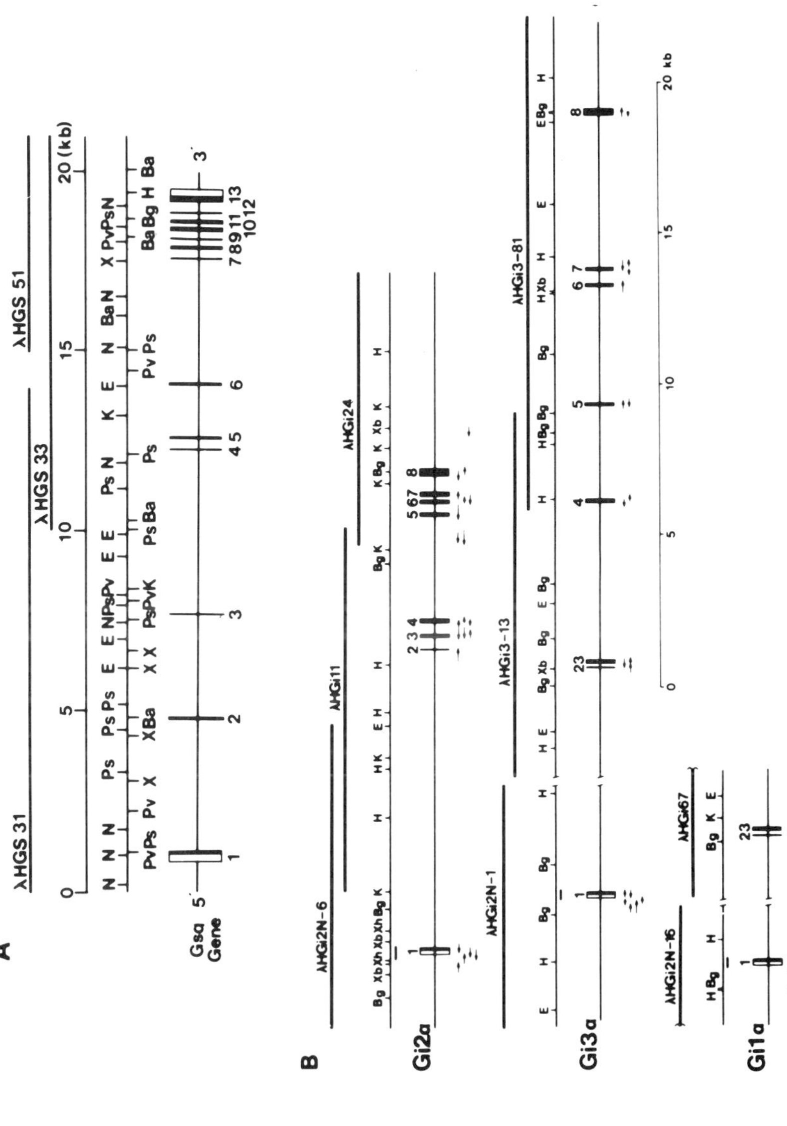

Figure 10.4. Organization of human genes for Gs1α (A) (from Kozasa et al., 1988) and three Giα subtypes (B) (from Itoh et al., 1988b).

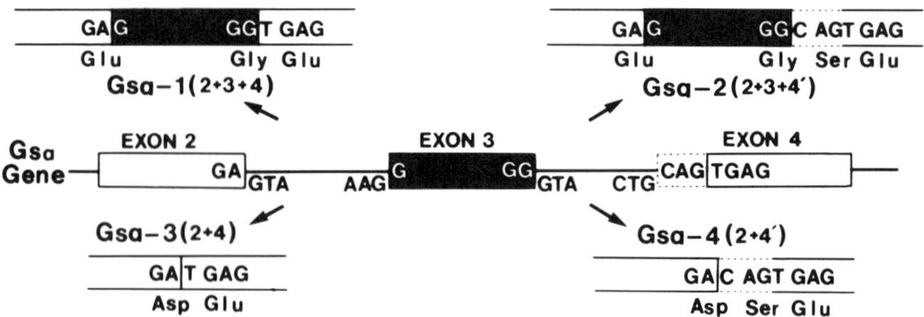

FIGURE 10.5. Generation of four different Gs1α mRNAs by alternative splicing. The Gsα gene is shown in the center. Gsα mRNAs are indicated by Gsα-1, -2, -3, and -4, which are now to be referred to as Gs1α-1, -2, -3, and -4. For details see Kozasa et al. (1988).

5. Organization of Human Gα Genes

The exon–intron organization of the Gsα, Gi2α, Gi3α, and Goα genes was compared with the predicted functional domain structure of proteins (Figure 10.6). The spatial orientation of each domain on the tertiary structure model will be discussed in a succeeding section (see later Figures 10.8 and 10.9). The NH$_2$-terminal domain encoded by exon 1 is hydrophilic and contains the site for limited tryptic digestions. Although this region

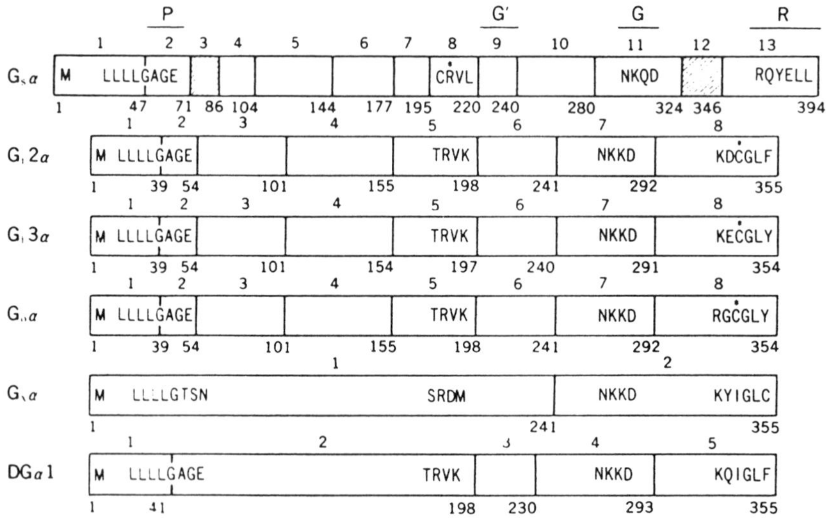

FIGURE 10.6. Organization of the exons of mammalian G-protein α subunits.

may be involved in interaction with $\beta\gamma$ subunits, its precise function has not yet been shown. Exon 2 encodes a short length region (24 and 14 amino acid residues, respectively, for Gsα and Giα's), which is the most conserved among all Gα proteins and responsible for GTP hydrolysis. Exon 3 of Gsα is the one which is unique to Gsα. This exon is lost by alternative splicing in some of the subtypes of Gsα. The domain encoded by exons 4–6 of Gsα and 3–4 of Giα is structurally divergent. Exon 8 of Gsα contains Arg[201], which is ADP-ribosylated in the presence of cholera toxin (Van Dop et al., 1984). ADP-ribosylation of Gsα by cholera toxin causes a decrease of affinity for $\beta\gamma$ subunits (Kahn and Gilman, 1984). Arg[179] in exon 5 of Gi2α corresponds to this arginine residue. The domain encoded by exons 9–11 of Gsα, and 6–7 of Giα is strongly conserved among all Gα proteins. This domain is involved in formation of a core structure for GTP binding together with that coded by exon 2. The sequence, Asn-Lys-Xaa-Asp, consensus to all guanine-nucleotide binding proteins, occurs in exon 11 of Gsα and exon 7 of Giα. The conserved Asp[223] in exon 9 of Gsα and Asp[201] in exon 6 of Gi2α may form a salt bridge to Mg^{2+}, which is linked to the β-phosphoryl group of GDP (Jurnak, 1985). The exchange of GDP to GTP may result in displacement of the surrounding region residues 230–238 in exon 9 of Gsα. A nonhydrolyzable GTP analog, but not GDP, prevents tryptic cleavage at Lys[210] in Goα or Lys[205] in Gt1α (Hurley et al., 1984).

Exons 12 of Gsα is unique to Gsα, and exon 13 of Gsα and exon 8 of Giα encode the COOH-terminus region. The domain may be involved in interaction with a receptor, since the Cys residue which is ADP-ribosylated by pertussis toxin is present in this region of Giα and also the structure of this region is heterogeneous. In Gxα, the Cys residue is replaced by Ile indicating that Gxα is refractory to modification by pertussis toxin. Gsα, which is also resistant to pertussis toxin, possessed Tyr instead of Cys in this position. It was shown that the replacement of Arg to Pro at -6 position of Gsα gives rise to a mutant protein which is uncoupled with β-adrenergic receptor in S49 cells (Sullivan et al., 1987).

Comparison of the exon organization of Giα subfamily and Goα with that of Gsα indicated that some of the exon junctions are conserved between Giα subfamily and Gsα. Thus, 3 out of 12 splice sites of the human Gsα gene are shared with the human Giα genes, and exon 1 and exons 7 and 8 of Gsα correspond to exon 1 and exon 5 of Giα, respectively. More recent work by Raport et al. (1989) revealed that the Gt1α and Gt2α genes possess the same organization as human Gi/Go genes.

Although not shown in Figure 10.6, human gene for Gxα consists of three exons, one for the 5'-noncoding region and two for coding regions (Matsuoka et al., 1990). Therefore, the gene organization of Gxα is quite different from other Gα's. The junction of exons 2 and 3 of Gxα is at the identical position as that of exons 6 and 7 of Giα and Goα.

6. Conservation of Primary Structure of Each Gα Among Mammalian Species

Table 10.3 shows that, in addition to the remarkable homologies of the overall structure, there is a strong conservation of the amino acid sequence in each subtype of G-protein α subunit. The amino acid sequence of Gsα is strongly conserved between human and rat; only 1 out of 394 amino acids are different. The sequence of Gi1α is completely identical between bovine and human. For Gi2α, Gi3α, Gxα, and Goα, over 98% identity of amino acid sequences is maintained among different mammalian species. The strong conservation of the amino acid sequence of each G-protein α subunit among distant mammalian species may reflect the presence of an evolutional pressure to maintain the specific physiological function of each G-protein gene product.

An evolutionary tree of G-protein α subunits can be drawn based on the homologies of the predicted amino acid sequences obtained from various mammalian sources (Figure 10.7). It is remarkable that the homologies among three Giα species are higher than that between rod (Gt1α) and cone (Gt2α) transducin α subunits.

7. Structural Model and Mutational Analysis

Figure 10.8 shows a structural model of Gα (Masters et al., 1988) constructed by adopting the X-ray crystal structure of *E. coli* EF-Tu at the GDP-binding domain (Jurnak, 1985; LaCour et al., 1985). As described in a previous section, the sequences conserved for all GTP-binding proteins, i.e., P site, G' site, and G site, are located in close proximity to the bound guanine nucleotide.

The NH$_2$-terminal region encoded by exon 1 is followed by exon 2 containing the P site which is responsible for GTP hydrolysis. Gly^{42} of Gi2α

TABLE 10.3. Conservation of G-protein α subunit sequences among different mammalian species

Species[a]	Amino acid sequences	Nucleotide sequences
rGsα vs. hGsα	393/394 (99.7%)	1128/1182 (95.4%)
bGi1α vs. hGi1α	354/354 (100 %)	998/1062 (94.0%)
rGi2α vs. hGi2α	350/355 (98.6%)	985/1065 (92.4%)
rGi3α vs. hGi3α	349/354 (98.6%)	981/1062 (92.4%)
rGoα vs. bGoα	348/354 (98.3%)	992/1062 (93.4%)
rGxα vs. hGxα	349/355 (98.3%)	977/1065 (91.7%)

[a]h, human; r, rat; b, bovine.

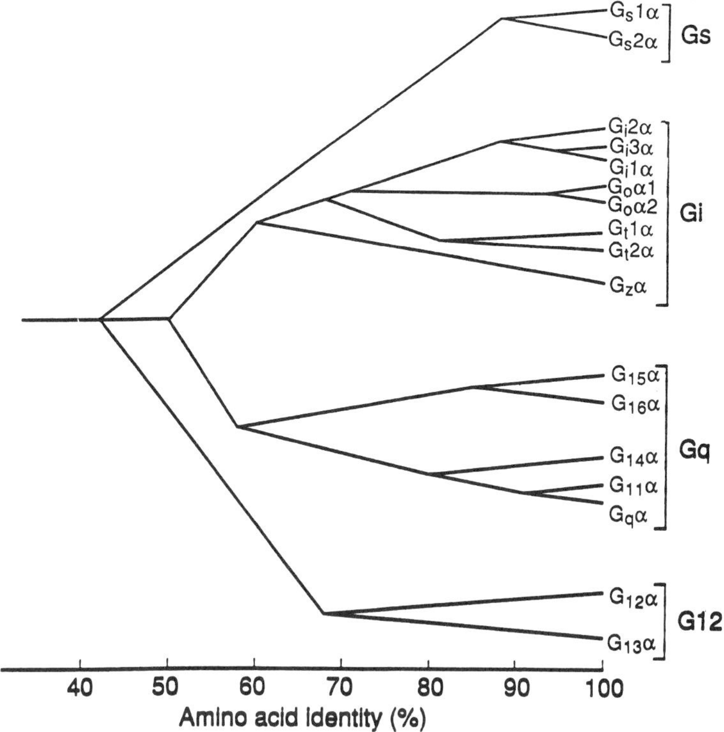

FIGURE 10.7. Relationships among mammalian Gα subunits. The α subunits are grouped by amino acid sequence identity into four distinct classes. Branch junctions approximate the values calculated for each pair of sequences. The splice variants of Gsα are not shown. Modified from Simon et al. Diversity of G proteins in signal transduction. *Science* 252:802–808 (1991).

in this site (shown in Figure 10.8 as Gly^{49} for Gsα) corresponds to Gly^{12} of Ha-ras p21. However, the replacement of Gly^{49} of Gsα by Val, using the site-directed mutagenesis, did not give the expected phenotype of persistently activated adenylyl cyclase in S49 cells (Bourne et al., 1988). However, a recent report by Woon et al. (1989a) showed the elevation of cAMP in the cells transfected with the Val^{49} mutant of Gsα cDNA. Lys^{46} of Gi2α (or Lys^{53} of Gsα) probably interacts electrostatically with the phosphoryl group of GDP.

The strongly conserved region of about 90 amino acids encoded by exons 6 and 7 of Giα/Goα forms a guanine nucleotide-binding pocket together with exon 2 containing the P site. Asp^{201} of Gi2α (shown as Asp^{225} in Figure 10.8) interacts with phosphoryl group through Mg^{2+}. The consensus sequence Asp-Xaa-Xaa-Gly-Gln of this region corresponds to the transforming mutation site of p21, that is, Ala^{59} and Gln^{61}. As will be described below, the activating mutation of Gsα was obtained by replacement of

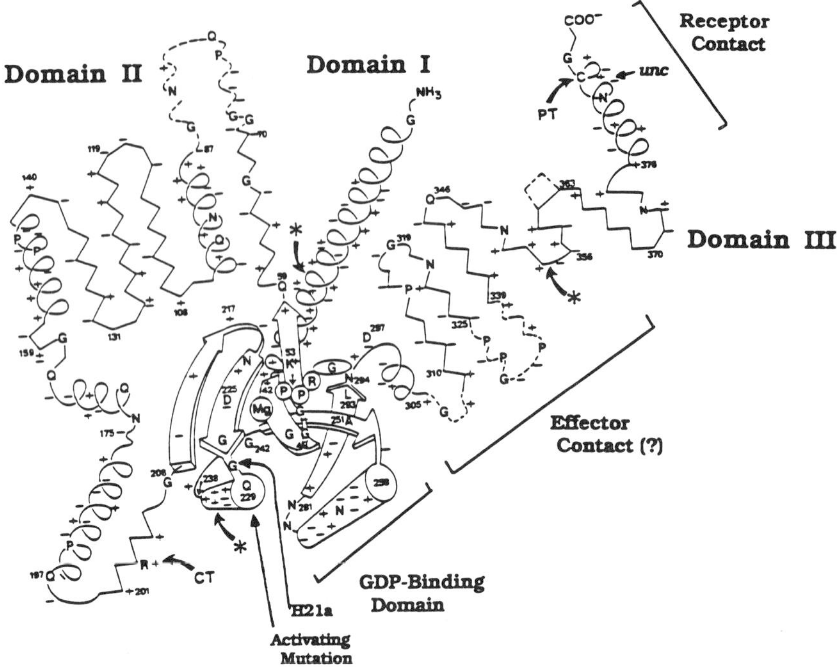

FIGURE 10.8. Structural model of Gα (from Masters et al., 1988). Numbers refer to positions of residues in α_{avg} (Masters et al., 1986). Asterisks with arrows are tryptic cleavage sites. Sites of point mutations and toxin modification sites are also indicated.

Gln[227] by Leu, and H21a mutant of Gsα was found to be the Gly[226] to Ala replacement. It must be noted that the corresponding region in *E. coli* EF-Tu is Asp[80]-Cys-Pro-Gly-His[84], where Cys[81] may be involved in interaction with aminoacyl-tRNA and ribosomes. A conformational change induced by the ligand change from GDP to GTP was detected in this region of EF-Tu (Kaziro, 1978). Presumably, β-strand-containing Asp[201] of Gi2α may be displaced by the γ-phosphoryl group of GTP. Lys[210] of Goα, which is located in an α-helix downstream of the DXXGQ sequence, may also be the region where the conformational alteration is taking place.

The receptor interaction site of G-protein α subunits is located in the C-terminal region. As described above, modification of Cys at − 4 position by ADP-ribosylation with pertussis toxin abolished the receptor coupling of the toxin-sensitive G-protein α subunits and the unc mutation of Gsα had a replacement of Lys by Pro at − 6 position (reviewed by Bourne et al., 1988). Furthermore, interaction of Gtα with rhodopsin was blocked by synthetic oligopeptides corresponding to the C-terminal sequences and also by monoclonal antibodies raised against this portion of Gtα (Deretic and Hamm, 1987; Hamm et al., 1988).

The effector binding region was previously assumed to be located in Domain II of Figure 10.8, the region encoded by exons 3 and 4 of Giα/Goα. This was based on the molecular heterogeneity of various G-protein α subunit species at this region and also on the location of the effector region of p21 around the NH_2-terminal region (residues 30–40 of p21, see Figure 10.9). However, the recent experiments in Bourne's lab (Bourne et al., 1988) using the chimeric constructs Gi2α/Gsα or Gtα/Gsα at the BamHl site (corresponding to amino acid residue 212 of Gi2α) revealed that these chimeras, when expressed in S49 cyc$^-$ cells, mediated stimulation of adenylyl cyclase in response to β-adrenoreceptor agonists. This implies that both receptor and effector binding sites of Gα may reside within the C-terminal 40% of the molecule (Domain III of Figure 10.8).

More recently, Holbrook and Kim (1989) proposed the molecular model of Gα based on the crystal structure of Ha-ras p21 (Figure 10.9). As can be seen, the schematic diagram of p21 and Gα are remarkably analogous. These authors defined three functional boxes, PO_4, G, and S, on the structure of p21 and Gα. PO_4- and G-Box correspond, respectively, to our P and G sites. S-Box, which they consider as the potential conformational switch region, corresponds to our G' site. Although this region in *E. coli* EF-Tu is one of the candidates for the potential conformational switch site induced by GTP/GDP, the conformation of the effector region of p21 (L2 in Figure 10.9) may also be modulated.

Therefore, we propose that the conformational change may take place in several regions. The potential conformation switch sites may include a region of Gα at residues 201–205 (in terms of Gi2α sequence) which is homologous to residues 57–61 of p21, a region surrounding Lys^{210} of which the sensitivity to trypsin was found to be modulated by the species of bound guanine nucleotide (L4 in Figure 10.9), a region around 167–177 where the sequence of Gα is homologous to the effector region (residues 31–41) of p21, and a certain sequence within a large region covering residues 212–315 in which the interaction with an effector was located by the use of two recombinant chimeras. Further studies including site-directed mutagenesis and construction of chimeric genes may shed more light on the structure–function relationship of Gα proteins (for more recent advances, see Update, pp. 258–263).

8. Oncogenic Mutation of Gα Proteins

As is well known, the oncogenic mutation of Ras p21 proteins often accompanies the decreased "intrinsic" GTPase activity. The mutations at Gly^{12} or Gln^{61}, which induce malignant transformations, result in the

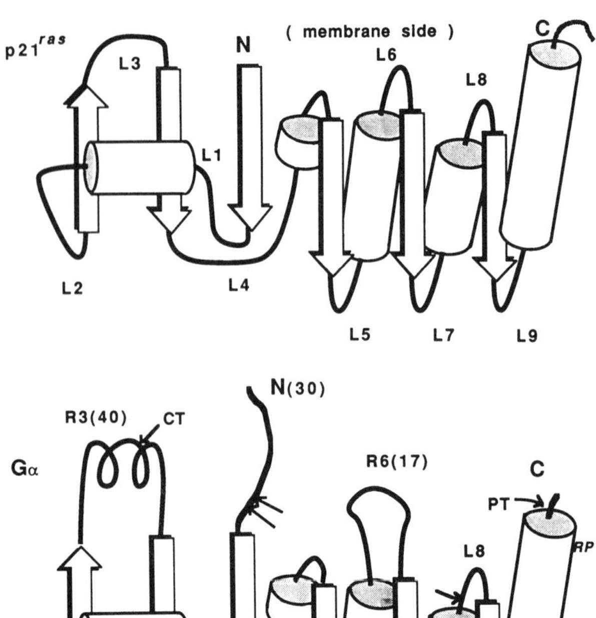

FIGURE 10.9. Schematic (topological) diagrams of Ha-ras p21 *(upper)* and the model of Gα *(lower)*. Modified from Holbrook and Kim (1989). The number of amino acid residues in the loops is indicated in parentheses, and the sites of tryptic cleavage by arrows. The point mutations GV, GA, QL, and RP indicate Gly → Val, Gly → Ala, Gln → Leu, and Arg → Pro substitutions, respectively. Hatched bars are junction sites for the chimeric proteins. The figure is the improved drawing [from correction in *Proc Natl Acad Sci USA* 86:7415 (1989)].

decrease in GTPase activity of the protein which can not be activated by the Ras GTPase-activating protein (GAP). The mutated protein exists in the cell in the GTP-bound form in contrast to the wild-type p21 which mostly exists in the GDP-bound form. Since the GTP-bound form is an active conformation which continues to transmit signals to the downstream effector molecule(s), the GTPase-defective mutations can be oncogenic.

Recently, Bourne and his co-workers made a remarkable observation that, in human pituitary tumors, such an activation mutation in fact had occurred in Gsα subunit (Landis et al., 1989). They found that a subset of growth hormone-secreting human pituitary tumors carries somatic muta-tions that inhibit GTPase activity of Gsα. The sequence analysis indicated that mutations in tumors replaced Arg^{201} with either Cys or His (R201C or R201H). In another tumor, Gln^{227} was replaced with Arg (mutation Q227R). In Gsα, Arg^{201} is the modification site of cholera toxin-catalyzed ADP-ribosylation, and GTPase activity is severely impaired by the modi-fication. Likewise, Gln^{227} is the counterpart of Gln^{61} of p21, the mutation of which also leads to the decrease in GTPase activity. These findings indicate that permanent activation of hormone or neurotransmitter signal transduction systems can be oncogenic, and G-proteins, as other compo-nents of the signaling pathways, can potentially be "proto-oncogenes." In our laboratory, K. Ahmed (unpublished observation) have shown that transfection of Gsα cDNA having a mutation of Gly^{49} to Val^{49} into PC12 cells can induce the differentiation to neuronal cells.

These observations are also in line with the previous reports that serotonin 1c receptor (Julius et al., 1989) and angiotensin/mas oncogene receptor (Jackson et al., 1988) have mitogenic potentials. It would be interesting to see whether mutations of other Gα, such as Giα and Goα, can also lead to the malignancy in certain cell types.

9. G-Proteins from *Saccharomyces cerevisiae*

A family of GTP-binding protein, the ras family, is widely distributed among eukaryotes (see Barbacid, 1987, for a review) including the yeasts *Saccharomyces cerevisiae* (DeFeo-Jones et al., 1983; Powers et al., 1984) and *Schizosaccharomyces pombe* (Fukui and Kaziro, 1985). It has been suggested that the RAS2 gene in *S. cerevisiae* is involved in the activation of adenylyl cyclase (Toda et al., 1985; Broek et al., 1985), and mimics the role of mammalian Gs.

However, in view of the strong conservation of the amino acid sequences of each G-protein species among different organisms (see Table 10.3), we speculated that G-protein may occur also in yeasts. We have searched for a G-protein homologous gene in *S. cerevisiae* and isolated two: GPA1 (Nakafuku et al., 1987) and GPA2 (Nakafuku et al., 1988) from *S. cerevisiae,* which are homologous with cDNAs for mammalian G-protein α subunits.

GPA1 and GPA2 code for the sequences of 472 and 449 amino acid residues, respectively, with calculated M_r's with 54,075 and 50,516. When aligned with the α subunit of mammalian G-proteins to obtain maximal homology, GP1α (GPA1-encoded protein) and GP2α (GPA2-encoded protein) were found to contain the stretches of 110 and 83 additional amino acid residues, respectively, near the NH$_2$-terminus (Figures 10.10 and 10.11).

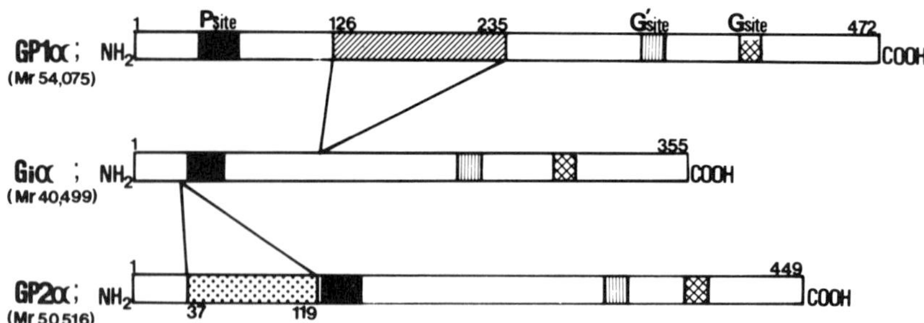

FIGURE 10.10. Schematic representation of the structure of yeast GP1α, yeast GP2α, and mammalian Gi2α.

10. Comparison of the Amino Acid Sequences of S. cerevisiae GP1α and GP2α with Those of Rat Brain Giα and Goα

The deduced amino acid sequence of *S. cerevisiae* GP1α and GP2α is highly homologous with those of rat brain Giα and Gsα (Fig. 10.11). The homology is most remarkable in the region of GTP hydrolysis (P site, see Figure 10.3A) (amino acid residues 43–58 of GP1α, 125–140 of GP2α, and 35–50 of Gi2α). As shown in Figure 10.3C, the region responsible for GTP binding (G site) (amino acid residues 381–392 of GP1α, 358–369 of GP2α, and 263–274 of Gi2α) was also highly homologous. Another region of homology (G' site, see Figure 10.3B) was found in amino acid residues 319–336 of GP1α, 296–313 of GP2α, and 201–218 of Gi2α where a sequence of 14 contiguous amino acids was completely identical between *S. cerevisiae* GP1α, GP2α, and rat Gi2α.

The overall homology in nucleotide and amino acid sequences of *S. cerevisiae* GP1α and GP2α and rat Gi2α and Gsα is remarkable. Disregarding the unique sequences present in GP1α (residues 126–235) and GP2α (residues 37–119), the proteins are 60% homologous if conservative amino acid substitutions are considered to be homologous. The homology is smaller than that between rat Gi2α and Goα (85%) but is comparable to that between rat Gi2α and Gsα (60%).

As is described elsewhere in detail (Miyajima et al., 1987; Dietzel and Kurjan, 1987), GPA1 is a haploid-specific gene and involved in the mating factor signal transduction. On the other hands, GPA2 is expressed both in haploid and diploid cells, and may be involved in the regulation of cAMP levels in *S. cerevisiae* (Nakafuku et al., 1988).

Recently, Whiteway et al. (1989) made a remarkable observation that the STE4 and STE18 genes of *S. cerevisiae* encode β and γ subunits, respectively, of GP1. Since both genes are required for the mating of *S. cerevisiae* cells, and expressed only in haploid cells, they are most likely the compo-

```
scGP1α:  1 - 40:  MG--CTVSTQTIGDESDPFLQNKRANDVIEQSLQLEKQR              DKN
scGP2α:  1 - 122: MG-LCASSEKNGSTPDTQTASAGSDN--VGKAKVPPKQE-(83aa)-      NDK
spGP1α:  1 - 73:  MG--CMSSKYADTSGGEVIQKKLSDTQTSNSSTTGSQNA-(33aa)-      GGN
ddGα1:   1 - 35:  MGNICGKPELGSP---EEIKANQH---INSLLKQARSK               LEG
ddGα2:   1 - 30:  MG-ICASSMEG------EKTNTDINLS-IE---KERKKK              -HN
atGPα1:  1 - 37:  MGLLCSRSRHHT-----EDTDENTQAAEIERRIEQEAKA              EKH
rGi2α:   1 - 32:  MG--CTVSA--------EDKAAAERSKMIDKNLREDGEK              AAR
rGsα:    1 - 39:  MG--CLGNSKTEDQRN-EEKAQREANKKIEKQLQKDKQV              YRA

scGP1α: 41 - 75:  EIKLLLLGAGESGKSTVLKQLKLLHQGGFSHQ-----------------ERL
scGP2α:123 - 157: ELKVLLLGAGESGKSTVLQQLKILHQNGFSEQ-----------------EIK
spGP1α: 74 - 108: DIKVLLLGAGDSGKTTIMKQMRLLYSPGFSQV-----------------VRK
ddGα1:  36 - 70:  EIKLLLLGAGESGKSTIAKQMKIIHLNGFNDE-----------------EKS
ddGα2:  31 - 65:  EVKLLLLGAGESGKSTISKQMKIIHQSGYSNE-----------------ERK
atGPα1: 38 - 72:  IRKLLLLGAGESGKSTIFKQIKLLFQTGFDEG-----------------ELK
rGi2α:  33 - 67:  EVKLLLLGAGESGKSTIVKQMKIIHEDGYSEE-----------------ECR
rGsα:   40 - 89:  THRLLLLGAGESGKSTIVKQMRILHVNGFNGEGGEEDPQAARSNSDGEKA

scGP1α: 76 - 125: QYAQVIWADAIQSMKILIIQARKLGIQLDCDDPINNKDLFACKRILLKAK
scGP2α:158 - 207: EYIPLIYQNLLEIGRNLIQARTRFNVNLEPECELTQQDLSRTMSYEMPNN
spGP1α:109 - 158: QYRVMIFENIISSLCLLEAMDNSNVSLLPENEKYRAVILRKHTSQPNEP
ddGα1:  71 - 119: SYKTIIYNNTVGSMRVLVNAAEELKIGISENNKEAASRISNDLGDHFNG-
ddGα2:  66 - 115: EFKPIITRNILDNMRVLLDGMGRLGMTIDPSNSDAAVMIKELTSLQASIV
atGPα1: 73 - 122: SYVPVIHANVYQTIKLLHDGTKEFAQNETDSAKYMLSSESIAIGEKLSEI
rGi2α:  68 - 117: QYRAVVYSNTIQSIMAIVKAMGNLQIDFADPQRADDARQLFALSCAAEEQ
rGsα:   90 - 139: TKVQDIKNNLKEAIETIVAAMSNLVPPVELANPENQFRVDYILSVMNVPN

scGP1α:126 - 268: ALDYINA-(103aa)-LIHEDIAKAIKQLWNNDKG-IKQCFARSNEF-QLE
scGP2α:208 - 244: YTG-----      QFPEDIAGVISTLWALPSTQDLVNGPNASKF-YLM
spGP1α:159 - 189: -------      -FSPEIYEAVHALTLDTK--LRTVQSCGTNL-SLL
ddGα1:120 - 151: ------       VLTAELAQDIKALWADPG--IQNTFQRSSEF-QLN
ddGα2:116 - 153: TDCWG--      ELNEDQGKKIKALWTDPG--VKQAMRRANEFSTLP
atGPα1:123 - 161: GGRLDYP      RLTKDIAEGIETLWKDPA--IQETCARGNEL-QVP
rGi2α:118 - 150: G------      MLPEDLSGVIRRLWADHG--VQACFGRSREY-QLN
rGsα:140 - 172: F------      DFPPEFYEHAKALWEDEG--VRACYERSNEY-QLI
```

FIGURE 10.11. (pp. 255–257). See p. 257 for legend.

scGP1α:269 - 314: G S A A Y Y F D N I E K F A S P N Y V C T D E D I L K G R I K T T G I T E T E F N I G S S K - - - -
scGP2α:245 - 291: D S T P Y F M E N F T R I T S P N Y R P T Q Q D I L R S R Q M T S G I F D T V I D M G S D I K - - -
spGP1α:190 - 235: D N F Y Y Y Q D H I D R I F D P Q Y I P S D Q D I L H C R I K T T G I S E E T F L L N R H H - - -
ddGα1:152 - 197:   D S A A Y Y F D S I D R I S Q P L Y L P S E N D V L R S R T K T T G I I E T V F E I Q N S T - - - -
ddGα2:154 - 199:   D S A P Y F F D S I D R M T S P V Y I P T D Q D I L H T R V M T R G V H E T N F E I G K I K - - - -
atGPα1:162 - 211: D C T K Y L M E N L K R L S D I N Y I P T K E D V L Y A R V R T T G V V E I Q F S P V G E N K K S G
rGi2α:151 - 196:   D S A A Y Y L N D L E R I A Q S D Y I P T Q Q D V L R T R V K T T G I V E T H F T F K D L H - - - -
rGsα:173 - 218:   D C A Q Y F L D K I D V I K Q A D Y V P S D Q D L L R C R V L T S G I F E T K F Q V D K V N - - - -

scGP1α:315 - 362: - - F K V L D A G G Q R S E R K K W I H C F E G I T A V L F V L A M S E Y D Q M L F E D E R V N R M
scGP2α:292 - 339: - - M H I Y D V G G Q R S E R K K W I H C F D N V T L V I F C V S L S E Y D Q T L M E D K N Q N R F
spGP1α:236 - 283: - - Y R F F D V G G Q R S E R R K W I H C F E N V T A L L F L V S L A G Y D Q C L V E D N S G N Q M
ddGα1:198 - 245: - - F R M V D V G G Q R S E R K K W M H C F Q E V T A V I F C V A L S E Y D L K L Y E D D T T N R M
ddGα2:200 - 247: - - F R L V D V G G Q R S E R K K W L S C F D D V T A V V F C V A L S E Y D L L L Y E D N S T N R M
atGPα1:212 - 261: E V Y R L F D V G G Q R N E R R K W I H L F E G V T A V I F C A A I S E Y D Q S V F E D E Q K N R M
rGi2α:197 - 244: - - F K M F D V G G Q R S E R K K W I H C F E G V T A I I F C V A L S A Y D L V L A E D E E M N R M
rGsα:219 - 266: - - F H M F D V G G Q R D E R R K W I Q C F N D V T A I I F V V A S S Y N M V I R E D N Q T N R L

scGP1α:363 - 408: H E S I M L F D T L L N S K W F K D T P F I L F L N K I D L F E E K V K - - S M P I R K Y - - F P D
scGP2α:340 - 385: Q E S L V L F D N I V N S R W F A R T S V V L F L N K I D L F A E K L R - - K V P M E N Y - - F P D
spGP1α:284 - 329: Q E A L L L W D S I C N S S W F S E S A M I L F L N K L D L F K R K G S - - H F P I Q K H - - F P D
ddGα1:246 - 291: Q E S L K L F K E I C N T K W F A N T A M I L F L N K R D I F S E K I T - - K T P I T V C - - F K E
ddGα2:248 - 292: L E S L R V F S D V C N S - W F V N T P I I L F L N K S D L F R E K I K - - H V D L S E T - - F P E
atGPα1:262 - 309: M E T K E L F D W V L K Q P C F E K T S F M L F L N K F D I F E K K V L - - D V P L N V C E W F R D
rGi2α:245 - 290: H E S M K L F D S I C N N K W F T D T S I I L F L N K K D L F E E K I T - - Q S P L T I C - - F P E
rGsα:267 - 314: Q E A L N L F K S I W N N R W L R T I S V I L F L N K Q D L L A E K V L A G K S K I E D Y - - F P E

256

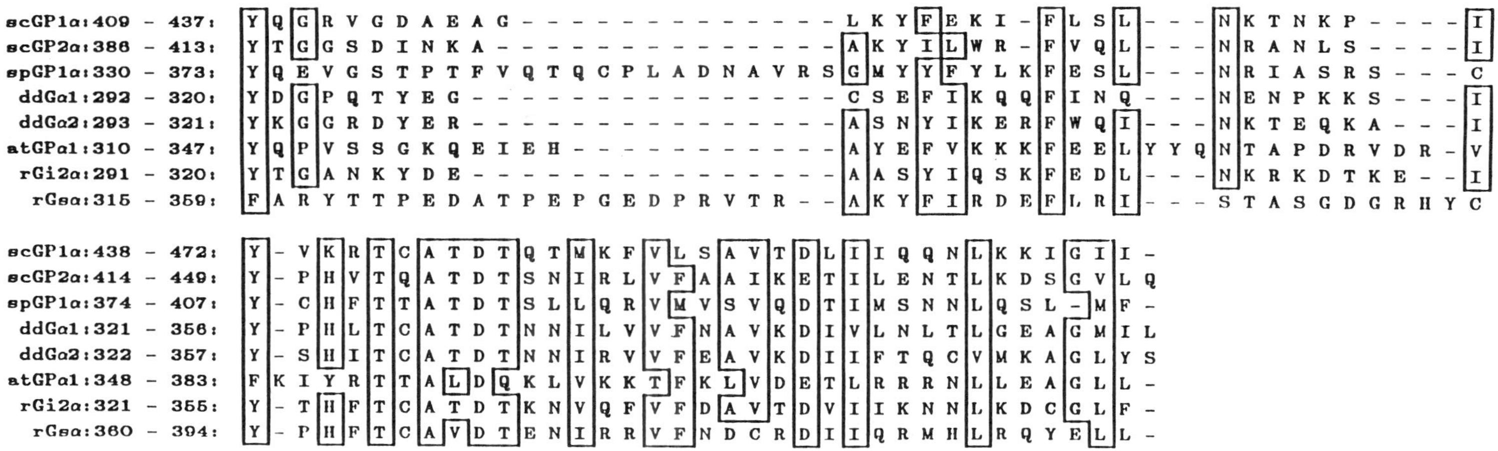

scGP1α:409 - 437: Y Q G R V G D A E A G - - - - - - - - - - - - - - - - L K Y F E K I - F L S L - - - N K T N K P - - - - - I
scGP2α:386 - 413: Y T G G S D I N K A - - - - - - - - - - - - - - - - A K Y I L W R - F V Q L - - - N R A N L S - - - - - I
spGP1α:330 - 373: Y Q E V G S T P T F V Q T Q C P L A D N A V R S G M Y Y F Y L K F E S L - - - N R I A S R S - - - - C
ddGα1:292 - 320: Y D G P Q T Y E G - - - - - - - - - - - - - - - - - C S E F I K Q Q F I N Q - - - N E N P K K S - - - - I
ddGα2:293 - 321: Y K G G R D Y E R - - - - - - - - - - - - - - - - - A S N Y I K E R F W Q I - - - N K T E Q K A - - - - I
atGPα1:310 - 347: Y Q P V S S G K Q E I E H - - - - - - - - - - - - - A Y E F V K K K F E E L Y Y Q N T A P D R V D R - V
rGi2α:291 - 320: Y T G A N K Y D E - - - - - - - - - - - - - - - - - A A S Y I Q S K F E D L - - - N K R K D T K E - - I
rGsα:315 - 359: F A R Y T T P E D A T P E P G E D P R V T R - - A K Y F I R D E F L R I - - - S T A S G D G R H Y C

scGP1α:438 - 472: Y - V K R T C A T D T Q T M K F V L S A V T D L I I Q Q N L K K I G I I -
scGP2α:414 - 449: Y - P H V T Q A T D T S N I R L V F A A I K E T I L E N T L K D S G V L Q
spGP1α:374 - 407: Y - C H F T T A T D T S L L Q R V M V S V Q D T I M S N N L Q S L - M F -
ddGα1:321 - 356: Y - P H L T C A T D T N N I L V V F N A V K D I V L N L T L G E A G M I L
ddGα2:322 - 357: Y - S H I T C A T D T N N I R V V F E A V K D I I F T Q C V M K A G L Y S
atGPα1:348 - 383: F K I Y R T T A L D Q K L V K K T F K L V D E T L R R R N L L E A G L L -
rGi2α:321 - 355: Y - T H F T C A T D T K N V Q F V F D A V T D V I I K N N L K D C G L F -
rGsα:360 - 394: Y - P H F T C A V D T E N I R R V F N D C R D I I Q R M H L R Q Y E L L -

FIGURE 10.11. Comparison of the predicted amino acid sequences of G-protein α subunits from different organisms. Identical or conservative amino acid residues are enclosed with solid lines. Aligned sequences are scGP1α and scGP2α, *Saccharomyces cerevisiae* GP1α and GP2α (Nakafuku et al., 1987, 1988); spGP1α, *Schizosaccharomyces pombe* GP1α (Obara et al., 1991); ddGα1 and ddGα2, *Dictyostelium discoideum* Gα-1 and Gα-2 (Pupillo et al., 1989); atGPα-1, *Arbidopsis thaliana* GPα-1 (Ma et al., 1990); rGi2α and rGsα, rat Gi2α and rat Gs1α (Itoh et al., 1986).

nents of the mating factor signaling system. The nucleotide sequence analysis revealed that STE4 codes for a protein of 423 amino acids highly homologous with mammalian $G\beta1$ and $G\beta2$. The deduced amino acid sequence of STE18 also exhibits a homology to bovine $Gt\gamma$, except that STE18 contains a poly Gln stretch near the NH_2-terminal region. From the genetic evidence, it appears that the products of STE4 and STE18 are involved, either directly or indirectly, in activation of the signals for the growth arrest at the late G_1 phase.

The fact that the expression of STE4 and STE18 is limited in haploid cells immediately predicts that *S. cerevisiae* cells may possess another set of genes coding for the potential β and γ subunits of G protein GP2 which is present in both haploid and diploid cells. This also implies that STE4 and STE18 are probably involved specifically in the mating factor signaling system, since they are not replaced by the putative β and γ subunits expressed both in the diploid and haploid cells.

Finally, it is expected that demonstration of the occurrence of G-proteins in *S. cerevisiae* may open the way for a detailed genetic analysis of the function of G-proteins in eukaryotic cells.

11. $G\alpha$'s in Other Organisms

The occurrence of G-protein α subunits has been found in *Drosophila melanogaster* and *Dictyostelium*. Provost et al. (1988) isolated a $G\alpha$ gene from a *D. melanogaster* genomic library using bovine $Gt1\alpha$ and $Gt2\alpha$ cDNAs as a probe. The gene, designated $DG\alpha1$, encodes a protein with an amino acid sequence 78% identical to bovine $Gi1\alpha$. However, $DG\alpha1$-encoded protein lacks a Cys residue at -4 position from the C-terminal, and therefore is not expected to be a pertussis toxin substrate. Northern blots of total cellular RNA revealed a major 2.3-kb transcript and a less abundant 1.7-kb transcript. These transcripts are most abundant in RNA from embryos and pupae.

In slime molds, the cDNAs for two $G\alpha$s, $G\alpha1$ and $G\alpha2$, were cloned and sequenced (Pupillo et al., 1989). The deduced amino acid sequences are 45% identical to each other and to $G\alpha$ proteins from mammals and yeast. $G\alpha1$ and $G\alpha2$ are differentially expressed during development. $G\alpha1$ is expressed in vegetative cells through aggregate stages while $G\alpha2$ is inducible by cAMP pulses and preferentially expressed in aggregation (Kumagai et al., 1989). $G\alpha2$ appears to associate with the cAMP receptor and mediates the signals for chemotaxis and gene regulation, while $G\alpha1$ may function in both the cell cycle and development.

12. Update

This review was completed at the end of 1989. Since then, considerable progress has been made in this area. To update this article, I have replaced

Table 10.1, Table 10.2, Figure 10.7 and Figure 10.11 with new versions and added a new table (Table 10.4). In addition, the readers are advised to refer several reviews which have appeared recently (Kaziro et al., 1991; Freissmuth et al., 1989; Ross, 1989; Bourne et al., 1990, 1991; Simon et al., 1991). Several monographs have also been published recently, which include those edited by Iyenger and Birnbaumer (1990), Houslay and Milligan (1990), and Moss and Vaughn (1990). The following is a brief summary of the recent progress.

12.1. Isolation of New cDNA Clones

Simon and his collaborators (Strathmann et al., 1989; Strathmann and Simon, 1990) have isolated cDNAs for several additional G-protein α subunits by using the polymerase chain reaction. The new $G\alpha$ cDNAs are classified into two groups by their sequence homology: the $Gq\alpha$ subgroup consisting of $Gq\alpha$, $G11\alpha$, $G14\alpha$, $G15\alpha$, and $G16\alpha$; and the $G12\alpha$ subgroup consisting of $G12\alpha$ and $G13\alpha$. The cDNAs for $G15\alpha$ and $G16\alpha$ were obtained from mouse and human, respectively, and they are probably coding for the counterparts of the same G-protein α subunits.

A remarkable finding by Exton and his group (Taylor et al., 1990, 1991; Taylor and Exton, 1991) is that $Gq\alpha$ stimulates the activity of phospholipase C β-subtype. The putative G_{PLC} (G-protein α-subunit involved in phospholipase C activation) was finally identified as $Gq\alpha$. It is likely that other members of the Gq subgroup might also be involved in phospholipase C β activation. However, since the members of $Gq\alpha$ subgroup have no cysteine at the -4 position from the C-terminal, Gq family represents only the pertussis-toxin-resistant phospholipase C activators. The search for another G_{PLC} species which mediate the pertussis-toxin-sensitive phospholipase C activation is still underway.

12.2. Alternative Splicing of the $Go\alpha$ Transcript

Recently two types of $Go\alpha$ cDNAs (designated here as $Go\alpha1$ and $Go\alpha2$; in other references as $G_oA\alpha$ and $G_oB\alpha$) were isolated from rat (Tsukamoto et al., 1991), mouse (Strathmann et al., 1990), and hamster (Hsu et al., 1990). Amino acid sequences of $Go\alpha1$ and $Go\alpha2$ are identical except for the C-terminal region. The human $Go\alpha$ gene spans more than 100 kb and contains 11 exons including one at the 3' noncoding region (Tsukamoto et al., 1991). It was found that exons 7 and 8 coding for amino acid residues 242–354 of $Go\alpha$ protein were duplicated (referred to as exons 7A and 7B, and 8 A and 8B, respectively). The amino acid sequence identity of exons 7A and 7B, and 8A and 8B, is 88% and 68%, respectively. It was found that exons 7A and 8A, and 7B and 8B, code for $Go\alpha1$ and $Go\alpha2$, respectively. Thus, two different $Go\alpha$ mRNAs are generated by alternative splicing of a single $Go\alpha$ gene. Recently, Kleuss et al. (1991) have demonstrated the functional difference of the two forms of $Go\alpha$ in terms of the receptor coupling. By intranuclear injection of antisense oligonucleotides into rat pituitary GH3 cells, they found that Goα1 and Goα2 subtypes mediate the inhibition of the Ca^{2+} channel through muscarinic and somatostatin receptors, respectively.

12.3. Mutational Analysis of the Function of G-Protein α Subunits

The availability of cDNA clones coding for various G-protein α subunits prompted the mutational analysis of the function of G-protein α subunits. Recent results along this line of research are summarized in Table 10.4.

As shown in the table, the amino acid replacements at the P, G', and G sites by point mutation were carried out by several groups in analogy to the transforming mutation of Ras proteins. In Gsα, replacement of Gly[49] by valine (G49V) reduced the rate of k_{cat} of GTP hydrolysis (Graziano and Gilman, 1989; Masters et al., 1989). The replacement of three amino acids at P site of Gsα from A-G-E to T-S-N caused a 50-fold reduction of $k_{cat \cdot GTP}$ compared with the wild-type Gsα (Casey et al., 1990). The T-S-N sequence occurs naturally in Gzα at the P site, and was thought to be responsible for the remarkably lower GTPase activity of Gzα compared with other G-protein α subunits. Conversion of Gln[61] to Leu at G' region also causes a dramatic reduction of GTPase activity.

The modification of Arg[201] of Gsα by cholera-toxin-catalyzed ADP-ribosylation reduces GTPase activity. Freissmuth and Gilman (1989) demonstrated that the replacement of this Arg by Ala, Glu, and Lys reduced $k_{cat \cdot GTP}$, and these mutant proteins activated adenylyl cyclase constitutively in the presence of GTP. As already discussed in Section 10, Bourne and his collaborators found that, in human pituitary tumors, mutations of Arg[201] to Cys or His, and of Gln[227] to His or Leu inhibit GTP hydrolysis and cause constitutive activation of adenylyl cyclase activity (Landis et al., 1989; Lyons et al., 1990; see also Clementi et al., 1990). In adrenal cortex tumors, mutations have been found in Gi2α that replace the conserved Arg[179] (corresponds to Arg[201] of Gsα) by Cys or His. In this case, the mutation causes the constitutive inhibition of adenylyl cyclase and reduces the level of cAMP which is required to suppress the proliferation of adrenal cells. In contrast, cAMP is probably required for growth of pituitary cells, and the elevation of its level may lead to transformation. It has been shown that the R179C mutation of Gi2α promotes neoplastic transformation of fibroblast (Gupta et al., 1992; Pace et al., 1991).

Analysis of the adenylyl cyclase activity in the transfected cells expressing the Gi2α(1–212)/Gsα(235–394) chimera protein and the Gsα(1–356)/Gi2α(320–355) chimera protein suggested that an adenylyl cyclase activation site may be located in residues 235–356 of Gsα. Itoh and Gilman (1991) found that the replacement of Leu[282]-Arg[283] of Gsα to Phe-The (the corresponding sequence in Giα) abolishes the adenylyl cyclase stimulating

TABLE 10.4. Genetic and biochemical studies of G-protein α subunits[a]

Type	Implication	Reference
Mutation		
unc (Gs1α, R389P)	Uncoupling of Gs with receptor	Sullivan et al. (1987); Rall and Harris (1987)
I121a (Gs1α, G226A)	Prevents GTP-dependent conformational change	Miller et al. (1988)
Phosphoryl-binding region	$\sim$5-fold reduction in k_{cat} . GTP hydrolysis	Graziano and Gilman (1989); Masters et al. (1989); Woon et al. (1989a)
(Gs1α, G49V)		
(Gs1α, AGE48–50TSN)	50-fold reduction in k_{cat} . GTP	Casey et al. (1990)
Conserved Gln	>100-fold reduction in k_{cat} .	Graziano and
(Gs1α, Q227L)	GTP. Constitutively activates adenylyl cyclase	Gilman (1989); Masters et al. (1989)
Conserved Arg	>100-fold reduction in k_{cat} .	Freissmuth and
(Gs1α, R201A/Q/K)[b]	GTP	Gilman (1989)
Pertussis toxin modification site (Gs1α, QYE390–392DCG)	Indirect enhancement of rate of GDP dissociation	Freissmuth and Gilman (1989)
Effector interaction (Gs1α, LR,282,283,FT)[b]	Reduction of ability to stimulate adenylyl cyclase	Itoh and Gilman (1991)
Oncogenic mutations		
gsp (Gs1α, Q227R/H/L)	Found in GH-secreting pituitary tumors and thyroid tumors	Landis et al. (1989); Lyons et al. (1990); Clementi et al. (1990)
gsp (Gs1α, R201C/H)	Found in GH-secreting pituitary tumors	Landis et al. (1989); Lyons et al. (1990)
gip2 (Gi2α, R179C/H)	Found in ovarian sex cord stromal tumors and adrenal cortical tumors	Lyons et al. (1990)
Chimera		
Gi2α (1–212)/Gs1α (235–394)	Receptor and effector interaction sites are in the carboxy-terminal 40% of Gs1α	Masters et al. (1988)
Gs1α (1–356)/Gi2α (320–355)	Constitutively activates adenylyl cyclase. Dominant negative effects on receptor-stimulated activation of phospholipase A_2	Woon et al. (1989b); Gupta et al. (1990)
Gi2α (1–54)/Gs1α (62–356)/Gi2α (320–355)	Dominant negative effects on receptor-stimulated activation of phospholipase A_2	Osawa et al. (1990a)
Gi2α (1–54)/Gs1α (62–394)	Constitutively activates adenylyl cyclase	Osawa et al. (1990a)

(continued)

TABLE 10.4. *(continued)*

Type	Implication	Reference
N-Terminal cleavage		
By trypsin (Goα, K21)	Involved in the association with βγ subunits	Neer et al. (1988)
By V8 protease (Gt1α, E21)	Involved in the association with βγ subunits	Navon and Fung (1987)
Internal cleavage		
By trypsin (Gt1α, R204) (Goα, R209)	GTP analog prevents cleavage at these sites	Hurley et al. (1984)
ADP-ribosylation		
By pertussis toxin (Gt2α, Gi3α, Goα, C351) (Gi2α, C352) (Gt1α, C347)	Impairs the interaction with receptor	Ui (1984); West et al. (1985)
By cholera toxin (Gt1α, R174), (Gs1α, R201)	Reduces GTPase activity and decreases affinity for βγ subunits	Van Dop et al. (1984); Kahn and Gilman (1984)
Myristoylation (Giα, Goα, Gtα, Gzα, G2)	Enhances interaction with βγ subunits and/or with membranes	Jones et al. (1990); Linder et al. (1991); Mumby et al. (1990)
Synthetic peptide		
C-terminal regions (Gt1α, 311–328), (Gt1α, 340–350), (Gs1α, 379–394)	Inhibits receptor–G-protein interaction	Hamm et al. (1988); Palm et al. (1990)

[a]Modified from Kaziro et al. (1991). See text for details.
[b]The exact number of these residues depends on the splice variants of α. In this table, we use the number of α, long form, which consists of 394 amino acid residues.

activity. The mutant protein was expressed and purified from *E. coli.* Neither the kinetic properties of GTP binding and hydrolysis nor the interacting properties with βγ-subunits and β-adrenergic receptor was affected by the mutation. Therefore, it was concluded that the amino acid residues Leu-Arg at positions 282 and 283 are essential for interaction with $G_s\alpha$ and adenylyl cyclase.

More recently, Berlot and Bourne (1992) carried out an extensive study of effector-recognition sites of Gsα using a scanning mutagenesis approach. Within the region comprising residues 236–356, they identified four clusters of residues in which substitutions prevented effector activation. According to them, in a three-dimensional Gα model based on the structure of Ras p21, the effector-activating residues of Gsα form a surface on the membrane-facing side of the molecule.

12.4. G-protein from *Schizosaccharomyces pombe*

Recently a G-protein homologous gene has been isolated from *S. pombe* (Obara et al., 1991). The gene, designated gpa1, encodes a protein (GP1α) of 407 amino acid residues (see Fig. 10.11). Genetic analyses suggest the involvement of the gpa1 gene in mating factor signal transduction of *S. pombe*. The activated mutant of gpa1 can abrogate the signal of mating factor. Thus, in contrast to the *S. cerevisiae* system where GPA1 apparently plays a negative role in the mating factor signaling, *S. pombe* gpa1 plays a positive function. It is interesting to note that the signal of nitrogen starvation, which is also required for the mating in *S. pombe,* is transmitted through *S. pombe ras1.*

References

Barbacid M (1987): *ras* genes. *Annu Rev Biochem* 56:779–827

Beals CR, Wilson CB, Perlmutter RM (1987): A small multigene family encodes Gi signal-transduction proteins. *Proc Natl Acad Sci USA* 84:7886–7890

Berlot CH, Bourne H (1992): Identification of effector-activating residues of $G_s\alpha$. *Cell* 68:911–922

Bourne HR, Masters SB, Miller RT, Sullivan KA, Heideman W (1988): Mutations probe structure and function of G-protein α chains. *Cold Spring Harbor Symp Quant Biol* 53:221–228

Bourne HR, Sanders DA, McCormick F (1990): The GTPase superfamily: A conserved switch for diverse cell functions. *Nature* 348:125–132

Bourne HR, Sanders DA, McCormick F (1991): The GTPase superfamily: Conserved structure and molecular mechanism. *Nature* 349:117–127

Bray P, Carter A, Simons C, Guo V, Puckett C, Kamholz J, Spiegel A, Nirenberg M (1986): Human cDNA clones for four species of Gαs signal transduction protein. *Proc Natl Acad Sci USA* 83:8893–8897

Bray P, Carter A, Guo V, Puckett C, Kamholz J, Spiegel A, Nirenberg M (1987): Human cDNA clones for an α subunit of Gi signal-transducing protein. *Proc Natl Acad Sci USA* 84:5115–5119

Broek D, Samiy N, Fasano O, Fujiyama A, Tamanoi F, Northup J, Wigler M (1985): Differential activation of yeast adenylate cyclase by wild-type and mutant *RAS* proteins. *Cell* 41:763–769

Casey PJ, Fong HK, Simon MI, Gilman AG (1990): Gz, a guanine nucleotide-binding protein with unique biochemical properties. *J Biol Chem* 265:2383–2390

Clementi E, Malgaretti N, Meldolesi J, Taramelli R (1990): A new constitutively activating mutation of the Gs protein alpha subunit-gsp oncogene is found in human pituitary tumours. *Oncogene* 5:1059–1061

Codina J, Olate J, Abramowitz J, Mattera R, Cook RG, Birnbaumer L (1988): αi-3 cDNA encodes the α subunit of Gk, the stimulatory G protein of receptor-regulated K^+ channels. *J Biol Chem* 263:6746–6750

DeFeo-Jones D, Scolnick EM, Koller R, Dhar R (1983): *ras*-Related gene sequences identified and isolated from *S. cerevisiae. Nature* 306:707–709

Deretic D, Hamm HE (1987): Topographic analysis of antigenic determinants recognized by monoclonal antibodies to the photoreceptor guanyl nucleotide-binding protein, transducin. *J Biol Chem* 262:10839–10847

De Vos AM, Tong L, Milburn MV, Matias PM, Jankark J, Noguchi S, Nishimura S, Miura K, Ohtsuka E, Kim S-H (1988): Three-dimensional structure of an oncogene protein: Catalytic domain of human c-H-*ras* p21. *Science* 239:888–893

Didsbury JR, Ho Y, Snyderman R (1987): Human Gi protein α-subunit: Deduction of amino acid structure from a cloned cDNA. *FEBS Lett* 211:160–164

Didsbury JR, Snyderman R (1987): Molecular cloning of a new human G protein. Evidence for two Giα-like protein families. *FEBS Lett* 219:259–263

Dietzel D, Kurjan J (1987): The yeast *SCG1* gene: A Gα-like protein implicated in the *a*- and α-factor response pathway. *Cell* 50:1001–1010

Fong, HKW, Yoshimoto KK, Eversole-Cire P, Simon MI (1988): Identification of a GTP-binding protein α subunit that lacks an apparent ADP-ribosylation site for pertussis toxin. *Proc Natl Acad Sci USA* 85:3066–3070

Freissmuth M, Gilman AG (1989): Mutations of Gsα designed to alter the reactivity of the protein with bacterial toxins: Substitutions at Arg 187 result in loss of GTPase activity. *J Biol Chem* 264:21907–21914

Freissmuth M, Casey PJ, Gilman AG (1989): G proteins control diverse pathways of transmembrane signaling. *FASEB J* 3:2125–2131

Fukui Y, Kaziro Y (1985): Molecular cloning and sequence analysis of a *ras* gene from *Schizosaccharomyces pombe*. EMBO J 4:687–691

Gilman AG (1987): G proteins: Transducers of receptor-generated signals. *Annu Rev Biochem* 56:615–649

Graziano MP, Gilman AG (1989): Synthesis in *Escherichia coli* of GTPase-deficient mutants of Gs alpha. *J Biol Chem* 264:15475–15482

Gupta SK, Diez E, Heasley LE, Osawa S, Johnson GL (1990): A G protein mutant that inhibits thrombin and purinergic receptor activation of phospholipase A2. *Science* 249:662–666

Gupta SK, Gallego C, Lowndes JM, Pleiman CM, Sable C, Eisfelder BJ, Johnson GJ (1992): Analysis of the fibroblast transformation potential of GTPase-deficient *gip2* oncogenes. *Mol Cell Biol* 12:190–197

Hamm HE, Deretic D, Arendt A, Hargrave PA, Koenig B, Hofmann KP (1988): Site of G protein binding to rhodopsin mapped with synthetic peptides from the α subunit. *Science* 241:832–835

Harris BA, Robishaw JD, Mumby SM, Gilman A (1985): Molecular cloning of complementary DNA for the alpha subunit of the G protein that stimulates adenylate cyclase. *Science* 229:1274–1277

Holbrook S, Kim S-H (1989): Molecular model of the G protein α subunit based on the crystal structure of the HRAS protein. *Proc Natl Acad Sci USA* 86:1751–1755

Houslay MD, Milligan G, eds (1990): *G-Proteins as Mediators of Cellular Signaling Processes*. New York: John Wiley & Sons, Inc., 232 pp

Hsu WH, Rudolph U, Sanford J, Bertrand P, Olate J, Nelson C, Moss LG, Boyd AE, Codina J, Birnbaumer L (1990): Molecular cloning of a novel splice variant of the alpha subunit of the mammalian Go protein. *J Biol Chem* 265, 11220–11226

Hurley JB, Simon MI, Teplow DB, Robishaw JD, Gilman AG (1984): Homologies between signal transducing G proteins and *ras* gene products. *Science* 226:860–862

Itoh H, Gilman AG (1991): Expression and analysis of $G_s\alpha$ mutants with decreased ability to activate adenylylcyclase. *J Biol Chem* 266:16226–16231

Itoh H, Kozasa T, Nagata S, Nakamura S, Katada T, Ui M, Iwai S, Ohtsuka E, Kawasaki H, Suzuki K, Kaziro Y (1986): Molecular cloning and sequence determination of cDNAs for α subunit of the guanine nucleotide-binding proteins Gs, Gi, and Go from rat brain. *Proc Natl Acad Sci USA* 83:3776–3780

Itoh H, Katada T, Ui M, Kawasaki H, Suzuki K, Kaziro Y (1988a): Identification of three pertussis toxin substrates (41, 40 and 39 kDa proteins) in mammalian brain. *FEBS Lett* 230:85–89

Itoh H, Toyama R, Kozasa T, Tsukamoto T, Matsuoka M, Kaziro Y (1988b): Presence of three distinct molecular species of Gi protein α subunit. *J Biol Chem* 263:6656–6664

Iyengar R, Birnbaumer L, eds. (1990): *G Proteins.* New York: Academic Press, 651 pp

Jackson TR, Blair LAC, Marshall J, Goedert M, Hanley MR (1988): The *mas* oncogene encodes an angiotensin receptor. *Nature* 335:437–440

Jones DT, Reed RR (1987): Molecular cloning of five GTP-binding protein cDNA species from rat olfactory neuroepithelium. *J Biol Chem* 262:14241–14249

Jones DT, Reed RR (1989): G_{olf}: Olfactory neuron specific-G protein involved in odorant signal transduction. *Science* 244:790–795

Jones TL, Simonds WF, Merendino JJ, Brann MR, Spiegel AM (1990): Myristoylation of an inhibitory GTP-binding protein alpha subunit is essential for its membrane attachment. *Proc Natl Acad Sci USA* 87, 568–572

Julius D, Jivelli TJ, Jessell TM, Axel R (1989): Ectopic expression of the serotonin 1c receptor and the triggering of malignant transformation. *Science* 244:1057–1062

Jurnak F (1985): Structure of the GDP domain of EF-Tu and location of the amino acids homologous to *ras* oncogene proteins. *Science* 230:32–36

Kahn RA, Gilman AG (1984): ADP-Ribosylation of Gs promoters the dissociation of its α and β subunits. *J Biol Chem* 259:6235–6240

Kaziro Y (1978): The role of guanosine 5′-triphosphate in polypeptide chain elongation. *Biochim Biophys Acta* 505:95–127

Kaziro Y, Itoh H, Kozasa T, Tsukamoto T, Matsuoka M, Nakafuku M, Obara T, Takagi T, Hernandez R (1988): Structure of the genes coding for G protein α subunits from mammalian and yeast cells. *Cold Spring Harbor Symposia Quant Biol* 53:209–220

Kaziro Y, Itoh H, Kozasa T, Nakafuku M, Satoh T (1991): Structure and function of signal-transducing GTP-binding proteins. *Ann Rev Biochem* 60:349–400

Kim SY, Ang SL, Bloch DB, Bloch KD, Kawahara Y, Tolman C, Lee R, Seidman JG, Neer EJ (1988): Identification of cDNA encoding an additional alpha subunit of a human GTP-binding protein: Expression of three alpha i subtypes in human tissues and cell lines. *Proc Natl Acad Sci USA* 85:4153–4157

Kleuss C, Hescheler J, Ewel C, Rosenthal W, Schultz G, Wittig B (1991): Assignment of G-protein subtypes to specific receptors inducing inhibition of calcium currents. *Nature* 353:43–48

Kozasa T, Itoh H, Tsukamoto T, Kaziro Y (1988): Isolation and characterization of human Gsα gene. *Proc Natl Acad Sci USA* 85:2081–2085

Kumagai A, Pupillo M, Gundersen R, Miake-Lye R, Devreotes PN, Firtel RA (1989): Regulation and function of Gα protein subunits in *Dictyostelium*. *Cell* 57:265–275

LaCour TFM, Nyborg J, Thirup S, Clark BFC (1985): Structural details of the binding of guanosine diphosphate to elongation factor Tu from *E. coli* as studied by X-ray crystallography. *EMBO J* 4:2385–2388

Landis CA, Masters SB, Spada A, Pace AM, Bourne HR, Vallar L (1989): GTPase inhibiting mutations activate the α chain of Gs and stimulate adenylyl cyclase in human pituitary tumours. *Nature* 340:692–696

Lavu S, Clark J, Swarup R, Matsushima K, Paturu K, Moss J, Kung H-F (1988): Molecular cloning and DNA sequence analysis of the human guanine nucleotide-binding protein Goα. *Biochem Biophys Res Commun* 150:811–815

Lerea CL, Somers DE, Hurley JB, Klock IB, Bunt-Milam AH (1986): Identification of specific transducin α subunits in retinal rod and cone photoreceptors. *Science* 324:77–80

Linder ME, Pang IH, Duronio RJ, Gordon JI, Sternweis PC, Gilman AG (1991): Lipid modifications of G protein subunits. Myristoylation of Go alpha increases its affinity for beta gamma. *J Biol Chem* 266:4654–4659

Lochrie MA, Hurley JB, Simon MI (1985): Sequence of the alpha subunit of photoreceptor G protein: Homologies between transducin, *ras,* and elongation factors. *Science* 228:96–99

Lyons J, Landis CA, Harsh G, Vallar L, Grunewald K, Feichtinger H, Duh QY, Clark OH, Kawasaki E, Bourne HR, McCormick F (1990): Two G proteins oncogenes in human endocrine tumors. *Science* 249:655–659

Ma H, Yanofsky MF, Meyerowitz EM (1990): Molecular cloning and characterization of GPA1, a G protein alpha subunit gene from *Arabidopsis thaliana. Proc Natl Acad Sci USA* 87:3821–3825

Masters SB, Stroud RM, Bourne HR (1986): Family of G protein α chains: Amphipathic analysis and predicted structure of functional domains. *Protein Eng.* 1:47–54

Masters SB, Sullivan KA, Miller RT, Beiderman B, Lopez NG, Ramachandran J, Bourne HR (1988): Carboxyl terminal domain of Gsα specifies coupling of receptors to stimulation of adenylyl cyclase. *Science* 241:448–451

Masters SB, Miller RT, Chi MH, Chang F-H, Beiderman B, Lopez NG, Bourne HR (1989): Mutations of the GTP-binding site of $G_s\alpha$ alter stimulation of adenylyl cyclase. *J Biol Chem* 264:15467–15474

Matsuoka M, Itoh H, Kozasa T, Kaziro Y (1988): Sequence analysis of cDNA and genomic DNA for a putative pertussis toxin-insensitive guanine nucleotide-binding regulatory protein α subunit. *Proc Natl Acad Sci USA* 85:5384–5388

Matsuoka M, Itoh H, Kaziro Y (1990): Characterization of the human gene for $G_x\alpha$, a pertussis toxin-insensitive regulatory GTP-binding protein. *J Biol Chem* 265:13215–13220

Mattera R, Codina J, Crozat A, Kidd V, Woo SLC, Birnbaumer L (1986): Identification by molecular cloning of two forms of the α-subunit of the human liver stimulatory (Gs) regulatory component of adenylate cyclase. *FEBS Lett* 206:36–41

Medynski DC, Sullivan K, Smith D, van Dop C, Chang F-H, Fung BK-K, Seeburg PH, Bourne HR (1985): Amino acid sequence of the α subunit of transducin deduced from the cDNA sequence. *Proc Natl Acad Sci USA* 82:4311–4315

Mercken L, Moras V, Tocque B, Mayaux JF (1990): The cDNA sequence of the alpha-subunit of the Chinese hamster adenylate cyclase-stimulatory G-protein. *Nucl Acids Res* 18:662

Michel T, Winslow JW, Smith JA, Seidman JG, Neer EJ (1986): Molecular cloning and characterization of cDNA encoding the GTP-binding protein αi and identification of a related protein, αh. *Proc Natl Acad Sci USA* 83:7663–7667

Miyajima I, Nakafuku M, Nakayama N, Brenner C, Miyajima A, Kaibuchi K, Arai K, Kaziro Y, Matsumoto K (1987): *GPA1,* a haploid-specific essential gene, encodes a yeast homolog of mammalian G protein which may be involved in mating factor signal transduction. *Cell* 50:1011–1019

Moss J, Vaughan M, eds (1990): *ADP-Ribosylating Toxins and G Proteins, Insight into Signal Transduction.* Washington, DC: *Am Soc Microbiol,* 567 pp

Mumby SM, Heukeroth RO, Gordon JI, Gilman AG (1990): G-protein alpha-subunit expression, myristoylation, and membrane association in COS cells. *Proc Natl Acad Sci USA* 87:728–732

Nakafuku M, Itoh H, Nakamura S, Kaziro Y (1987): Occurrence in *Saccharomyces cerevisiae* of a gene homologous to the cDNA coding for the α subunit of mammalian G proteins. *Proc Natl Acad Sci USA* 84:2140–2144

Nakafuku M, Obara T, Kaibuchi K, Miyajima I, Miyajima A, Itoh H, Nakamura S, Arai K, Matsumoto K, Kaziro Y (1988): Isolation of a second yeast *Saccharomyces cerevisiae* gene *(GPA2)* coding for guanine nucleotide-binding regulatory protein: Studies on its structure and possible functions. *Proc Natl Acad Sci USA* 85:1374–1378

Navon SE, Fung BK (1987): Characterization of transducin from bovine retinal rod outer segments. Participation of the amino-terminal region of T alpha in subunit interaction. *J Biol Chem* 262:15746–15751

Neer EJ, Pulsifer L, Wolf LG (1988): The amino terminus of G protein alpha subunits is required for interaction with beta gamma. *J Biol Chem* 263:8996–8970

Northup JK, Sternweise PC, Smigel MD, Shleifer LS, Ross EM, Gilman AG (1980): Purification of the regulatory component of adenylate cyclase. *Proc Natl Acad Sci USA* 77:6516–6520

Nukada T, Tanabe T, Takahashi H, Noda M, Hirose T, Inayama S, Numa S (1986a): Primary structure of the α-subunit of bovine adenylate cyclase-stimulating G-protein deduced from the cDNA sequence. *FEBS Lett* 195:220–224

Nukada T, Tanabe T, Takahashi H, Noda M, Haga K, Haga T, Ichi-yama A, Kangawa K, Hiranaga M, Matsuo H, Numa S (1986b): Primary structure of the α subunit of bovine adenylate cyclase-inhibiting protein deduced from the cDNA sequence. *FEBS Lett* 197:305–310

Obara T, Nakafuku M, Yamamoto M, Kaziro Y (1991): Isolation and characterization of a gene encoding a G-protein α subunit from *Schizosaccharomyces pombe:* Involvement in mating and sporulation pathways. *Proc Natl Acad Sci USA* 88:5877–5881

Osawa S, Dhanasekaran N, Woon CW, Johnson GL (1990a): $G_{\alpha i}$-$G_{\alpha s}$ chimeras define the function of α chain domains in control of G protein activation and $\beta\gamma$ subunit complex interactions. *Cell* 63:697–706

Osawa S, Heasley, LE, Dhanasekaran N, Gupta SK, Woon CW, Berlot C, Johnson GL (1990b): Mutation of the Gs protein alpha subunit NH2 terminus relieves an attenuator function, resulting in constitutive adenylyl cyclase stimulation. *Mol Cell Biol* 10:2931–2940

Ovchinnikov YA, Slepak VZ, Pronin AN, Shlensky AB, Levina NB, Voeikov VL, Lipkin VM (1987): Primary structure of bovine cerebellum GTP-binding protein G_{39} and its effect on the adenylate cyclase system. *FEBS Lett* 226:91–95

Pace AM, Wong YH, Bourne HR (1991): A mutant α subunit of G_i2 induces neoplastic transformation of Rat-1 cells. *Proc Natl Acad Sci USA* 88:7031–7035

Palm D, Munch G, Malek D, Dees C, Hekman M (1990): Identification of a Gs-protein coupling domain to the beta-adrenoceptor using site-specific synthetic peptides. Carboxyl terminus of Gs alpha is involved in coupling to beta-adrenoceptors. *FEBS Lett* 261:294–298

Powers S, Kataoka T, Fasano O, Goldfarb M, Strathern J, Broach J, Wigler M (1984): Genes in *S. cerevisiae* encoding proteins with domains homologous to the mammalian *ras* proteins. *Cell* 36:607–612

Price SR, Murtagh JJ, Tsuchiya M, Serventi IM, Van MK, Angus CW, Moss J, Vaughan M (1990): Multiple forms of Go alpha mRNA: Analysis of the 3'-untranslated regions. *Biochemistry* 29:5069–5076

Provost NM, Somers DE, Hurley JB (1988): A *Drosophila melanogaster* G protein α subunit gene is expressed primarily in embryos and pupae. *J Biol Chem* 263:12070–12076

Pupillo M, Kumagai A, Pitt GS, Firtel RA, Devreotes PN (1989): Multiple α subunits of guanine nucleotide-binding proteins in *Dictyostelium*. *Proc Natl Acad Sci USA* 86:4892–4896

Rall T, Harris BA (1987): Identification of the lesion in the stimulatory GTP-binding protein of the uncoupled S49 lymphoma. *FEBS Lett* 224:365–371

Raport CJ, Dere B, Hurley JB (1989): Characterization of the mouse rod transducin α subunit gene. *J Biol Chem* 264:7122–7128

Robishaw JD, Russell DW, Harris BA, Smigel MD, Gilman AG (1986a): Deduced primary structure of the α subunit of the GTP-binding stimulatory protein of adenylate cyclase. *Proc Natl Acad Sci USA* 83:1251–1255

Robishaw JD, Smigel MD, Gilman AG (1986b): Molecular basis for two forms of the G protein that stimulates adenylate cyclase. *J Biol Chem* 261:9587–9590

Ross EM (1989): Signal sorting and amplification through G protein-coupled receptors. *Neuron* 3:141–152

Sewell JL, Kahn RA (1988): Sequences of the bovine and yeast ADP-ribosylation factor and comparison to other GTP-binding proteins. *Proc Natl Acad Sci USA* 85:4620–4624

Simon MI, Strathmann MP, Gautam N (1991): Diversity of G proteins in signal transduction. *Science* 252:802–808

Strathmann M, Simon MI (1990): G protein diversity: A distinct class of α subunits is present in vertebrate and invertebrates. *Proc Natl Acad Sci USA* 87:9113–9117

Strathmann M, Wilkie TM, Simon MI (1989): Diversity of the G-protein family: Sequences from five additional α subunits in the mouse. *Proc Natl Acad Sci USA* 86:7407–7409

Strathmann M, Wilkie TM, Simon MI (1990): Alternative splicing produces transcripts encoding two forms of the α subunit of GTP-binding protein G$_o$. *Proc Natl Acad Sci USA* 87:6477–6481

Suki WN, Abramowitz J, Mattera R, Codina J, Birnbaumer L (1987): The human genome encodes at least three non-allellic G proteins with αi-type subunits. *FEBS Lett* 220:187–192

Sullivan KA, Liao Y-C, Alborzi A, Beiderman B, Chang F-H, Masters SB, Levinson AD, Bourne HR (1986): Inhibitory and stimulatory G proteins of adenylate cyclase: cDNA and amino acid sequences of the α chains. *Proc Natl Acad Sci USA* 83:6687–6691

Sullivan KA, Miller RT, Masters SB, Beiderman B, Heideman W, Bourne HR

(1987): Identification of receptor contact site involved in receptor-G protein coupling. *Nature* 330:758–760

Tanabe T, Nukada T, Nishikawa Y, Sugimoto K, Suzuki H, Takahashi H, Noda M, Haga T, Ichiyama A, Kanagawa K, Minamino N, Matsuo H, Numa S (1985): Primary structure of the α-subunit of transducin and its relationship to *ras* proteins. *Nature* 315:242–245

Taylor SJ, Exton JH (1991): Two alpha subunits of the Gq class of G proteins stimulate phosphoinositide phospholipase C-beta 1 activity. *FEBS Lett* 286:214–216

Taylor SJ, Smith JA, Exton JH (1990): Purification from bovine liver membranes of a guanine nucleotide-dependent activator of phosphoinositide-specific phospholipase C. Immunologic identification as a novel G-protein alpha subunit. *J Biol Chem* 265, 17150–17156

Taylor SJ, Chae HZ, Rhee SG, Exton JH (1991): Activation of the beta 1 isozyme of phospholipase C by alpha subunits of the Gq class of G proteins. *Nature* 350:516–518

Toda T, Uno I, Ishikawa T, Powers S, Kataoka T, Broek D, Cameron S, Broach J, Matsumoto K, Wigler M (1985): In yeast, *RAS* proteins are controlling elements of adenylate cyclase. *Cell* 40:27–36

Tsukamoto T, Toyama R, Itoh H, Kozasa T, Matsuoka M, Kaziro Y (1991): Structure of the human gene and two rat cDNAs encoding the α chain of GTP-binding regulatory protein G_o: Two different mRNAs are generated by alternative splicing. *Proc Natl Acad Sci USA* 88:2974–2978

Van Dop C, Tsubokawa M, Bourne HR, Ramachandran J (1984): Amino acid sequences of retinal transducin at the site ADP-ribosylated by cholera toxin. *J Biol Chem* 259:696–698

Van Meurs KP, Angus CW, Lavu S, Kung H-F, Czarnecki SK, Moss J, Vughan M (1987): Deduced amino acid sequence of bovine retinal Goα: Similarities to other guanine nucleotide-binding proteins. *Proc Natl Acad Sci USA* 84:3107–3111

Weinstein LS, Spiegel AM, Carter AD (1988): Cloning and characterization of the human gene for the α-subunit of Gi2, a GTP-binding signal transduction protein. *FEBS Lett* 232:333–340

West RJ, Moss J, Vaughan M, Liu T, Liu TY (1985): Pertussis toxin-catalyzed ADP-ribosylation of transducin. Cysteine 347 is the ADP-ribose acceptor site. *J Biol Chem* 260:14428–14430

Whiteway M, Hougan L, Dignard D, Thomas DY, Bell L, Saari GC, Grant FJ, O'Hara P, MacKay VL (1989): The STE4 and STE18 genes of yeast encode potential β and γ subunits of the mating factor receptor-coupled G protein. *Cell* 56:467–477

Woon CW, Heasley L, Osawa S, Johnson GL (1989a): Mutation of glycine 49 to valine in the α subunit of G_s results in the constitutive elevation of cyclic AMP synthesis. *Biochemistry* 28:4547–4551

Woon CW, Soparkar S, Heasley L, Johnson GL (1989b): Expression of a G alpha s/G alpha i chimera that constitutively activates cyclic AMP synthesis. *J Biol Chem* 264:5687–5693

Yatsunami K, Khorana HG (1985): GTPase of bovine rod outer segments: The amino acid sequence of the α subunit as derived from the cDNA sequence. *Proc Natl Acad Sci USA* 82:4316–4320

11

Immunological Probes of the Structure, Function, and Expression of Heterotrimeric GTP-Binding Proteins

ALLEN M. SPIEGEL, PAUL K. GOLDSMITH,
WILLIAM F. SIMONDS, TERESA JONES, KEVIN ROSSITER, AND
CECILIA G. UNSON

1. Introduction

G-proteins convey extracellular signals from cell surface receptors to diverse effectors such as enzymes of second messenger metabolism and ion channels. Receptor–effector coupling G-proteins are composed of three subunits, α, β, and γ, each separate gene products. Multiple forms of each type of subunit have been identified at the DNA and protein level (Gilman, 1987; Lochrie and Simon, 1988; Spiegel, 1987). Historically, α subunits were first identified by GTP affinity labeling and bacterial toxin-catalyzed ADP-ribosylation. These methods lack requisite specificity. Pertussis toxin, for example, is now known to covalently modify as many as six different Gα subunits (Gilman, 1987; Lochrie and Simon, 1988; Spiegel, 1987). Antibodies specific for G-protein subunits have proved to be powerful tools for studies of G-proteins. Both polyclonal (Gierschik et al., 1985, 1986c; Huff et al., 1985; Katada et al., 1987; Pines et al., 1985a; Roof et al., 1985) and monoclonal (Halpern et al., 1986; Hamm and Bownds, 1984; Navon and Fung, 1987, 1988; Tsai et al., 1987) antibodies have been raised against purified G-proteins. Antisera raised against purified G-holoproteins and subunits provided initial evidence for similarities (β subunits) and differences (α and γ subunits) between members of the G-protein family (Figure 11.1). Immunochemical studies with such antisera suggested diversity in pertussis toxin substrates (Gierschik et al., 1986b), subsequently confirmed by cloning of cDNAs encoding multiple types of putative pertussis toxin substrate (Gilman, 1987; Lochrie and Simon, 1988; Spiegel, 1987). Antisera raised against synthetic peptides corresponding to amino acid sequences predicted by cloned cDNAs (Tables 11.1 and 11.2) have proved even more

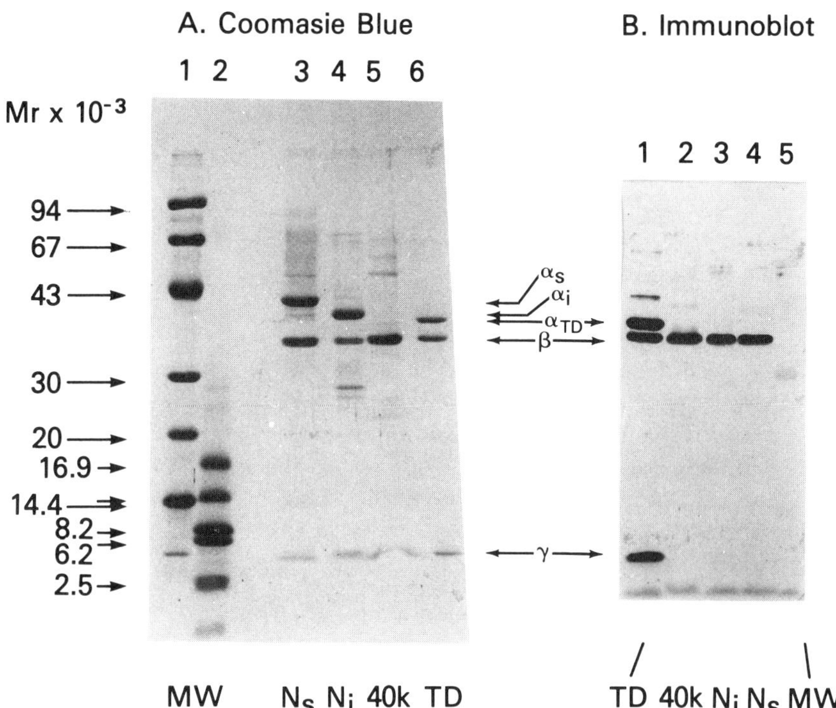

FIGURE 11.1. Analysis of purified G-proteins by protein staining and immunoblotting. Purified Gs and Gi (Ns and Ni) from human erythrocytes and bovine Gt (TD) from retina were separated by urea-sodium dodecyl sulfate-polyacrylamide gel electrophoresis (SDS-PAGE) and either stained with Coomassie blue (A) or immunoblotted with an antiserum raised against holoprotein Gt (B). Lanes 1 and 2 in panel A contain molecular weight markers. Lane marked "40k" contains the $\beta\gamma$ complex purified from human erythrocytes. Positions of the subunits are marked by arrows. The immunoblot shows reactivity of the antiserum with all three Gt subunits but only with the β subunits of Gs and Gi. (From Gierschik et al., 1985.)

useful, making possible identification of the protein products of several Gα cDNAs (Goldsmith et al., 1987, 1988a; Harris et al., 1985; Lerea et al., 1986), and facilitating discrimination between Gα subtypes (Goldsmith et al., 1988a,b; Mumby et al., 1988).

In addition to providing insights into G-protein structure, antibodies have offered unique insights into G-protein function, including the specificity of receptor and effector interactions. "Site-directed" G-protein antibodies have been used as probes to determine the domains involved in various G-protein subunit interactions. The ability to immunoprecipitate G-protein subunits, particularly in their native form, offers the possibility of identifying closely associated proteins (e.g., effectors). Posttranslational modification of G-proteins, such as phosphorylation or fatty acid acylation, can also be studied in this way. Antibodies also provide a convenient method for monitoring G-protein expression by recombinant DNA methods. Finally, antibodies have been extensively used to quantitate G-proteins, and to study their cellular and tissue distribution. Antibodies are superior to bacterial toxin-catalyzed ADP-ribosylation in such studies, since they are more specific, allow measurement of β and γ, not just α, subunits, and are less susceptible to interference by extraneous factors (see Milligan et al., 1987a, for example).

The p21 products of ras protooncogenes and an expanding list of related low-molecular-weight GTP-binding proteins share significant sequence homology with the α subunit of heterotrimeric G-proteins. As yet, there is no evidence that members of the family of G-protein-coupled receptors interact directly with any of the low-molecular-weight GTP-binding proteins. Antibodies directed against members of the low-molecular-weight GTP-binding protein family have provided important structural and functional information, but a discussion of the data obtained with these reagents is beyond the scope of this chapter.

2. Immunochemical Studies of G-Protein Structure

2.1. γ Subunits

The γ subunits are low-molecular-weight (~ 8–11 kDa) proteins tightly, but noncovalently, associated with β subunits to form the $\beta\gamma$ complex of G-proteins. The amino acid sequence of Gtγ, a very hydrophilic 8.4-kDa protein, has been defined (Hurley et al., 1984). cDNA cloning indicates that brain contains at least two distinct forms of γ subunit, each different from Gtγ (Robishaw et al., 1989; Gautam et al., 1989).

Antisera raised against Gt holoprotein or its $\beta\gamma$ subunit readily react with Gtγ on immunoblots but fail to recognize the γ subunits of other G-proteins (Figure 11.1; Gierschik et al., 1985). Immunochemical distinction between Gtγ and other G-protein γ subunits has been confirmed (Audigier et al., 1985; Evans et al., 1987; Roof et al., 1985). Silver staining of purified G-proteins suggests diversity among G-protein γ subunits other than Gt (Sternweis and Robishaw, 1984). Antisera raised against a $\beta\gamma$ complex purified from human placenta recognized placental γ subunits on immunoblot, but failed to react with either Gtγ or the γ subunits of other G-proteins purified from platelet, liver, and brain (Evans et al., 1987). Species-specific differences could not account for these findings since placental and platelet proteins were both human. Peptide antisera raised against distinct sequences predicted by two different brain γ cDNAs identify two distinct low-molecular-weight proteins as brain γ subunits (Robishaw et al., 1989). Thus, both immunochemical studies and cDNA cloning suggest substantial diversity among G-protein γ subunits. The functional significance of this diversity with respect to interaction with β subunits and with $\beta\gamma$ subunit function is not clear. Gt $\beta\gamma$, unlike brain G-protein $\beta\gamma$ subunits, is soluble in aqueous buffers, and can be released from rod outer segment membranes in hypotonic buffers without detergents. This may reflect differences in primary sequence and/or differences in posttranslational modifications. γ Subunits share carboxy-terminal sequence homology ("CAAX Box") with p21 products of ras protooncogenes. The latter undergo a complex series of posttranslational modifications, including proteolysis of the carboxy-terminal tripeptide, carboxymethylation, and farnesylation of the resultant cysteine residue (Hancock et al., 1989). Brain G-protein γ subunits appear to undergo similar, if not identical, modifications which may account for their relative hydrophobicity and tight membrane attachment (Backlund et al., 1990; Simonds et al., 1991).

2.2. β Subunits

Two forms of G-protein β subunit, β_{36} and β_{35}, have been purified (Sternweis and Robishaw, 1984). Gtβ contains only the 36-kDa form; in brain, β_{36} predominates over the 35-kDa form, and in placenta, β_{35} predominates (Evans et al., 1987). Antisera raised against Gtβ recognize the corresponding subunit in all G-proteins (Figure 11.1; Gierschik et al., 1985; Roof et al., 1985), and antisera raised against β subunits of other G-proteins, e.g., brain, recognize Gtβ (Evans et al., 1987; Gierschik et al., 1986c). cDNA clones encoding the 36-kDa form of β (β_1) predict identical 340 amino acid sequences for Gt- and other Gβs (Codina et al., 1986; Fong et al., 1986; Sugimoto et al., 1985), thus accounting for immunologic cross-reactivity.

Immunochemical differences have been noted between β_{36} and β_{35}. Some Gtβ antisera react preferentially with β_{36} (Evans et al., 1987; Roof et al., 1985); an antiserum raised against a synthetic peptide representing residues 130–145 of β_{36} reportedly recognized β_{36} but not β_{35} (Mumby et al., 1986). cDNAs (β_2 and β_3) clearly related to but distinct from β_1 have recently been cloned (Fong et al., 1987; Gao et al., 1987b; Levine et al., 1990). The purified β_{35} protein has not yet been sequenced, but antisera to unique peptide sequences predicted by the β_2 cDNA suggest that the latter corresponds to β_{35} (Amatruda et al., 1988; Gao et al., 1987a). A protein corresponding to the β_3 cDNA (Levine et al., 1990) has not yet been identified.

Trypsin cleaves soluble native $\beta\gamma$ complexes preferentially at Arg[129] of the β subunit (Mumby et al., 1986; Sugimoto et al., 1985). This results in two fragments of β, 14 kDa amino-terminal and 25 kDa carboxy-terminal, as well as the intact γ subunit. An identical pattern is observed on immunoblots after tryptic cleavage of retinal rod outer segment membranes and brain plasma membranes (Pines et al., 1985b). This implies that the tryptic cleavage site between residues 129–130 of β is exposed in the native membrane-bound protein.

The majority of antisera raised against β subunits recognize epitope(s) within the 14-kDa amino-terminal fragment (Zaremba et al., 1988). This epitope was mapped to the amino-terminal decapeptide which is the most hydrophilic region of β and may be exposed at the surface of the native $\beta\gamma$ complex (Zaremba et al., 1988). A synthetic decapeptide corresponding to the amino-terminal sequence of β_{36} effectively blocks binding of most β antisera to the 14- but not 25-kDa tryptic fragment of β. The matrix-bound peptide could also be used to affinity-purify from crude β antisera antibodies capable of binding both the 36- and 35-kDa forms of β. Antibodies (termed MS) raised against this peptide, moreover, react with both forms of β (Zaremba et al., 1988). This is consistent with the identity in sequence, with the exception of a single conserved residue of the 36- and 35-kDa forms of β in the amino-terminal decapeptide.

Immunochemical studies show that β subunits are highly conserved among vertebrates (Audigier et al., 1985; Gierschik et al., 1986c). Immunoblots of invertebrate tissues with polyclonal β antisera failed to detect β subunits (Audigier et al., 1985), but recently, blots performed with peptide antisera MS, KT (directed against residues 127–137), and SW (directed against residues 330–340) revealed cross-reacting β subunits of approximately 36 kDa in membranes from adult *Drosophila* (A. M. Spiegel, unpublished observations). This is consistent with the identity (KT) or close homology (MS, SW) in sequence of the *Drosophila* β subunit to the mammalian sequence in the corresponding peptide regions (Yarfitz et al., 1988).

2.3. α Subunits

2.3.1. GENERAL COMMENTS

Nine distinct mammalian genes encoding α subunits have been identified by isolation of genomic and cDNA clones. These include Gt1 (Tanabe et al., 1985) and Gt2 (Lochrie et al., 1985), Gs (Harris et al., 1985; Nukada et al., 1986a), Golf (Jones and Reed, 1989), Gi1, Gi2, and Gi3 (Jones and Reed, 1987), Go (Itoh et al., 1986; Van Meurs et al., 1987), and Gz(x) (Fong et al., 1988; Matsuoka et al., 1988). Based on cDNA cloning using a polymerase chain reaction (PCR) technique, at least four additional α subtypes may exist (Strathmann et al., 1989). G-protein α subunits range from 350 to 395 amino acids in length. All show at least 40% homology, corresponding largely to regions identified in EF-Tu to be involved in binding guanine nucleotide (Jurnak, 1985). Besides these regions that are highly conserved in all Gα subunits, additional regions of homology are apparent in subsets of the family. Gi1 and Gi3, for example, share 94% homology. Given the degree of homology between Gα subunits, it is perhaps surprising that most polyclonal antisera raised against purified G-proteins are quite specific for a given Gα subtype. Of the 18 antisera we have raised against Gt, 16 fail to cross-react with other Gα subunits (see, for example, Figure 11.1). An antiserum, RV/3 (Gierschik et al., 1986c), raised against purified brain G-proteins (a mixture of Go and several Gi subtypes) cross-reacts with all G-protein β subunits, but is specific for Goα. The highly conserved regions of α subunits show low immunogenicity, but this cannot be explained simply by the fact that these regions are highly conserved. Thus, a peptide corresponding to a highly conserved region involved in binding the phosphate of GDP is highly immunogenic, and yields antisera reactive with all Gα subunits (Mumby et al., 1986). The amino acid sequence of the β subunit is highly conserved in mammals (Codina et al., 1986) and even invertebrates (Yarfitz et al., 1988), yet is highly immunogenic and yields antisera that react with β subunits of other species (Audigier et al., 1985; Gierschik et al., 1986c). It is possible that the highly conserved regions of α subunits are not exposed at the surface and, hence, are less immunogenic.

2.3.2. Gt

Two distinct cDNA clones corresponding to Gtα have been isolated. One (Tanabe et al., 1985) corresponded perfectly to amino acid sequence derived from the α_{39} protein purified from bovine rod outer segment membranes, whereas the other (Lochrie et al., 1985) showed several discrepancies. Antisera raised against α_{39} purified from bovine rod outer segment membranes showed specific staining of rod outer segments but failed to stain cone outer segments in chick retina (Grunwald et al., 1986). Antisera raised against synthetic peptides corresponding to sequences predicted by the two forms of Gt cDNA (Table 11.1) proved capable of distinguishing between rod α_{39} and an α_{41} present in retinal homogenates (Lerea et al., 1986). The peptide antiserum corresponding to the rod-type cDNA-predicted sequence stained only rod outer segments on immunohistochemical analysis, consistent with previous results (Grunwald et al., 1986). The peptide antiserum based on the alternative Gt cDNA-predicted sequence specifically stained cone outer segments. Therefore, this cDNA likely corresponds to a cone-specific form of Gt (Lerea et al., 1986).

2.3.3. Gs AND Golf

Cholera toxin-catalyzed ADP-ribosylation of Gsα, and purification of the protein on the basis of a complementation assay (restoration of adenylyl cyclase activation to the Gsα deficient cyc$^-$ cell line), reveal α subunits variously referred to as 42 and 47, or 45 and 52 kDa. cDNAs cloned from brain libraries (Harris et al., 1985; Nukada et al., 1986a) predicted a protein sequence unrelated to those sequenced to date. Synthetic peptide antisera identified the protein encoded by the cDNA as Gsα (Harris et al., 1985). Subsequently, it was found (Bray et al., 1986; Kozasa et al., 1988) that four

TABLE 11.1. Comparison of the amino acid sequences of residues 159–168 in Gi1 and corresponding residues in other Gα subunits[a]

G-protein	Peptide	Antiserum
Gt (rod)[b]	L D R I T A P D Y L	T-I[d]
Gt (cone)[b]	L E R L V T P G Y V	T-II[d]
Gi1	L D R I A Q P N Y I	LD[e]
Gi2	L E R I A Q S D Y I	LE[e]
Gi3[c]	L D R I S Q S N Y I	SQ[e]
Go	L D R I G A A D Y Q	
Gz(x)	L E R I A A A D Y I	
Gs	I D V I K Q A D Y V	

[a]Single-letter amino acid code. Unless indicated, the sequences shown are conserved in all mammals for which sequence data is available.
[b]Bovine.
[c]Human.
[d]Lerea et al. (1986).
[e]Goldsmith et al. (1988a).

forms of Gsα mRNA could be derived by alternative splicing from a single gene. The major difference between these forms is the presence or absence of a 15-amino acid peptide (between residues 71–87). Immunoblots, using a synthetic peptide antiserum corresponding to the 15-amino acid peptide (Robishaw et al., 1986), showed that the "long" forms of mRNA encode 52-kDa α subunits and the "short" mRNAs encode 45-kDa α subunits. A novel cDNA was cloned from a rat olfactory mucosa library (Jones and Reed, 1989). It encodes a protein termed Golf that is highly homologous to Gs and may link olfactory receptors to adenylyl cyclase. Based on its high homology to Gs, it is likely that many Gs-specific antibodies would cross-react with Golf, but antisera raised against peptides specific to Golf allow discrimination from Gs. Immunohistochemical studies with such antisera identify Golf in cilia of olfactory neurons (Jones and Reed, 1989).

2.3.4. Go

A novel G-protein termed Go, with a 39-kDa α subunit, was discovered upon purification of pertussis toxin substrates from bovine brain (Neer et al., 1984; Sternweis and Robishaw, 1984). cDNAs cloned from rat (Itch et al., 1986; Jones and Reed, 1987) and bovine (Van Meurs et al., 1987) libraries predict a protein sequence that corresponds precisely to amino acid sequence derived for the purified 39-kDa protein. Antisera prepared by immunization with purified brain G-holoprotein (Gierschik et al., 1986c) or with resolved α_{39} subunit (Huff et al., 1985; Tsai et al., 1987) distinguish Goα from other Gα subunits. Antisera prepared against synthetic peptides with sequences predicted by Goα cDNAs also show unique reactivity with Goα (Eide et al., 1987; Goldsmith et al., 1988b; Mumby et al., 1986). A novel G-protein purified from brain differs in charge from the predominant form of Go, but shows an identical pattern of immunoreactivity with a series of peptide antisera. This suggests that the novel protein may represent a Go variant, differing in either primary sequence or in posttranslational modification (Goldsmith et al., 1988b). Others have confirmed the presence of variant forms of Goα in brain (Lang, 1989; Kobayashi et al., 1989). Goα antisera have been used to identify putatively novel forms of Goα in adipocytes (Rouot et al., 1989) and myometrium (Milligan et al., 1989). Further elucidation of the structure of these proteins will be required to confirm these observations.

2.3.5. Gi

Gi was initially defined functionally as the G-protein-coupling receptors to inhibition of adenylyl cyclase. Gi is uncoupled from receptors by pertussis toxin-catalyzed ADP-ribosylation (Ui, 1984). Subsequently, it was found that G-protein modulation of other effectors (e.g., regulation of certain ion channels and activation of phospholipase C in neutrophils) is also pertussis toxin-sensitive. Since inhibition of agonist-stimulated phospholipase C activity by pertussis toxin correlated with ADP-ribosylation of a 40- to 41-kDa protein in neutrophil membranes, it was assumed that this protein is "Gi."

Neutrophil membranes contain an abundant (>20 pmol/mg membrane protein) pertussis toxin substrate(s). Immunoblots with Gt- and Goα-specific antisera show that these pertussis toxin substrates are undetectable in neutrophil membranes (Gierschik et al., 1986b). A unique Gt antiserum, CW/6, was found to cross-react with the 41 kDa form of Giα in brain, but not with Goα (Pines et al., 1985a). Immunoblots with CW/6 showed that the amount of "Gi" in neutrophil membranes also could not account for the abundant pertussis toxin substrate (Gierschik et al., 1986b). These studies provided immunochemical evidence for a novel pertussis toxin substrate in neutrophils.

cDNA cloning likewise suggested heterogeneity of pertussis toxin substrates. In addition to Goα, three closely related forms of "Giα" were cloned (Bray et al., 1987; Didsbury et al., 1987; Didsbury and Snyderman, 1987; Itoh et al., 1986; Jones and Reed, 1987; Nukada et al., 1986b; Suki et al., 1987; Sullivan et al., 1986). The three types, arbitrarily designated Gi1, Gi2, and Gi3 in order of cloning, all contain a cysteine, the putative pertussis toxin ADP-ribosylation site (West et al., 1985), as the fourth residue from the carboxy-terminus. The predicted amino acid sequence of Gi1α cDNA matches the amino acid sequence obtained for the α_{41} purified from bovine and rat brain (Itoh et al., 1986; Nukada et al., 1986b). The Gi2- and Gi3α cDNAs, however, were not initially correlated with the sequence of a purified protein.

278   Allen M. Spiegel et al.

Antisera directed against peptides with sequence based on residues 159–168 in Gi1 and Gi3 and 160–169 in Gi2 (Table 11.1) can differentiate these three Gi subtypes on immunoblots (Figure 11.2). These antisera were used to identify the product of Gi2 cDNA as the predominant pertussis toxin substrate in neutrophils (Gierschik et al., 1987; Goldsmith et al., 1987; Goldsmith et al., 1988a). This 40-kDa protein is also found in brain, and may well be ubiquitously expressed (Goldsmith et al., 1987, 1988a,b; Katada et al., 1987; Mumby et al., 1988). Amino acid sequencing of the 40-kDa protein purified from brain confirms its identity as Gi2 (Itoh et al., 1988). The product of Gi3 cDNA was identified as a 41-kDa protein expressed in HL-60 cells, as well as many other cell types, but poorly expressed in brain (Goldsmith et al., 1988a,b).

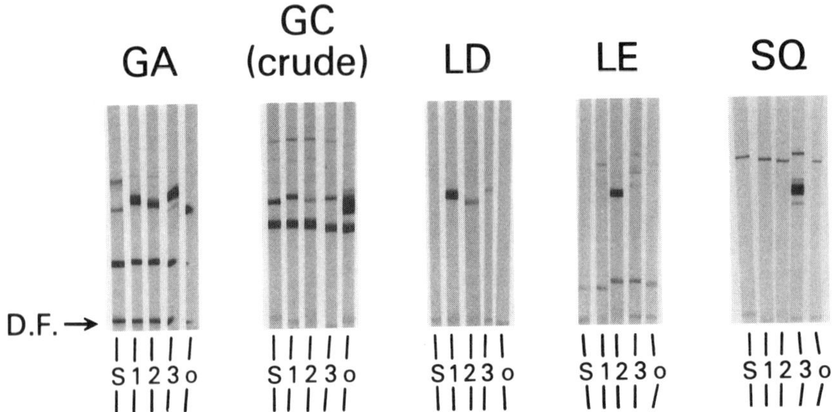

FIGURE 11.2. Assessment of specificity of Gα peptide antisera. cDNAs encoding the indicated Gα subunits were expressed in *E. coli*. (Kindly provided by J. Codina.) After lysing the bacteria, a soluble fraction was collected and subjected to SDS-PAGE. The expressed Gs protein is the approximately 42-kDa protein encoded by the "short" form of mRNA. About 0.5 mL of lysate was loaded on laneless blocks of a 10% SDS gel, transblotted to nitrocellulose paper, and individual strips cut for incubation with antisera. The dye front (D.F.) is indicated. Antibodies used were affinity purified, except where "crude" is indicated. Antibodies used were LD, LE, and SQ (see Table 11.1), GC (raised against the amino-terminus of Go, and GA (raised against a sequence beginning at about residue 40 and conserved in all Gα subunits except for Gz). Each antibody was incubated with strips containing α subunits of Gs ("S"), Gi1 ("1"), Gi2 ("2"), Gi3 ("3"), and Go ("o"). Note that GA reacts with all Gα subunits (weakly with Gs); GC reacts primarily with Go, and the Gi subtype-selective antibodies (LD, LE, and SQ) are reasonably specific (LD shows weak cross-reactivity with Gi2). Several antibodies, particularly the crude antisera, in addition to reacting with expressed Gα subunits, react with endogenous *E. coli* proteins (e.g., lower bands in GA and GC).

2.3.6. OTHER Gα SUBUNITS

A novel cDNA encoding a 355-amino acid Gα subunit was cloned from human retinal (Fong et al., 1988) and rat brain (Matsuoka et al., 1988) libraries. The predicted protein product, termed Gz (Fong et al., 1988) or Gx (Matsuoka et al., 1988), shows greatest homology to Gi2, but lacks a cysteine residue in the position characteristic of pertussis toxin substrates. Antisera have successfully been raised against peptides predicted by the cDNA (e.g., QN in Table 11.2; see also Figure 11.3), and these react with recombinant-expressed protein (Figure 11.4). Such antisera have identified Gzα as a 40-kDa protein in human platelet membranes (Simonds et al., 1989a; Carlson et al., 1989), as well as in brain and adrenal medulla (Casey et al., 1990).

It is likely that additional Gα subunits remain to be identified. A 43-kDa pertussis toxin substrate of unknown primary structure, but reactive with a G-"common" antiserum, has been purified (Iyengar et al., 1987). Immunochemical studies suggest that this protein may be a variant form of Gi3α (Carty and Iyengar, 1990). One or more pertussis-toxin insensitive G-proteins coupling to phospholipase C remain to be identified. Fragments of several novel α subunit cDNAs have been cloned (Strathmann et al., 1989) using PCR. It will be necessary to define the full sequence of such cDNAs which will enable generation of antibodies against cDNA-predicted peptide sequences and identification of the proteins encoded by novel Gα subunit cDNAs.

2.3.7. SPECIFICITY OF Gα SUBUNIT ANTISERA

Given the degree of homology between certain Gα subtypes, and the difficulty in resolving individual subtypes through conventional purification methods, evaluation of the specificity of G-protein antisera is not straightforward. Results obtained on immunoblots of "purified" proteins

TABLE 11.2. Carboxy-terminal decapeptides and corresponding antisera

Peptide	Sequence[a]	Antiserum[b]	Gα subunit[c]
RM	R M H L R Q Y E L L	RM	Gs
QN	Q N N L K Y I G L C	QN	Gx(z)
GO	A N N L R G C G L Y	GO	Go
EC	K N N L K E C G L Y	EC	Gi3
KE	K E N L K D C G L F	AS	Gt1, Gt2
KN	K N N L K D C G L F	−	Gi1, Gi2

[a]Single-letter amino acid code.
[b]Antiserum raised by immunization with designated peptide.
[c]Designated peptide represents carboxy-terminus of corresponding Gα subunit (Lochrie and Simon, 1988).

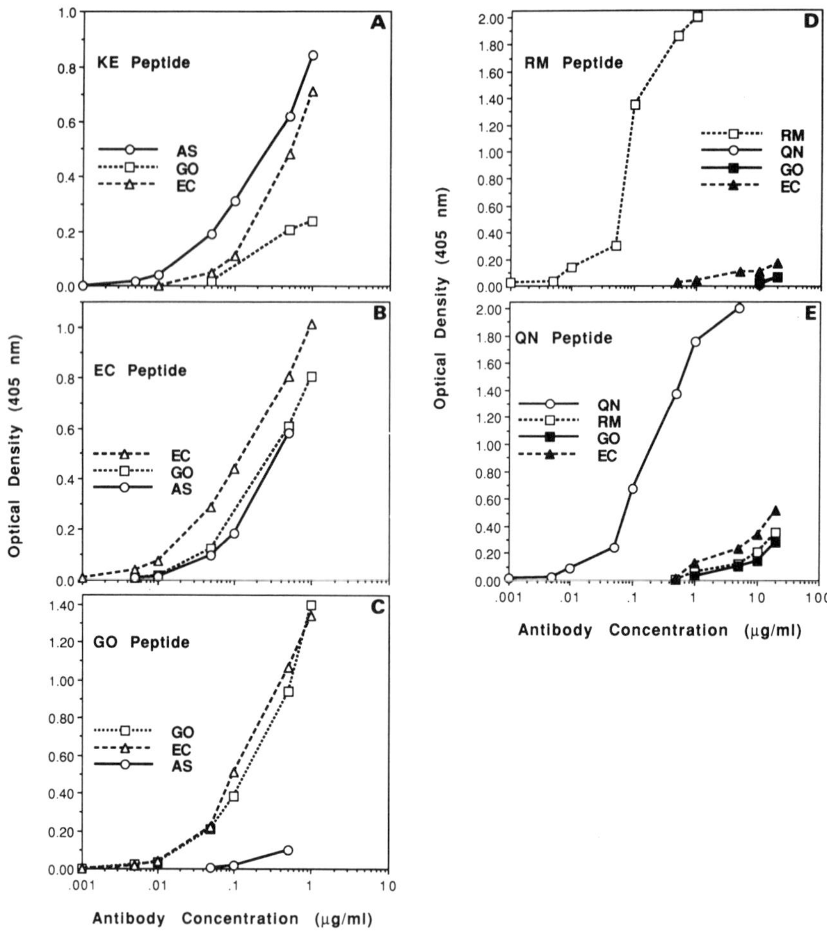

FIGURE 11.3. Characterization of specificity of Gα carboxy-terminal decapeptide antisera by ELISA. ELISA was performed with plates coated with the indicated peptides (see Table 11.2), and then incubated with increasing concentrations of the affinity-purified antibodies. For plates coated with the KE, EC, and GO peptides, affinity-purified RM and QN antibodies (up to 10 μg/mL) gave no detectable reactivity. The values are the means of quadruplicate determinations.

must be considered tentative. A more rigorous analysis is possible with individual, defined G-protein subtypes expressed by recombinant techniques, e.g., in *E. coli* (see Figures 11.2 and 11.4). Thus, one can show that three Gi subtype antisera are at least relatively (LD for Gi1), if not absolutely (LE for Gi2 and SQ for Gi3), specific.

Antisera directed against the carboxy-terminal decapeptide of Gα subunits illustrate the unexpected patterns of cross-reactivity that may develop.

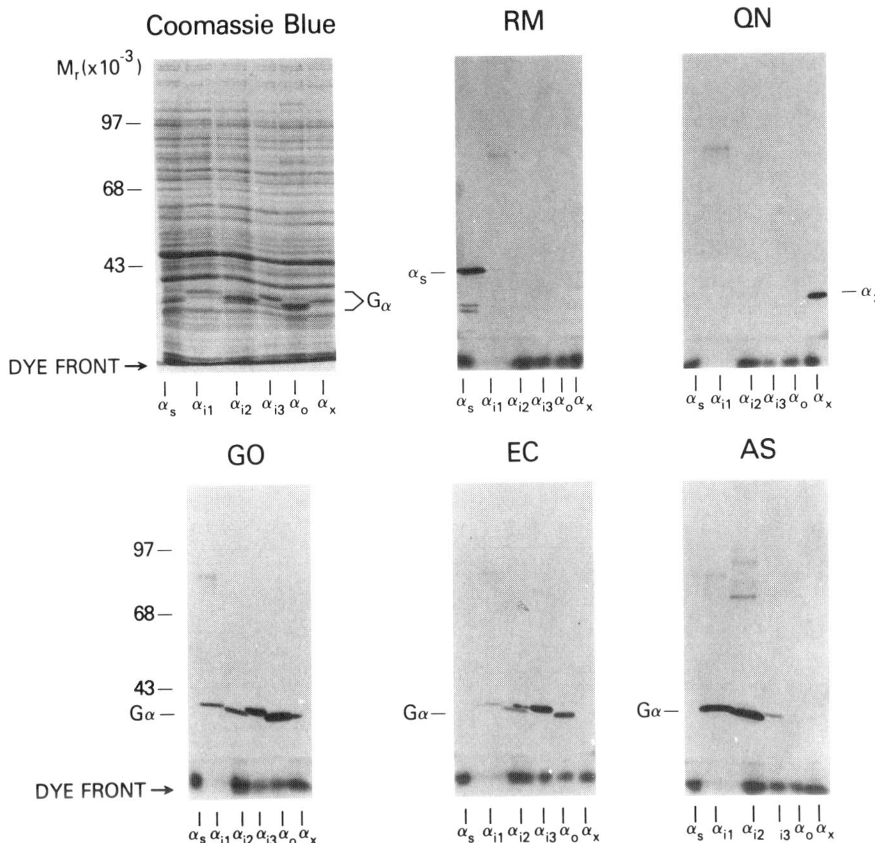

FIGURE 11.4. Characterization of specificity of Gα carboxy-terminal decapeptide antisera by immunoblot of Gα subunits expressed in *E. coli*. Total *E. coli* lysates containing the indicated Gα subunits were applied to a 10% SDS-polyacrylamide gel. One gel was stained for protein with Coomassie blue. Five other replica gels containing one-half as much protein were transblotted to nitrocellulose and the blots incubated with the indicated affinity-purified antibodies. The positions of molecular size markers and of the dye front are indicated. The expressed Gα subunits are visible as Coomassie blue-stained bands with the exception of αs which is obscured by an abundant, endogenous *E. coli* protein beneath the 43-kDa marker. (From Simonds et al., 1989a.)

Both ELISA (Figure 11.3) and immunoblot methods (Figure 11.4) have been used to assess the specificity of these antisera. Antisera RM and QN (Table 11.2) directed against Gs and Gz, respectively, show absolute specificity, as might be predicted by the relatively unique carboxy-terminal decapeptide sequences. The data in Figure 11.4 confirm earlier observations (Goldsmith et al., 1987, 1988a,b) showing that antisera (AS/6 and 7) raised against the carboxy-terminal decapeptide of Gtα react strongly with Gi1 and Gi2, poorly with Gi3, and are virtually unreactive with Go. Antisera against the carboxy-terminal decapeptides of Gi3- and Goα, EC and GO, respectively, both show weak cross-reactivity with Gi1 and Gi2, and strong reciprocal cross-reactivity. This pattern is somewhat surprising given the closer homology of the Gi3 peptide to the Gi1/Gi2 sequence rather than that of Go (Table 11.2). The explanation, based on experiments with "hybrid" peptides such as KNNLKDCGLY, appears to be that the carboxy-terminal residue, phenylalanine (Gt, Gi1, Gi2) vs. tyrosine (Gi3, Go), appears to be a major determinant of antigenicity. The specificity studies done with *E. coli-* expressed G-proteins (Figure 11.4) explain the otherwise puzzling results of immunoblots of crude brain membranes (Figure 11.5). Immunoreactive bands detected with the Gi3 carboxy-terminus antiserum, EC, reflect cross-reactivity with Gi1 and Go, rather than authentic Gi3, which is poorly expressed in brain.

G-proteins expressed in mammalian cells by transfection (see Figure 11.6) may also be useful in assessing antibody specificity, but the presence of endogenous G-proteins in untransfected cells may limit the conclusions that can be drawn. Antibodies, in turn, offer an excellent method for monitoring expression after transfection (e.g., Masters et al., 1988). Immunoreactivity is independent of biologic activity which might, for example, be adversely affected by in vitro mutagenesis.

2.3.8. CONSERVATION OF Gα SUBUNIT STRUCTURE

Polyclonal antisera raised against purified mammalian G-proteins, including Gt (Grunwald et al., 1986; van Veen et al., 1986) and Go (Gierschik et al., 1986c; Homburger et al., 1987), cross-react with G-proteins in

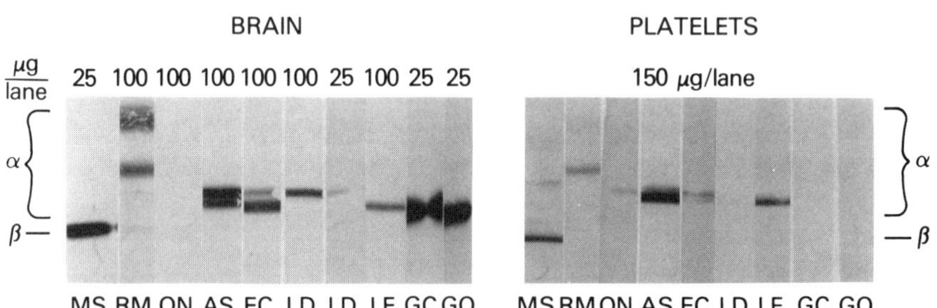

FIGURE 11.5. Immunoblot analysis of G-protein subunit expression in bovine brain and human platelet membranes. Aliquots of a cholate extract of bovine brain membranes containing the indicated amounts of protein, and of human platelet membranes (150 μg/lane) were separated on a 10% SDS-polyacrylamide gel, and immunoblotting performed with the antipeptide antisera indicated. (From Simonds et al., 1989a.)

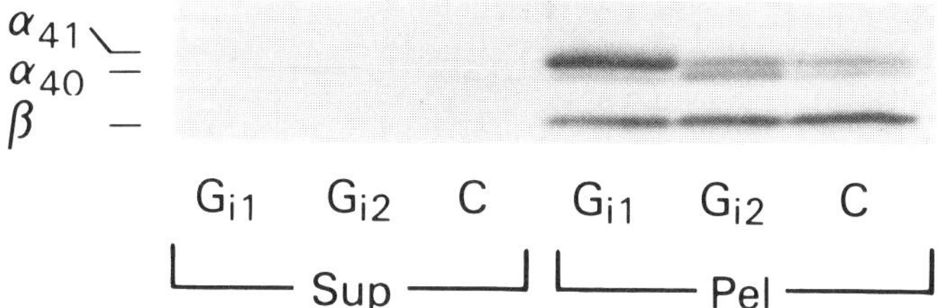

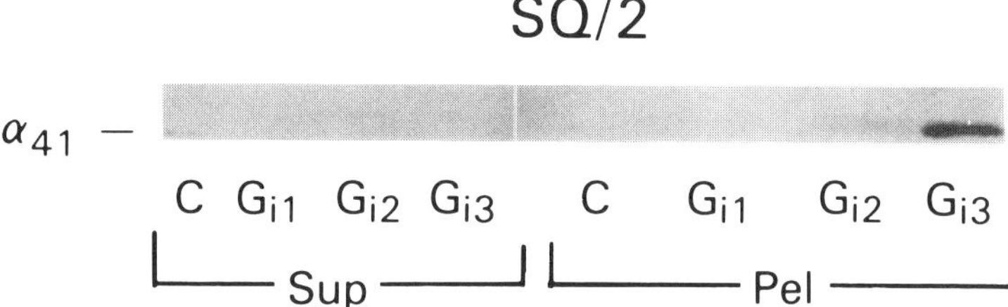

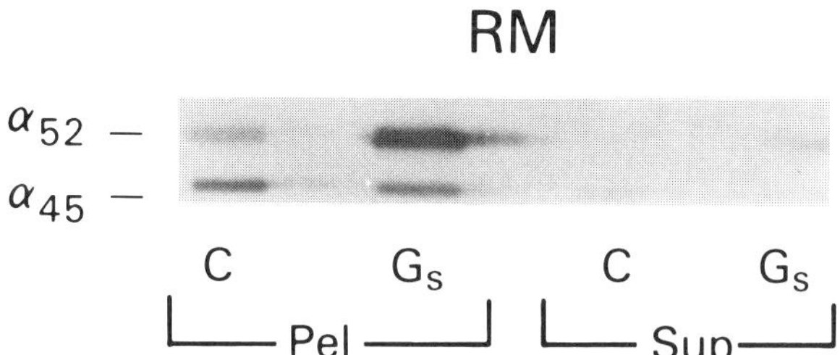

FIGURE 11.6. Subcellular distribution of transiently expressed Gα subunits detected by immunoblotting. Samples of the high speed supernatant (Sup) or crude membrane pellet (Pel) of COS cell lysates were analyzed by immunoblotting with the indicated antisera. [Note that in the top panel AS (αi-specific) and MS (β-specific) were combined.] Below each lane the Gα cDNA employed for transfection is shown (C, sham-transfected control). (From Simonds et al., 1989b.)

submammalian vertebrates. A monoclonal antibody raised against frog Gt (Hamm and Bownds, 1984) recognizes mammalian G-proteins (Yatani et al., 1988). Recently, antisera raised against peptides based on mammalian cDNA-predicted sequences have been used to detect G-proteins in invertebrate species, including slime mold (Snaar-Jagalska et al., 1988), *Drosophila* (Thambi et al., 1989), aplysia (Vogel et al., 1989), and plants (Blum et al., 1988). These results reflect a high degree of evolutionary conservation of primary structure of G-proteins, which is not surprising given the constraints arising from the multiple interactions in which G-proteins participate.

3. Immunochemical Studies of Gα Subunit Posttranslational Modifications

3.1. Phosphorylation

Purified Gα subunits serve as in vitro substrates for phosphorylation by protein kinase C and the insulin receptor tyrosine kinase (Zick et al., 1986). The preferred substrate is the dissociated, GDP-bound α subunit. Demonstration that Gα subunits serve as in vivo substrates for phosphorylation by protein kinases has proved more difficult. Immunoprecipitation with specific antisera has provided evidence for phosphorylation in intact cells of α subunits of Gi2 (Daniel-Issakani et al., 1989; Pyne et al., 1989) and Gz (Carlson et al., 1989) by protein kinase C, but not by the insulin receptor tyrosine kinase (Rothenberg and Kahn, 1988). The functional significance of such phosphorylation is unclear. Speculations include effects on association with $\beta\gamma$ subunits, and inhibition of receptor coupling, but proof is lacking. Likewise, residues subject to phosphorylation have not yet been mapped.

3.2. Myristoylation and Membrane Association of Gα Subunits

Purified Gα subunits behave as hydrophilic proteins in aqueous buffers. This has led to the suggestion that $\beta\gamma$ subunits serve to anchor α subunits to the cytoplasmic side of the plasma membrane (Sternweis, 1986). Under activating conditions (0.1 mM GTP-γ-S), however, most α subunits remain associated with plasma membranes in brain and neutrophil (Eide et al., 1987). Tryptic cleavage of the amino-terminus of Go-, Gi1-, or Gi2α releases these proteins from the membrane, as visualized by immunoblot. This suggests a role for the amino-terminus of α subunits in membrane attachment. When Gα subunits are transiently expressed in COS cells, the majority of immunoreactive protein is membrane-associated, despite the fact that some α subunits (e.g., Gi3) can be expressed in amounts 20-fold greater than endogenous $\beta\gamma$ subunits (Figure 11.6) (Simonds et al., 1989b). This suggests that α subunits can bind to membranes independent of $\beta\gamma$. Immunoprecipitation of cells labeled with [³H]myristate had shown that several forms of Gi- and Goα are myristoylated (Buss et al., 1987). Subsequent studies confirm this, but show that Gsα is not myristoylated (Figure 11.7) (Jones et al., 1990; Mumby et al., 1990). Mutation of Gly² (to alanine) of either Gi1α (Jones et al., 1990) or Goα (Mumby et al., 1990) prevents myristoylation, and subsequent membrane attachment of the mutant proteins (Figure 11.8). Thus, myristoylation is critical for membrane attachment of Gα subunits with the exception of Gsα.

Lack of Gsα myristoylation suggests that this subunit may be more loosely attached to membranes than others. Using an ELISA, one group has claimed that receptor agonists and other activators release Gsα from the membrane to the cytosol (Ransnäs et al., 1989). This observation has not been confirmed; the nature of Gsα interaction with the membrane requires further study. Likewise, the concept that Gα subunits are exclusively localized on the cytoplasmic surface of the cell membrane must be

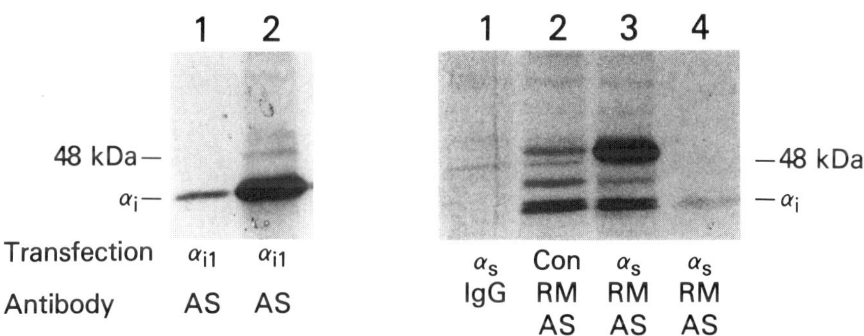

FIGURE 11.7. Immunoprecipitation of G-protein α subunits in transfected COS cells. COS cells were transfected using the diethylaminoethyl (DEAE)-dextran method with the cDNA for αs or αi1 inserted into the pCD-PS expression vector or sham-transfected (Con) and radiolabeled after 48 hr with [³⁵S]methionine or [³H]myristic acid. A particulate fraction was prepared after cell lysis, homogenization, low-speed centrifugation (1000 g for 3 min) and ultracentrifugation of the supernatant (400,000 g for 30 min). Equivalent amounts of protein were immunoprecipitated with rabbit IgG, affinity-purified RM antibodies specific for the αs subunit, or affinity-purified AS antibodies specific for αi1 and αi2. Samples were analyzed by SDS-PAGE, fixed, treated with En³Hance and exposed to XAR-2 film at $-70°C$ for 7–21 days. An endogenous [³⁵S]methionine radiolabeled protein of molecular weight 48 kDa, determined by molecular weight markers, is labeled for reference in the figures. Left panel: [³H]myristic acid- (lane 1) and [³⁵S]methionine- (lane 2) labeled proteins (13 μg protein). Right panel: [³⁵S]methionine- (lanes 1–3) and [³H]myristic acid- (lane 4) labeled proteins (13 μg protein). (From Jones et al., 1990.)

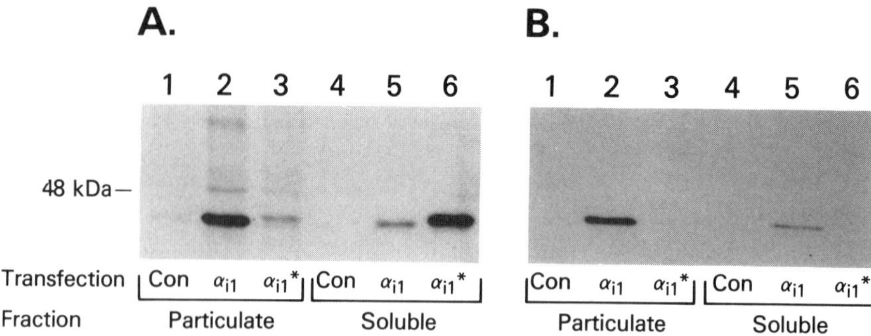

FIGURE 11.8. Immunoprecipitation of cellular fractions after transfection with the mutant αi1. COS cells were transfected with the pCD-PS vector without inserts (Con) or with cDNA for the wild-type αi1 (αi1) or the mutant αi1 (αi1*) and radiolabeled after 48 hr with [^{35}S]methionine or [^{3}H]myristic acid. The cells were lysed, homogenized, and centrifuged at low-speed (1000 g for 3 min). The supernatant was centrifuged at 400,000 g for 30 min. The pellet was resuspended (particulate fraction) and the supernatant recentrifuged at 400,000 g. The final supernatant was concentrated with a centricon-10 microcentrator (soluble fraction). Equivalent amounts of protein were immunoprecipitated with affinity-purified AS antibodies and analyzed by SDS-PAGE and fluorography as described in Figure 11.3. (A) [^{35}S]methionine-labeled proteins (40 μg protein); (B) [^{3}H]myristic acid-labeled proteins (40 μg protein). (From Jones et al., 1990.)

reevaluated. For example, subcellular fractionation of adipocytes, followed by immunoblotting, suggests that both Gs and Gi may move from the plasma membrane to a low-density microsomal fraction upon agonist stimulation (Haraguchi and Rodbell, 1990).

4. Immunochemical Studies of G-Protein Function

G-protein subunits interact with each other, with guanine nucleotides, receptors, effectors, membranes, and perhaps with other, as yet unidentified, entities. Antibodies have been used to define the domains involved in G-protein subunit interactions (Hingorani et al., 1988; Navon and Fung, 1988). In particular, this approach indicates the involvement of the Gα carboxy-terminus in coupling to receptors. Polyclonal antisera raised against Gα carboxy-terminal decapeptides (Simonds et al., 1989c) and a monoclonal antibody against Gtα whose epitope was mapped near the carboxy-terminus (Deretic and Hamm, 1987) both block receptor–G-protein coupling. The monoclonal antibody not only blocks rhodopsin–Gt interaction (Hamm et al., 1987), but also blocks muscarinic stimulation of a pertussis toxin-sensitive G-protein that modulates atrial K$^+$ channel activity (Yatani et al., 1988). This presumably reflects the conservation of the antibody's epitope in various Gα subunits. This also limits the utility of this reagent in defining the specificity of receptor–G-protein coupling. For this purpose, a panel of antisera directed against synthetic decapeptides corresponding to carboxy-termini of Gα subunits (Table 11.2) has proved quite useful. Gtα-specific antibodies block rhodopsin–Gt interaction without interfering with G-protein–effector interaction (Cerione et al., 1988). The comparable Gsα-specific antibodies block agonist-stimulated adenylyl cyclase activation, but are far less effective against fluoride stimulation (Figure 11.9) which bypasses the receptor. The same antibodies immunoprecipitate an activated Gsα–adenylyl cyclase complex (Simonds et al., 1989c). This indicates that the carboxy-terminal decapeptide is not involved in G-protein–effector interaction. Antibodies against the Gtα carboxy-terminus (AS, Table 11.2) cross-react strongly with Gi1- and Gi2α

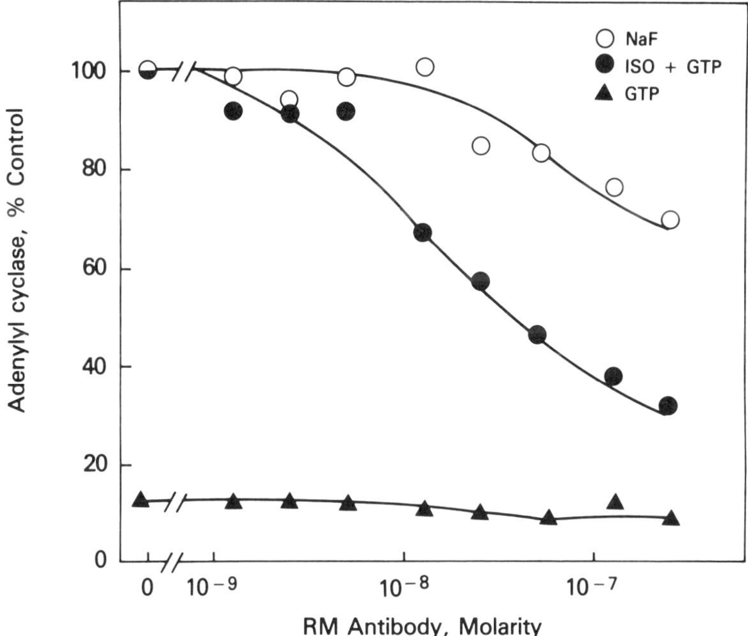

FIGURE 11.9. RM antibody inhibition of Gs activation in S49 mouse lymphoma cell membranes. Membranes were incubated with increasing concentrations (as shown) of RM affinity-purified antibody for 2 hr at 4°C, after which aliquots were assayed for adenylyl cyclase activity with 10 mM NaF, or with 100 μM GTP with or without 500 μM 1-isoproterenol (as shown). Values (in pmol cAMP/min/mg) are the mean of triplicate determinations and are expressed as percentage of control (84 for NaF and 60 for isoproterenol). (From Simonds et al., 1989c.)

(Figure 11.4). Human platelet membranes contain Gi2-, Gi3-, and Gz-, in addition to Gsα (Figure 11.5). AS, but not EC or QN, antibodies (Table 11.2 and Figure 11.10) block α₂-adrenergic-mediated adenylyl cyclase inhibition in human platelet membranes (Simonds et al., 1989a). This indicates that the α₂-adrenergic receptor is coupled to Gi2, and that the latter, at least, can function as a true "Gi." AS antibodies also block opiate-stimulated GTPase in NG 108 cells (McKenzie et al., 1988). This effect presumably reflects antibody-induced uncoupling of opiate receptors from Gi rather than Go, since AS antibodies cross-react strongly with Gi but not with Go (Figure 11.4). In contrast, antibodies directed against the carboxy-terminal decapeptide of Goα block α₂-adrenergic inhibition of a calcium channel in NG 108 cells, indicating a role for Go in this aspect of receptor–effector coupling (McFadzean et al., 1989). Similarly, a polyclonal antiserum raised against purified Go was shown to react with an endogenous pertussis toxin substrate in snail neurons, and upon microinjection to block the activation of this G-protein by dopamine (Harris-Warrick et al., 1988).

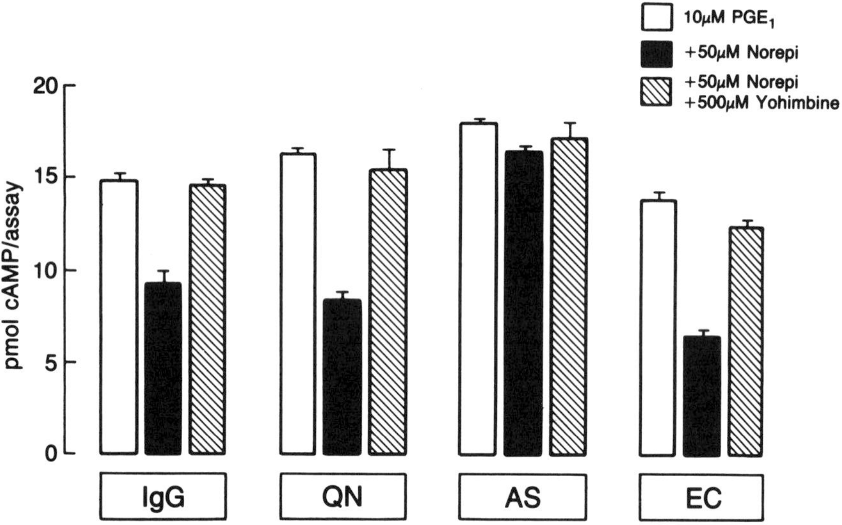

FIGURE 11.10. Effect of carboxy-terminal peptide antisera on α₂-adrenergic-inhibition of adenylyl cyclase in human platelet membranes. Aliquots of membranes were incubated with control rabbit immunoglobulin (IgG), or affinity-purified antibodies QN, AS, or EC at a final antibody concentration of 50 μg/mL. After 1 hr at 4°C, PGE₁-stimulated adenylyl cyclase activity was determined with or without norepinephrine and yohimbine as indicated by the symbols. Values are the mean ± SE of triplicate determinations. PGE₁, prostaglandin E₁; Norepi, norepinephrine. (From Simonds et al., 1989a.)

Antibodies have been used to define specificity of receptor–G-protein coupling in other ways. Under certain conditions, particularly after agonist pretreatment, receptors copurify with G-protein(s). Antibodies to G-proteins have not yet been used to immunoprecipitate receptor–G-protein complexes, but have been used on immunoblots to identify G-proteins copurifying (e.g., after ligand affinity chromatography) with receptors. The anterior pituitary D2-dopamine receptor copurifies with a pertussis toxin substrate. Immunochemical studies suggest the latter may include both Go as well as a form of Gi (Senogles et al., 1987). Similar studies suggest that both Go as well as a form of Gi copurify with muscarinic cholinergic (M2) receptors in heart (Matesic et al., 1989). In liver, the vasopressin receptor couples to a pertussis toxin-insensitive G-protein (Gp) that stimulates phospholipase C. The receptor copurifies with a G-protein shown to contain the immunoreactive β subunit (Fitzgerald et al., 1986). This suggests that the pertussis toxin-insensitive Gp is a heterotrimeric G-protein.

5. Immunochemical Studies of G-Protein Expression

Antibodies have been used extensively in studies of G-protein expression. Limitations of space preclude a detailed review of these applications, but selected examples will be cited. Immunohistochemistry has been successfully performed with antisera directed against α subunits of Gt (Brann and Cohen, 1987; Grunwald et al., 1986; Lad et al., 1987; Lerea et al., 1986; van Veen et al., 1986), Go (Lad et al., 1987; Terashima et al., 1987a,b; Worley et al., 1986), and Gi3 (Cortès et al., 1988). Immunoassays developed to quantitate G-protein subunits include several based on immunoblot techniques (Gierschik et al., 1986c; Homburger et al., 1987; Watkins et al., 1987), and on ELISA techniques (Asano et al., 1987, 1989; Ransnäs, and Insel, 1989). Developmental (Luetje et al., 1987; Milligan et al., 1987a) and differentiation-dependent (Falloon et al., 1986; Gierschik et al., 1986a; Murphy et al., 1987; Watkins et al., 1987) changes in G-protein expression have been demonstrated using both techniques. Modulation of G-protein expression by factors such as thyroid hormone (Milligan et al., 1987b), glucocorticoids (Saito et al., 1989), and ethanol (Charness et al., 1988) has been studied with G-protein antisera. In some instances, e.g., adenosine-mediated desensitization in fat cells, changes in G-protein expression are not accompanied by changes in mRNA (Longabaugh et al., 1989). Such results indicate that regulation may occur at the level of G-protein subunit mRNA translation and protein degradation. Initial studies of Goα turnover confirm this idea (Silbert et al., 1990). Antibodies have also proved invaluable in studies of altered G-protein expression in pathologic states such as pituitary tumors (Collu et al., 1988), heart failure (Longabaugh et al., 1988), and retinitis pigmentosa (Navon et al., 1987). Antibodies defined Gi2α as the pertussis toxin substrate differentially expressed in high vs. low metastatic melanoma cell lines (Lester et al., 1989). Evidence that a pertussis toxin substrate mediates the chemotactic response of tumor cells to extracellular matrix factors (Aznavoorian et al., 1990) emphasizes the potential importance of differences in Gi2α expression on the metastatic behavior of melanoma cells. With the identification of novel G-proteins and G-protein-mediated functions, one can anticipate further examples of dysregulated G-protein expression as a cause of disease.

References

Amatruda TT 3rd, Gautam N, Fong HK, Northup JK, Simon MI (1988): The 35- and 36-kDa β subunits of GTP-binding regulatory proteins are products of separate genes. *J Biol Chem* 263:5008–5011

Asano T, Semba R, Ogasawara N, Kato K (1987): Highly sensitive immunoassay for the α subunit of the GTP-binding protein Go and its regional distribution in bovine brain. *J Neurochem* 48:1617–1623

Asano T, Morishita R, Semba R, Itoh H, Kaziro Y, Kato K (1989): Identification of lung major GTP-binding protcin as Gi2 and its distribution in various rat tissues determined by immunoassay. *Biochemistry* 28:4749–4754

Audigier Y, Pantaloni C, Bigay J, Deterre P, Bockaert J, Homburger V (1985): Tissue expression and phylogenetic appearance of the β and γ subunits of GTP binding proteins. *FEBS Lett* 189:1–7

Aznavoorian S, Stracke ML, Krutzsch H, Schiffman E, Liotta LA (1990): Signal transduction for chemotaxis and haptotaxis by matrix molecules in tumor cells. *J Cell Biol* 110:1427–1438

Backlund PS Jr, Simonds WF, Spiegel AM (1990): Carboxyl-methylation and C-terminal processing of the brain G protein γ subunit. *J Biol Chem* 265:15572–15577

Blum W, Hinsch KD, Schultz G, Weiler EW (1988): Identification of GTP-binding proteins in the plasma membrane of higher plants. *Biochem Biophys Res Commun* 156:954–959

Brann MR, Cohen LV (1987): Diurnal expression of transducin mRNA and translocation of transducin in rods of rat retina. *Science* 235:585–587

Bray P, Carter A, Simons C, Guo V, Puckett C, Kamholz J, Spiegel A, Nirenberg M (1986): Human cDNA clones for four species of G α s signal transduction protein. *Proc Natl Acad Sci USA* 83:8893–8897

Bray P, Carter A, Guo V, Puckett C, Kamholz J, Spiegel A, Nirenberg M (1987): Human cDNA clones for an α subunit of Gi signal-transduction protein. *Proc Natl Acad Sci USA* 84:5115–5119

Buss JE, Mumby SM, Casey PJ, Gilman AG, Sefton BM (1987): Myristoylated α subunits of guanine nucleotide-binding regulatory proteins. *Proc Natl Acad Sci USA* 84:7493–7497

Carlson KE, Brass LF, Manning DR (1989): Thrombin and phorbol esters cause the selective phosphorylation of a G protein other than Gi in human platelets. *J Biol Chem* 264:13298–13305

Carty DJ, Iyengar R (1990): A 43kDa form of the GTP-binding protein Gi3 in human erythrocytes. *FEBS Lett* 262:101–103

Casey PJ, Fong HKW, Simon MI, Gilman AG (1990): G$_z$, a guanine nucleotide-binding protein with unique biochemical properties. *J Biol Chem* 265:2383–2390

Cerione RA, Kroll S, Rajaram R, Unson C, Goldsmith P, Spiegel AM (1988): An antibody directed against the carboxyl-terminal decapeptide of the α subunit of the retinal GTP-binding protein, transducin effects on transducin function. *J Biol Chem* 263:9345–9352

Charness ME, Querimit LA, Henteleff M (1988): Ethanol differentially regulates G proteins in neural cells. *Biochem Biophys Res Commun* 155:138–143

Codina J, Stengel D, Woo SL, Birnbaumer L (1986): β-subunits of the human liver Gs/Gi signal-transducing proteins and those of bovine retinal rod cell transducin are identical. *FEBS Lett* 207:187–192

Collu R, Bouvier C, Lagacé G, Unson CG, Milligan G, Goldsmith P, Spiegel AM (1988): Selective deficiency of guanine nucleotide-binding protein Go in two dopamine-resistant pituitary tumors. *Endocrinology* 122:1176–1178

Cortès R, Hökfelt T, Schalling M, Goldstein M, Goldsmith P, Spiegel A, Unson C, Walsh J (1988): Antiserum raised against residues 159–168 of the guanine nucleotide-binding protein Gi3-α reacts with ependymal cells and some neurons in the rat brain containing cholecystokinin- or cholecystokinin- and tyrosine 3-hydroxylase-like immunoreactivities. *Proc Natl Acad Sci USA* 85:9351–9355

Daniel-Issakani S, Spiegel AM, Strulovici B (1989): Lipopolysaccharide response is linked to the GTP binding protein, Gi$_2$, in the promonocytic cell line U937. *J Biol Chem* 264:20240–20247

Deretic D, Hamm HE (1987): Topographic analysis of antigenic determinants recognized by monoclonal antibodies to the photoreceptor guanyl nucleotide-binding protein, transducin. *J Biol Chem* 262:10839–10847

Didsbury JR, Ho YS, Snyderman R (1987): Human Gi protein α-subunit: Deduction of amino acid structure from a cloned cDNA. *FEBS Lett* 211:160–164

Didsbury JR, Snyderman R (1987): Molecular cloning of a new human G protein. Evidence for two Gi α-like protein families. *FEBS Lett* 219:259–263

Eide B, Gierschik P, Milligan G, Mullaney I, Unson C, Goldsmith P, Spiegel A (1987): GTP-binding proteins in brain and neutrophil are tethered to the plasma membrane via their amino termini. *Biochem Biophys Res Commun* 148:1398–1405

Evans T, Fawzi A, Fraser ED, Brown ML, Northup JK (1987): Purification of a β 35 form of the $\beta\gamma$ complex common to G-proteins from human placental membranes. *J Biol Chem* 262:176–181

Falloon J, Malech H, Milligan G, Unson C, Kahn R, Goldsmith P, Spiegel A (1986): Detection of the major pertussis toxin substrate of human leukocytes with antisera raised against synthetic peptides. *FEBS Lett* 209:352–356

Fitzgerald TJ, Uhing RJ, Exton JH (1986): Solubilization of the vasopressin receptor from rat liver plasma membranes. Evidence for a receptor X GTP-binding protein complex. *J Biol Chem* 261:16871–16877

Fong HK, Hurley JB, Hopkins RS, Miake-Lye R, Johnson MS, Doolittle RF, and Simon MI (1986): Repetitive segmental structure of the transducin beta subunit: Homology with the CDC4 gene and identification of related mRNAs. *Proc Natl Acad Sci USA* 83:2162–2166

Fong HK, Amatruda TT 3rd, Birren BW, Simon MI (1987): Distinct forms of the β subunit of GTP-binding regulatory proteins identified by molecular cloning. *Proc Natl Acad Sci USA* 84:3792–3796

Fong HK, Yoshimoto KK, Eversole-Cire P, and Simon MI (1988): Identification of a GTP-binding protein α subunit that lacks an apparent ADP-ribosylation site for pertussis toxin. *Proc Natl Acad Sci USA* 85:3066–3070

Gao B, Mumby S, Gilman AG (1987a): The G protein β 2 complementary DNA encodes the beta 35 subunit. *J Biol Chem* 262:17254–17257

Gao B, Gilman AG, Robishaw JD (1987b): A second form of the β subunit of signal-transducing G proteins. *Proc Natl Acad Sci USA* 84:6122–6125

Gautam N, Baetscher M, Aebersold R, Simon MI (1989): A G protein γ subunit shares homology with ras proteins. *Science* 244:971–974

Gierschik P, Codina J, Simons C, Birnbaumer L, Spiegel A (1985): Antisera against a guanine nucleotide binding protein from retina cross-react with the β subunit of the adenylyl cyclase-associated guanine nucleotide binding proteins, Ns and Ni. *Proc Natl Acad Sci USA* 82:727–731

Gierschik P, Morrow B, Milligan G, Rubin C, Spiegel A (1986a): Changes in the guanine nucleotide-binding proteins, Gi and Go, during differentiation of 3T3-L1 cells. *FEBS Lett* 199:103–106

Gierschik P, Falloon J, Milligan G, Pines M, Gallin JI, Spiegel A (1986b): Immunochemical evidence for a novel pertussis toxin substrate in human neutrophils. *J Biol Chem* 261:8058–8062

Gierschik P, Milligan G, Pines M, Goldsmith P, Codina J, Klee W, Spiegel A (1986c): Use of specific antibodies to quantitate the guanine nucleotide-binding protein Go in brain. *Proc Natl Acad Sci USA* 83:2258–2262

Gierschik P, Sidiropoulos D, Spiegel A, Jakobs KH (1987): Purification and immunochemical characterization of the major pertussis-toxin-sensitive guanine-nucleotide-binding protein of bovine-neutrophil membranes. *Eur J Biochem* 165:185–194

Gilman AG (1987): G proteins: Transducers of receptor-generated signals. *Annu Rev Biochem* 56:615–649

Goldsmith P, Gierschik P, Milligan G, Unson CG, Vinitsky R, Malech HL, Spiegel AM (1987): Antibodies directed against synthetic peptides distinguish between GTP-binding proteins in neutrophil and brain. *J Biol Chem* 262:14683–14688

Goldsmith P, Rossiter K, Carter A, Simonds W, Unson CG, Vinitsky R, Spiegel AM (1988a): Identification of the GTP-binding protein encoded by Gi3 complementary DNA. *J Biol Chem* 263:6476–6479

Goldsmith P, Backlund PS Jr, Rossiter K, Carter A, Milligan G, Unson CG, Spiegel A (1988b): Purification of heterotrimeric GTP-binding proteins from brain: Identification of a novel form of Go. *Biochemistry* 27:7085–7090

Grunwald GB, Gierschik P, Nirenberg M, Spiegel A (1986): Detection of α-transducin in retinal rods but not cones. *Science* 231:856–859

Halpern JL, Tsai SC, Adamik R, Kanaho Y, Bekesi E, Kung HF, Moss J, Vaughan M (1986): Structural and functional characterization of guanyl nucleotide-binding proteins using monoclonal antibodies to the α-subunit of transducin. *Mol Pharmacol* 29:515–519

Hamm HE, Bownds MD (1984): A monoclonal antibody to guanine nucleotide binding protein inhibits the light-activated cyclic GMP pathway in frog rod outer segments. *J Gen Physiol* 84:265–280

Hamm HE, Deretic D, Hofmann KP, Schleicher A, Kohl B (1987): Mechanism of action of monoclonal antibodies that block the light activation of the guanyl nucleotide-binding protein, transducin. *J Biol Chem* 262:10831–10838

Hancock JF, Magee AI, Childs JE, Marshall CJ (1989): All ras proteins are polyisoprenylated but only some are palmitoylated. *Cell* 57:1167–1177

Haraguchi K, Rodbell M (1990): Isoproterenol stimulates shift of G proteins from plasma membrane to pinocytotic vesicles in rat adipocytes: A possible means of signal dissemination. *Proc Natl Acad Sci USA* 87:1208–1212

Harris BA, Robishaw JD, Mumby SM, Gilman AG (1985): Molecular cloning of complementary DNA for the α subunit of the G protein that stimulates adenylate cyclase. *Science* 229:1274–1277

Harris-Warrick RM, Hammond C, Paupardin-Tritsch D, Homburger V, Rouot B, Bockaert J, Gerschenfeld HM (1988): An α-40 subunit of a GTP-binding protein immunologically related to Go mediates a dopamine-induced decrease of calcium current in snail neurons. *Neuron* 1:27–32

Hingorani VN, Tobias DT, Henderson JT, Ho YK (1988): Chemical cross-linking of bovine retinal transducin and cGMP phosphodiesterase. *J Biol Chem* 263:6916–6926

Homburger V, Brabet P, Audigier Y, Pantaloni C, Bockaert J, and Rouot B (1987): Immunological localization of the GTP-binding protein Go in different tissues of vertebrates and invertebrates. *Mol Pharmacol* 31:313–319

Huff RM, Axton JM, Neer EJ (1985): Physical and immunological characterization of a guanine nucleotide-binding protein purified from bovine cerebral cortex. *J Biol Chem* 260:10864–10871

Hurley JB, Fong HK, Teplow DB, Dreyer WJ, Simon MI (1984): Isolation and characterization of a cDNA clone for the γ subunit of bovine retinal transducin. *Proc Natl Acad Sci USA* 81:6948–6952

Itoh H, Kozasa T, Nagata S, Nakamura S, Katada T, Ui M, Iwai S, Ohtsuka E, Kawasaki H, Suzuki K, Kaziro Y (1986): Molecular cloning and sequence determination of cDNAs for alpha subunits of the guanine nucleotide-binding proteins Gs, Gi, and Go from rat brain. *Proc Natl Acad Sci USA* 83:3776–3780

Itoh H, Katada T, Ui M, Kawasaki H, Suzuki K, Kaziro Y (1988): Identification of three pertussis toxin substrates (41, 40 and 39 kDa proteins) in mammalian brain. Comparison of predicted amino acid sequences from G-protein α-subunit genes and cDNAs with partial amino acid sequences from purified proteins. *FEBS Lett* 230:85–89

Iyengar R, Rich KA, Herberg JT, Grenet D, Mumby S, Codina J (1987): Identification of a new GTP-binding protein. A Mr = 43,000 substrate for pertussis toxin. *J Biol Chem* 262:9239–9245

Jones DT, Reed RR (1987): Molecular cloning of five GTP-binding protein cDNA species from rat olfactory neuroepithelium. *J Biol Chem* 262:14241–14249

Jones DT, Reed RR (1989): Golf: An olfactory neuron specific-G protein involved in odorant signal transduction. *Science* 244:790–795

Jones TLZ, Simonds WF, Merendino JJ Jr, Brann MR, Spiegel AM (1990): Myristoylation of an inhibitory GTP-binding protein a subunit is essential for its membrane attachment. *Proc Natl Acad Sci USA* 87:568–572

Jurnak F (1985): Structure of the GDP domain of EF-Tu and location of the amino acids homologous to ras oncogene proteins. *Science* 230:32–36

Katada T, Oinuma M, Kusakabe K, Ui M (1987): A new GTP-binding protein in brain tissues serving as the specific substrate of islet-activating protein, pertussis toxin. *FEBS Lett* 213:353–358

Kobayashi I, Shibasaki H, Takahashi K, Kikkawa S, Ui M, Katada T (1989):

Purification of GTP-binding proteins from bovine brain membranes: Identification of heterogeneity of the α-subunit of G$_o$ proteins. *FEBS Lett* 257:177–180

Kozasa T, Itoh H, Tsukamoto T, Kaziro Y (1988): Isolation and characterization of the human Gs α gene. *Proc Natl Acad Sci USA* 85:2081–2085

Lad RP, Simons C, Gierschik P, Milligan G, Woodard C, Griffo M, Goldsmith P, Ornberg R, Gerfen CR, Spiegel A (1987): Differential distribution of signal-transducing G-proteins in retina. *Brain Res* 423:237–246

Lang J (1989): Purification and characterization of subforms of the guanine-nucleotide-binding proteins G α i and G α o. *Eur J Biochem* 183:687–692

Lerea CL, Somers DE, Hurley JB, Klock IB, Bunt-Milam AH (1986): Identification of specific transducin α subunits in retinal rod and cone photoreceptors. *Science* 234:77–80

Lester BR, McCarthy JB, Sun ZQ, Smith RS, Furcht LT, Spiegel AM (1989): G-protein involvement in matrix-mediated motility and invasion of high and low experimental metastatic B16 melanoma clones. *Cancer Res* 49:5940–5948

Levine MA, Smallwood PM, Moen PT Jr, Helman LJ, Ahn TG (1990): Molecular cloning of β3 subunit, a third form of the G protein β-subunit polypeptide. *Proc Natl Acad Sci USA* 87:2329–2333

Lochrie MA, Hurley JB, Simon MI (1985): Sequence of the α subunit of photoreceptor G protein: Homologies between transducin, ras, and elongation factors. *Science* 228:96–99

Lochrie MA, Simon MI (1988): G protein multiplicity in eukaryotic signal transduction systems. *Biochemistry* 27:4957–4965

Longabaugh JP, Vatner DE, Vatner SF, Homcy CJ (1988): Decreased stimulatory guanosine triphosphate binding protein in dogs with pressure-overload left ventricular failure. *J Clin Invest* 81:420–424

Longabaugh JP, Didsbury J, Spiegel A, Stiles GL (1989): Modification of the rat adipocyte A$_1$ adenosine receptor-adenylate cyclase system during chronic exposure to an A$_1$ adenosine receptor agonist: Alterations in the quantity of G$_S\alpha$ and G$_i\alpha$ are not associated with changes in their mRNAs. *Mol Pharmacol* 36:681–688

Luetje CW, Gierschik P, Milligan G, Unson C, Spiegel A, Nathanson NM (1987): Tissue-specific regulation of GTP-binding protein and muscarinic acetylcholine receptor levels during cardiac development. *Biochemistry* 26:4876–4884

Masters SB, Sullivan KA, Miller RT, Beiderman B, Lopez NG, Ramachandran J, Bourne HR (1988): Carboxyl terminal domain of Gs α specifies coupling of receptors to stimulation of adenylyl cyclase. *Science* 241:448–451

Matesic DF, Manning DR, Wolfe BB, Luthin GR (1989): Pharmacological and biochemical characterization of complexes of muscarinic acetylcholine receptor and guanine nucleotide-binding protein. *J Biol Chem* 264:21638–21645

Matsuoka M, Itoh H, Kozasa T, Kaziro Y (1988): Sequence analysis of cDNA and genomic DNA for a putative pertussis toxin-insensitive guanine nucleotide-binding regulatory protein alpha subunit. *Proc Natl Acad Sci USA* 85:5384–5388

McFadzean I, Mullaney I, Brown DA, Milligan G (1989): Antibodies to the GTP binding protein, Go, antagonize noradrenaline-induced calcium current inhibition in NG108-15 hybrid cells. *Neuron* 3:177–182

McKenzie FR, Kelly EC, Unson CG, Spiegel AM, Milligan G (1988): Antibodies which recognize the C-terminus of the inhibitory guanine-nucleotide-binding protein (Gi) demonstrate that opioid peptides and foetal-calf serum stimulate the high-affinity GTPase activity of two separate pertussis-toxin substrates. *Biochem J* 249:653–659

Milligan G, Streaty RA, Gierschik P, Spiegel AM, Klee WA (1987a): Development of opiate receptors and GTP-binding regulatory proteins in neonatal rat brain. *J Biol Chem* 262:8626–8630

Milligan G, Spiegel AM, Unson CG, Saggerson ED (1987b): Chemically induced hypothyroidism produces elevated amounts of the α subunit of the inhibitory guanine nucleotide binding protein (Gi) and the β subunit common to all G-proteins. *Biochem J* 247:223–227

Milligan G, Tanfin Z, Goureau O, Unson C, Harbon S (1989): Identification of both Gi2 and a novel, immunologically distinct, form of Go in rat myometrial membranes. *FEBS Lett* 244:411–416

Mumby SM, Kahn RA, Manning DR, Gilman AG (1986): Antisera of designed specificity for subunits of guanine nucleotide-binding regulatory proteins. *Proc Natl Acad Sci USA* 83:265–269

Mumby SM, Pang IH, Gilman AG, Sternweis PC (1988): Chromatographic resolution and immunologic identification of the α 40 and α 41 subunits of guanine nucleotide-binding regulatory proteins from bovine brain. *J Biol Chem* 263:2020–2026

Mumby SM, Heukeroth RO, Gordon JI, Gilman AG (1990): G-protein α-subunit expression, myristoylation, and membrane association in COS cells. *Proc Natl Acad Sci USA* 87:728–732

Murphy PM, Eide B, Goldsmith P, Brann M, Gierschik P, Spiegel A, Malech HL (1987): Detection of multiple forms of Gi α in HL60 cells. *FEBS Lett* 221:81–86

Navon SE, Fung BK (1987): Characterization of transducin from bovine retinal rod outer segments. Participation of the amino-terminal region of T α in subunit interaction. *J Biol Chem* 262:15746–15751

Navon SE, Fung BK (1988): Characterization of transducin from bovine retinal rod outer segments. Use of monoclonal antibodies to probe the structure and function of the subunit. *J Biol Chem* 263:489–496

Navon SE, Lee RH, Lolley RN, Fung BK (1987): Immunological determination of transducin content in retinas exhibiting inherited degeneration. *Exp Eye Res* 44:115–125

Neer EJ, Lok JM, Wolf LG (1984): Purification and properties of the inhibitory guanine nucleotide regulatory unit of brain adenylate cyclase. *J Biol Chem* 259:14222–14229

Nukada T, Tanabe T, Takahashi H, Noda M, Hirose T, Inayama S, Numa S (1986a): Primary structure of the α-subunit of bovine adenylate cyclase-stimulating G-protein deduced from the cDNA sequence. *FEBS Lett* 195:220–224

Nukada T, Tanabe T, Takahashi H, Noda M, Haga K, Haga T, Ichiyama A, Kanagawa K, Hiranaga M, Matsuo, Numa S (1986b): Primary structure of the α-subunit of bovine adenylate cyclase-inhibiting G-protein deduced from the cDNA sequence. *FEBS Lett* 197:305–310

Pines M, Gierschik P, Milligan G, Klee W, Spiegel A (1985a): Antibodies against the carboxyl-terminal 5-kDa peptide of the α subunit of transducin crossreact with the 40-kDa but not the 39-kDa guanine nucleotide binding protein from brain. *Proc Natl Acad Sci USA* 82:4095–4099

Pines M, Gierschik P, Spiegel A (1985b): The tryptic and chymotryptic fragments of the β-subunit of guanine nucleotide binding proteins in brain are identical to those of retinal transducin. *FEBS Lett* 182:355–359

Pyne NJ, Murphy GJ, Milligan G, Houslay MD (1989): Treatment of intact hepatocytes with either the phorbol ester TPA or glucagon elicits the phosphorylation and functional inactivation of the inhibitory guanine nucleotide regulatory protein Gi. *FEBS Lett* 243:77–82

Ransnäs LA, Insel PA (1989): Quantitation of a guanine nucleotide binding regulatory protein by an enzyme-linked immunosorbent competition assay. *Anal Biochem* 176:185–190

Ransnäs LA, Svoboda P, Jasper JR, Insel PA (1989): Stimulation of β-adrenergic receptors of S49 lymphoma cells redistributes the α subunit of the stimulatory G protein between cytosol and membranes. *Proc Natl Acad Sci USA* 86:7900–7903

Robishaw JD, Smigel MD, Gilman AG (1986): Molecular basis for two forms of the G protein that stimulates adenylate cyclase. *J Biol Chem* 261:9587–9590

Robishaw JD, Kalman VK, Moomaw CR, Slaughter CA (1989): Existence of two γ subunits of the G proteins in brain. *J Biol Chem* 264:15758–15761

Roof DJ, Applebury ML, Sternweis PC (1985): Relationships within the family of GTP-binding proteins isolated from bovine central nervous system. *J Biol Chem* 260:16242–16249

Rothenberg PL, Kahn CR (1988): Insulin inhibits pertussis toxin-catalyzed ADP-ribosylation of G-proteins. Evidence for a novel interaction between insulin receptors and G-proteins. *J Biol Chem* 263:15546–15552

Rouot B, Carrette J, Lafontan M, Lan Tran P, Fehrentz JA, Bockaert J, Toutant M (1989): The adipocyte Go α-immunoreactive polypeptide is different from the α subunit of the brain Go protein. *Biochem J* 260:307–310

Saito N, Guitart X, Hayward M, Tallman JF, Duman RS, Nestler EJ (1989): Corticosterone differentially regulates the expression of Gs α and Gi α messenger RNA and protein in rat cerebral cortex. *Proc Natl Acad Sci USA* 86:3906–3910

Senogles SE, Benovic JL, Amlaiky N, Unson C, Milligan G, Vinitsky R, Spiegel AM, Caron MG (1987): The D2-dopamine receptor of anterior pituitary is functionally associated with a pertussis toxin-sensitive guanine nucleotide binding protein. *J Biol Chem* 262:4860–4867

Silbert S, Michel T, Lee R, Neer EJ (1990): Differential degradation rates of the G protein α o in cultured cardiac and pituitary cells. *J Biol Chem* 265:3102–3105

Simonds WF, Goldsmith PK, Codina J, Unson CG, Spiegel AM (1989a): G_{i2} mediates α_2-adrenergic inhibition of adenylyl cyclase in platelet membranes: *In situ* identification with Gα C-terminal antibodies. *Proc Natl Acad Sci USA* 86:7809–7813

Simonds WF, Collins RM, Spiegel AM, Brann MR (1989b): Membrane attachment of recombinant G-protein α subunits in excess of βγ subunits in a eukaryotic expression system. *Biochem Biophys Res Commun* 164:46–53

Simonds WF, Goldsmith PK, Woodard CJ, Unson CG, Spiegel AM (1989c): Receptor and effector interactions of Gs. Functional studies with antibodies to the α s carboxyl-terminal decapeptide. *FEBS Lett* 249:189–194

Simonds WF, Butrynski JE, Gautam N, Unson CG, Spiegel AM (1991): G protein βγ dimers: Membrane targeting requires subunit coexpression and intact γ CAAX domain. *J Biol Chem* 266:5363–5366

Snaar-Jagalska BE, Kesbeke F, Pupillo M, Van Haastert PJ (1988): Immunological detection of G-protein alpha-subunits in *Dictyostelium discoideum*. *Biochem Biophys Res Commun* 156:757–761

Spiegel AM (1987): Signal transduction by guanine nucleotide binding proteins. *Mol Cell Endocrinol* 49:1–16

Sternweis PC, Robishaw JD (1984): Isolation of two proteins with high affinity for guanine nucleotides from membranes of bovine brain. *J Biol Chem* 259:13806–13813

Sternweis PC (1986): The purified α subunits of Go and Gi from bovine brain require beta gamma for association with phospholipid vesicles. *J Biol Chem* 261:631–637

Strathmann M, Wilkie TM, Simon MI (1989): Diversity of the G-protein family: Sequences from five additional α subunits in the mouse. *Proc Natl Acad Sci USA* 86:7407–7409

Sugimoto K, Nukada T, Tanabe T, Takahashi H, Noda M, Minamino N, Kangawa K, Matsuo H, Hirose T, Inayama S, Numa S (1985): Primary structure of the β-subunit of bovine transducin deduced from the cDNA sequence. *FEBS Lett* 191:235–240

Suki WN, Abramowitz J, Mattera R, Codina J, Birnbaumer L (1987): The human genome encodes at least three non-allellic G proteins with α i-type subunits. *FEBS Lett* 220:187–192

Sullivan KA, Liao YC, Alborzi A, Beiderman B, Chang FH, Masters SB, Levinson AD, Bourne HR (1986): Inhibitory and stimulatory G proteins of adenylate cyclase: cDNA and amino acid sequences of the α chains. *Proc Natl Acad Sci USA* 83:6687–6691

Tanabe T, Nukada T, Nishikawa Y, Sugimoto K, Suzuki H, Takahashi H, Noda M, Haga T, Ichiyama A, Kangawa K, Minamino N, Matsuo H, Numa S (1985): Primary structure of the α-subunit of transducin and its relationship to ras proteins. *Nature* 315:242–245

Terashima T, Katada T, Oinuma M, Inoue Y, Ui M (1987a): Immunohistochemical localization of guanine nucleotide-binding protein in rat retina. *Brain Res* 410:97–100

Terashima T, Katada T, Oinuma M, Inoue Y, Ui M (1987b): Endocrine cells in pancreatic islets of Langerhans are immunoreactive to antibody against guanine nucleotide-binding protein (Go) purified from rat brain. *Brain Res* 417:190–194

Thambi NC, Quan F, Wolfgang WJ, Spiegel A, Forte M (1989): Immunological and molecular characterization of Go α-like proteins in the *Drosophila* central nervous system. *J Biol Chem* 264:18552–18560

Tsai SC, Adamik R, Kanaho Y, Halpern JL, Moss J (1987): Immunological and biochemical differentiation of guanyl nucleotide binding proteins: Interaction of Go α with rhodopsin, anti-Go α polyclonal antibodies, and a monoclonal antibody against transducin α subunit and Gi α. *Biochemistry* 26: 4728–4733

Ui M (1984): Islet-activating protein, pertussis toxin: A probe for functions of the inhibitory component of adenylate cyclase. *Trends Pharmacol Sci* 5:277–279

Van Meurs KP, Angus CW, Lavu S, Kung HF, Czarnecki SK, Moss J, Vaughan M (1987): Deduced amino acid sequence of bovine retinal Go α: Similarities to other guanine nucleotide-binding proteins. *Proc Natl Acad Sci USA* 84:3107–3111

van Veen T, Ostholm T, Gierschik P, Spiegel A, Somers R, Korf HW, Klein DC (1986): α-Transducin immunoreactivity in retinae and sensory pineal organs of adult vertebrates. *Proc Natl Acad Sci USA* 83:912–916

Vogel SS, Chin GJ, Mumby SM, Schonberg M, Schwartz JH (1989): G proteins in aplysia: Biochemical characterization and regional and subcellular distribution. *Brain Res* 478:281–292

Watkins DC, Northup JK, Malbon CC (1987): Regulation of G-proteins in differentiation. Altered ratio of α- to β-subunits in 3T3-L1 cells. *J Biol Chem* 262:10651–10657

West RE Jr, Moss J, Vaughan M, Liu T, Liu TY (1985): Pertussis toxin-catalyzed ADP-ribosylation of transducin cysteine 347 is the ADP-ribose acceptor site. *J Biol Chem* 260:14428–14430

Worley PF, Baraban JM, Van Dop C, Neer EJ, Snyder SH (1986): Go, a guanine nucleotide-binding protein: Immunohistochemical localization in rat brain resembles distribution of second messenger systems. *Proc Natl Acad Sci USA* 83:4561–4565

Yarfitz S, Provost NM, Hurley JB (1988): Cloning of a *Drosophila melanogaster* guanine nucleotide regulatory protein β-subunit gene and characterization of its expression during development. *Proc Natl Acad Sci USA* 85:7134–7138

Yatani A, Hamm H, Codina J, Mazzoni MR, Birnbaumer L, Brown AM (1988): A monoclonal antibody to the α subunit of Gk blocks muscarinic activation of atrial K^+ channels. *Science* 241:828–831

Zaremba T, Gierschik P, Pines M, Bray P, Carter A, Kahn R, Simons C, Vinitsky R, Goldsmith P, Spiegel A (1988): Immunochemical studies of the 36-kDa common β subunit of guanine nucleotide-binding proteins: Identification of a major epitope. *Mol Pharmacol* 33:257–264

Zick Y, Sagi-Eisenberg R, Pines M, Gierschik P, Spiegel AM (1986): Multisite phosphorylation of the α subunit of transducin by the insulin receptor kinase and protein kinase C. *Proc Natl Acad Sci USA* 83:9294–9297

Index

A23187, as initiator of arachidonic acid release, 123
a cells
  response to α factor, 220
  STE2 gene transcript in, 222
A431 cells, receptor desensitization studies on, 45, 46
Acetylcholine
  5-HT$_{1A}$ receptor role in release of, 124
  muscarine and nicotine responses to, 170
  as muscarinic receptor agonist, 174, 180, 183
Actin, mRNA of, 51, 52
A-current, muscarinic agent inhibition of, 179
Address-message theory, of peptide-receptor interactions, 200–201, 202
Adenylyl cyclase
  α-adrenergic receptor inhibition of, 95, 96, 97, 98
  desensitization of, 42, 46
  dopamine D1 receptor activation of, 142, 149, 154, 160
  dopamine D2 receptor inhibition of, 142, 160, 163, 166
  5-HT$_{1AD}$ receptor inhibition of, 130
  muscarinic receptor inhibition of, 174, 175, 183
  possible linkage of 5-HT$_{1A}$ receptor to, 121
  stimulation by α subunits, 251, 261
  stimulation by glycoprotein hormone receptors, 211, 212, 214, 215
  stimulation by G-proteins, 31, 39, 41, 44, 45, 62, 67, 71, 82, 129, 233, 253
ADP-ribosylation, of α subunits, 247, 250, 262, 271, 275, 277
Adrenal cortex tumors, Gi2α mutations in, 260, 261
β-Adrenergic blockers, 5-HT$_{1A}$ receptor affinity for, 130
α$_2$,β$_2$-Adrenergic chimeric receptors, ligand binding in, 225
Adrenergic receptors, 36
  amino acids in, 37
  G-protein interaction with, 83–84
  ligand binding site of, 164
α-Adrenergic receptors, 76–112
  affinity labeling of, 78
  cloning and localization of, 101–103
  differentiation from β-adrenergic receptors, 77
  general structure of, 83–85
  ligand binding by, 68, 84
  molecular identification of, 78–85
  original classification and pharmacology of, 77–78
  pharmacological studies on, 85–86
  purification of, 78–80
  schematic model of, 84
  structural studies on, 98–101
  subtypes of, 76, 77, 85–95, 101–103
α$_1$-Adrenergic receptors
  activation of, 95–96
  affinity labeling studies on, 78
  antagonists selective for, 77

cloning of, 79–80, 82–83
purification of, 78–79
second messenger systems of, 95–98
spiroxatrine binding to, 116
structure-function linkages in, 129
subtypes of, 86–87, 93–95
   activation, 95
   biochemical characteristics, 99
α_1-A-Adrenergic receptors
   biochemical characteristics of, 99
   characteristics of, 87
   cloning studies on, 90, 94, 95
   ligand binding by, 90
   Northern blot analysis of, 94
   tissue distribution of, 94, 95, 96
α_1-B-Adrenergic receptors
   characteristics of, 87, 99
   cloning studies on, 90, 94, 95
   ligand binding by, 90
   Northern blot analysis of, 94
   tissue distribution of, 94, 95, 96
α_1-C-Adrenergic receptors
   characteristics of, 87, 99
   cloning studies on, 90, 95, 114
   ligand binding by, 90
   Northern blot analysis of, 94
   tissue distribution of, 94
α_2-Adrenergic receptors
   affinity labeling studies on, 78
   binding site of, 127
   calcium channel inhibition by, 288
   chimeric (with β_2-adrenergic receptors)
      G-protein coupling studies on, 40, 98,
         100
      ligand binding by, 209
   cloning of, 80–82, 101–103
   comparison among, 103
   coupling to Gi2 protein, 86
   coupling to Na^+/H^+ exchanger, 96
   dopamine D2 receptor similarity to, 167
   functional expression of, 103–105
   human, 77, 86
   ligand binding by, 105–106
   mutagenesis/structural studies on,
      105–106
   Northern blot analysis of, 104
   phosphorylation of, 37, 40, 45, 106
   pre- and postsynaptic, 86
   purification of, 79–80
   second messenger systems of, 95–98
   sodium effects on, 19
   subtypes of, 85–86
      biochemical characteristics, 99
      cloning studies, 88–93

   proposed classification of, 86
   wild-type, 105
α_2-A-Adrenergic receptors, 85, 86, 90, 102,
      103, 104
   as α_2-C10-adrenergic receptor subtype,
      98
   proteolysis of, 105
α_2-B-Adrenergic receptors
   cloning studies on, 85, 86, 89, 92, 102,
      103
α_2-C-Adrenergic receptors, cloning of, 102,
      103
α_2-C2-Adrenergic receptors
   amino acid sequence of, 91
   characteristics of, 86, 99
   cloning studies on, 86, 90, 91, 92, 93,
      102, 103
   ligand binding by, 90
   localization on chromosome 2, 80
   Northern blot analysis of, 92
   similarity to α_2-B-Adrenergic receptor, 92
   structure of, 84, 98
   tissue distribution of, 92
α_2-C4-Adrenergic receptors
   activation of, 96–97
   amino acid sequence of, 91
   biochemical characteristics, 99
   characteristics of, 86
   cloning studies on, 86, 88–89, 90, 92, 93,
      96, 102, 103
   glycosylation of, 126
   ligand binding by, 90
   localization on chromosome 4, 80
   Northern blot analysis of, 92, 103
   similarity to other subtypes, 103
   structure-function linkages in, 129
   tissue distribution of, 92
α_2-C10-Adrenergic receptors
   activation of, 96–97
   amino acid sequence of, 91, 93
   biochemical characteristics, 99
   characteristics of, 86
   cloning studies on, 81, 86, 88, 89, 90,
      92, 93, 96, 102, 103
   5-HT_{1A} receptor homology with, 127
   ligand binding by, 90
   localization on chromosome 10, 80
   Northern blot analysis of, 92
   structure-function linkages in, 129
   tissue distribution of, 92
α_2-pre-Adrenergic receptors, cloning studies
      on, 93
α_2-R_s-Adrenergic receptors, cloning studies
      on, 89

β-Adrenergic receptor kinase
  dopamine D1 receptor as possible substrate for, 149
  receptor phosphorylation by, 40, 43, 44, 45, 55, 105–106
  role in receptor desensitization, 45, 46, 211
β-Adrenergic receptors, 14, 22
  acidic residue in TM 3 of, 209
  binding pocket in, 65, 66, 67
  cloning of, 198
  desensitization control by, 224
  differentiation from α-adrenergic receptors, 73
  genetic analysis of, 62–75
  glycosylation in, 63
  G-protein coupling by, 62, 63
  hydrophilic regions of, 64
  hydrophobic core of, 64–65
  i3 loop of, role in G-protein interactions, 71, 72
  ligand binding by, 62, 63, 64–70, 73, 105, 205, 209
  model for agonist activation of, 73
  mutational analysis of, 62–63, 262
  phosphorylation of, 106
  RS1 dopamine receptor clone similarity to, 144
  subtypes of, 70, 222
  wild-type, 64, 67
β_1-Adrenergic receptors
  affinity for β-adrenergic blocking agents, 116
  amino acids in, comparison with dopamine D1 receptors, 147
  chimeric (with β_2-adrenergic receptors), ligand binding by, 70, 98–99
  cloning of, 80, 90, 93
  5-HT$_{1A}$ receptor homology with, 127, 128
  ligand binding by, 65, 70
  structure-function linkages in, 129
β_2-Adrenergic receptors
  affinity for β-adrenergic blocking agents of, 116
  amino acids in, comparison with dopamine D1 receptors, 147
  binding site of, 127
  for catecholamines, 31, 62
  characterization of, 127
  cloning of, 80, 82, 114, 161
  desensitization of, 41–49
  disulfide bond formation in, 39
  downregulation of, 42, 47–49, 53
  functional uncoupling from adenylyl cyclase activation, 42–46
  gene expression of, 51–52, 54
  glycosylation of, 126
  G-protein coupling with, 40–41, 71–72
  5-HT$_{1A}$ receptor homology with, 127, 128
  human, 35, 39, 126
  ligand binding by, 65, 67, 70
  mRNA of, 49–52
  mutants of, 45, 47
  open reading frame in, 144
  palmitoylation of, 39, 98, 145
  phosphorylation of, 37, 40, 43, 45, 47, 54–55
  regulation of transmembrane signaling by, 41–49
  sequence responsible for cAMP regulation in, 53
  sequestration of, 46–47
  structure-function linkages in, 129
  transmembrane domains of, 114
a-factor receptors, encoding by STE genes, 220
a factors, discrimination from α factors, 225
AF-DX 116
  as muscarinic receptor antagonist, 171, 175
  use in bradycardia therapy, 171
AF-DX derivatives, binding by muscarinic m1–m5 receptors, 172, 173
Affinity chromatography
  α_1-adrenergic receptor purification by, 78–79
  of α_2-adrenergic receptors, 79, 80
Affinity labeling, of α-adrenergic receptors, 78
After hyperpolarization (AHP), mediation by muscarinic M1 receptors, 179
AG sequence, in human Gsα proteins, 243
Alanine residues
  in α_2-adrenergic receptors, 44
  in β-adrenergic receptors, 67, 68
  in rhodopsins, 10, 11, 13, 16, 18
A9 L cells, muscarinic receptor expression in, 172, 174, 176, 179
Alcoholism, dopamine receptor role in, 142
α cells
  α factor secretion by, 220
  STE3 expression in, 224
α factor
  discrimination from a factor, 225
  secretion by yeast cells, 220
  STE2 protein mutant binding of, 222

α-factor receptors, 220, 222
  encoding by STE genes, 220
α1 protein, properties of, 239
α2 protein, as control of STE2 expression, 222
α2 protein, properties of, 239
α receptor, 220
α subunits, 223, 233–269, 271, 274–284
  ADP-ribosylation of, 262
  antisera against, 289
  cDNA clones for, 234–235, 236–237, 258, 259, 272, 274
  chimeric, 261
    structural studies on, 249
  conservation of primary structure in, 248
  in *Dictyostelium*, 239, 258
  in *Drosophila*, 238–239, 258
  ELISA studies on, 280
  evolutionary tree of, 248, 249
  function of, 238–239, 260–262, 284–285, 286–289
  genetic and biochemical studies of, 235, 261–262
  homology in, 274–275
  human genes for, 235, 244–247
  immunological studies on, 270, 274–284, 286–289
  mutational analysis of, 248–251, 260–262
  myristoylation of, 262, 284–286
  oncogenic mutation of, 251–253
  phosphorylation of, 284
  point-mutation and toxin-modification sites of, 248–251
  posttranslational modifications of, 284–286
  properties of, 234, 236–237, 238–239
  receptor interaction site of, 250–251
  relationships among, 249
  role in binding, 21
  SCG1 relationship to, 220
  schematic model of, 252
  sera specificity of, 279–282
  structural comparison from different organisms, 255–257
  structural model and mutational analysis of, 248–251
  structure conservation of, 282, 284
  subtype discrimination of, 271
  of thyroid-stimulating hormone receptor, 215
  transient expression of, 283
  of yeast. *See* under type of yeast
α39 subunits, 275, 276
α41 subunits, 275, 277

Alprenolol, receptor binding by, 105
Alternative splicing, in dopamine receptors, 98
Alzheimer's disease
  dopamine receptor role in, 142
  muscarinic receptors as drug targets for, 171, 186
Amino acid side chains, of rhodopsins, 6
p-Aminoclonidone
  α1-adrenergic receptor binding of, 90
  α2-adrenergic receptor binding of, 89, 90
1-[2-(4-Aminophenyl)ethyl]-4-(3- trifluoro-methylphenyl)piperazine. *See* PAPP
Amygdala, dopamine receptor mRNA in, 150–151
Angiotensin III, receptor docking site for, 219
Angiotensin binding protein receptor, peptide-receptor interaction in, 202
Angiotensin dysfunctions, mas role in, 218
Angiotensinogen, predicted complementary peptides of, 203
Angiotensin receptors. *See also* mas/angiotensin receptor
  glycosylation site of, 38
  mas oncogene product as, 35
  mechanism of, 199
  peptide-receptor interaction in, 203
Antibodies
  use in G-protein subunit studies, 270, 271–272
  use in receptor-G-protein coupling studies, 286–289
Anxiety, role of 5-HT receptors in, 118
APDQ
  chemical name for, 77
  use in of α-adrenergic receptor subclassification, 78
Aplysia, G-proteins in, 284
AQ-RA 741, binding by muscarinic m1–m5 receptors, 173, 183
Arachidonic acid, receptor effects on metabolism of, 123, 170, 175
Arbidopsis thaliana, α subunits of, 239
  comparison with other organisms, 255–257
Arecoline, as muscarinic receptor agonist, 174
ARF proteins, 233
  structure of, 235
Arginine residues
  in β-adrenergic receptors, 67
  in β2-adrenergic receptors, 40
  in Gs1α subunit, 261

in luteinizing hormone-chorionic gonadotropin hormone receptor, 215
in rhodopsins, 10, 11
in STE2 receptor, 221
Arrestin
protein analogous to, role in phosphorylation by β_2-adrenergic kinase, 44
role in rhodopsin deactivation, 22–23
β-Arrestin gene, 44
Asparagine residues
in β-adrenergic receptors, 67
in dopamine D2 receptor model, 164
in rhodopsins, 10, 11, 15, 19, 20, 24
Aspartate, muscarinic receptor effect on
release of, 180
Aspartic acid residues
in α-adrenergic receptor subtypes, 91, 105
in β-adrenergic receptors, 67, 68, 73, 205,
225
in β_2-adrenergic receptors, 35, 36, 127
in dopamine D1 receptors, 145, 146, 147
in dopamine D2 receptor model, 164
in 5-HT$_{1A}$ receptors, 127
in luteinizing hormone-chorionic gonadotropin hormone receptor, 214
in muscarinic receptors, 182
in peptide receptors, 224
in rhodopsins, 10, 11, 14, 23, 127
possible function of, 14, 15, 19
in STE2 receptor, 221
in STE3 receptor, 223
in substance K receptor, 205
Astaxanthin, crustacyanin binding to, 15
Astrocytoma cells, β_2-adrenergic receptor
studies on, 42, 46
Astyanax fasciatus. See Cave fish
AT1-angiotensin AII receptor, cloning of,
218
ATP, as initiator of arachidonic acid release, 123
Atrial peptide family, 198
Atropine, binding by muscarinic m1–m5
receptors, 173
AtT20 cells, muscarinic m4 receptor expression in, 178
Azido-iodobenzene reagent, use in
rhodopsin structure studies, 6
p-Azido-PAPP, as photoaffinity label for
5-HT$_{1A}$ receptor, 125

Bacteriorhodopsin
chromophore-protein interaction in, 16
structure of, 5, 6, 12, 24, 31, 37–38
BamHI endonuclease

use in cloning studies of α_2-adrenergic
receptors, 88
use in dopamine D2 receptors studies, 165
use in studies of RFLP in dopamine D1
receptors, 151
BamHI fragment, of α_2-adrenergic receptor,
80
Basal ganglia, muscarinic receptors in, 185
BE 2254
as α_1-adrenergic receptor antagonist, 77,
83
chemical name for, 77
β-Blockers, as 5-HT$_{1A}$ receptor antagonists,
118
$\beta\gamma$ complex, 272
effector action of, 220, 223, 261
expression of, 284
function of, 273, 284
immunochemical studies on, 272–273
interaction of, 247
structure of, 273, 274
β-sheet elements, in membrane proteins, 4
β subunits, 270
amino acid sequence conservation in, 275
forms of, 273
of GP2 protein, 258
in G-protein coupled to vasopressin receptor, 289
immunological studies on, 270, 273–274
RV/3 serum cross reaction with, 275
STE4 relationship to, 220
of thyroid-stimulating hormone receptor,
215
β_1 subunits, 273
β_2 subunits, 273
β_3 subunits, 273
β_{35} subunits, 273
β_{36} subunits, 273, 274
BgIII endonuclease
use in cloning studies of dopamine D1
receptor, 144
use in studies of RFLP in dopamine D1
receptors, 151
BH-8-MeO-*N*-PAT
as 5-HT$_{1A}$ receptor radioligand, 116
structure of, 117
Binding pocket, in β-adrenergic receptors,
65
Blood pressure, serotonin effects on, 118
Blue color pigment, of humans, structure
of, 3, 11, 14, 19
BM-5, muscarinic receptor binding of, 174
BMY 7378, as 5-HT$_{1A}$ receptor partial antagonist, 118

Bolton Hunter reagent
  as 5-HT$_{1A}$ receptor radioligand, 116
  use in α-adrenergic receptor purification, 79, 83
Bradycardia, AF-DX 116 therapy of, 171
Bradykinin receptor, mechanism of, 199
Brain
  dopamine D1 receptors in, 142, 143, 150–151, 152, 154, 160
  dopamine D2 receptors in, 160, 162
  γ subunits in, 272, 273
  5-HT$_{1A}$ receptors in, 119–121
  mas/angiotensin receptors in, 218
  membranes, dopamine D1 receptor binding studies on, 150
  muscarinic receptors in, 178, 185, 187–188
  muscarinic receptor subtypes in, 174
  novel tachykinin receptors in, 207
Brainstem, muscarinic receptors in, 185
Bsu361 restriction endonuclease, use in cloning studies of α_2-adrenergic receptors, 88
Bufotenine, 5-HT$_{1A}$ receptor affinity for, 116
Buspirone
  as anxiolytic drug, 118
  D2-dopamine receptor binding to, 116
  as 5-HT$_{1A}$ receptor radioligand, 116, 118
(+)-Butaclamol
  effect on dopamine D1 and D5 receptor binding, 149, 150
  5-HT$_{1A}$ receptor affinity for, 116, 118

cA-7 adrenergic receptor clone, amino acid sequence of, 101
CAAX box, of γ subunits, 273
CAG sequence, in human Gsα proteins, 243
Calcitonin-gene-related protein receptor, mechanism of, 199
Calcitonin receptor, mechanism of, 199
Calcium channels
  α_1-adrenergic receptor A subtype linkage to, 87
  Go protein role in regulation of, 233
  muscarinic receptor effects on, 179–180
  voltage-sensitive, as second messengers, 96
Calcium ion
  increase of
    dopamine D1 receptors effects on, 154
    5-HT$_{1A}$ receptor mediation of, 123
  intracellular, receptor-mediated release of, 95–96, 175, 177

Calliphora vicini opsins, structure of, 3, 11
cAMP
  α_2-adrenergic receptor effect on levels of, 97–98
  dopamine receptor effects on levels of, 160, 163
  membrane-permeable analogues of, effects on β_2-adrenergic receptors, 43
  receptor inhibition of synthesis of, 122–123, 170
  regulatory action of, 52, 53–54
  role in receptor downregulation, 48
cAMP response elements, 52, 104
Carbachol, muscarinic receptor binding of, 174
Carbene-generating reagents, use in rhodopsin structure studies, 6
Carbohydrate group, attachment point in opsin, 4
5-Carboxyamidotryptamine, 5-HT$_{1A}$ receptor affinity for, 116, 119
Carboxypeptidase, 44
Carboxy-terminal decapeptide antisera, of Gα subunits, 279–282, 286
Carotenoid binding proteins, ligand binding to, 15
Catecholaminergic agents, binding by dopamine D1 receptors, 147
Catecholamines, β_2-adrenergic receptors for, 31, 66, 68, 127
Caudate membranes, dopamine D1 receptor binding studies on, 150
Caudate-putamen
  dopamine receptor mRNA in, 150, 151
  muscarinic receptors in, 185, 186
Cave fish
  green pigment structure of, 3
  red pigment structure of, 2, 18
C127 cells, α_2-A-adrenergic receptor subcloned into, 104
Central nervous system
  5-HT$_{1A}$ receptor distribution in, 119–121, 130
  muscarinic receptor activity in, 180
  role of 5-HT$_{1A}$ receptors in disorders of, 118
Cerebral cortex, muscarinic receptors in, 171, 175, 185, 186
c-fos, sequence responsible for cAMP regulation in, 53
C6 glioma cells, β_2-adrenergic receptor studies on, 42, 46, 53, 54
cGMP phosphodiesterase, transducin role in regulation of, 233

cGMP phosphodiesterase transduction pathway, rhodopsin stimulation of, 44
CGP 12177, use in receptor sequestration studies, 46, 47
Chemoattractant receptors, G-protein interaction with, 84
Chemotaxis, $G\alpha 1$ as mediator of, 258
Chicken
    red color pigment structure of, 2
    rhodopsin structure of, 2
Chimeric receptors, in studies of structure-function relationships, 40, 63, 70–71, 98–99
Chloramphenicol acetyltransferase, coding sequences of, in promoter-reporter gene fusions, 52–53
Chlorethylclonidine, α_1-adrenergic receptor binding by, 87
Chloride conductance, activation by muscarinic stimulation, 178, 179
CHO cells
    α_2-A-adrenergic receptor transfection of, 104
    α_2-C10-adrenergic receptor activation in, 97–98
    5-HT$_{1A}$ receptor expression in, 122, 123, 129
    thyroid-stimulating hormone receptor expression in, 214
CHO-K1 cells, muscarinic receptor expression in, 172, 176, 179, 184
Cholera toxin
    downregulation promotion by, 49
    effects on α subunits, 262
Cholinergic cells, muscarinic receptors in, 186
Cholinergic systems, role in psychomotor diseases, 151
Chorionic gonadotropin, sequence responsible for cAMP regulation in, 53
Chromaffin cells, muscarinic receptor activity in, 179
Chromosome 2, α_2-C2-adrenergic receptor gene on, 80
Chromosome 4, α_2-C4-adrenergic receptor gene on, 80
Chromosome 5
    α_1-adrenergic receptor gene on, 93–94, 153
    β_2-adrenergic receptor gene on, 153
    dopamine D1 receptor gene on, 153, 155
    gene for schizophrenia susceptibility on, 153
    5-HT$_{1A}$ receptor gene on, 153
Chromosome 6, mas gene on, 218

Chromosome 8, α_1-adrenergic receptor gene on, 93–94
Chromosome 10, α_2-C10-adrenergic receptor on, 80
Chromosome 11, dopamine D2 receptor gene on, 165, 166
Clonidone, as α_2-adrenergic receptor agonist, 77
Cloning, of α-adrenergic receptors, 79–85
Concanavalin A, use in receptor sequestration studies, 46
Coomassie stain, use in α-adrenergic receptor identification, 79
Cortex
    dopamine receptor mRNA in, 150
    5-HT$_{1A}$ receptors in, 119, 120
Coryanthine, α_1- and α_2-adrenergic receptor binding of, 77, 90
COS cells, α subunit expression in, 284
COS-1 cells, α_2-adrenergic receptor expression in, 103
COS-7 cells
    α_2-adrenergic receptor expression in, 82, 103
    α_2-adrenergic receptor subtype expression in, 88–89, 90
    dopamine D1 receptor expression in, 147, 149, 150
    5-HT$_{1A}$ receptor expression in, 122, 125, 126
    luteinizing hormone-chorionic gonadotropin hormone receptor expression in, 212
    muscarinic receptor expression in, 172
    substance P and neuromedin K receptor expression in, 208
    thyroid-stimulating hormone receptor expression in, 214
COS-m6 cells, $Gs1\alpha$ and $Gs4\alpha$ expression in, 243
CRE binding protein, sequence motif recognized by, 52, 53
CREB transcription factor, 53
Crustacyanin, astaxanthin binding to, 15
c-src protein, of protein kinase, 217
C-termini
    of β_2-adrenergic receptors, 41, 44, 72, 147–148, 211
    of α subunits, 262
    of dopamine D1 receptors, 144, 145
    of dopamine D2 receptors, 148, 164
    of dopamine D3 receptors, 148
    of G-protein-coupled receptors, 6, 20, 21, 23, 36–37, 38, 39, 40, 63

of 5-HT$_{1A}$ receptors, 129
of muscarinic receptors, 181
of neuromedin K receptor, 208
of peptide receptors, 224, 226, 227
phosphorylation of, 45
of pituitary glycoprotein hormone receptors, 211, 214, 215, 217
of STE2 receptor, 221–222
of STE3 receptor, 224
of substance K receptor, 205
of tachykinin receptors, 210–211
Cyanogen bromide cleavage, of
    α_1-adrenergic receptor, 82, 83
Cyanopindolol
    receptor binding by, 100, 114
    use in receptor downregulation studies, 50
cyc$^-$ cells, Gsα deficiency in, 275
Cyproheptadine, 5-HT$_{1A}$ receptor affinity
    for, 116
Cysteine residues
    in β-adrenergic receptors, 67
    in dopamine D1 receptors, 145, 148, 149
    in dopamine D2 receptor model, 164
    in EF-Tu protein, 234
    in Giα proteins, 278
    in 5-HT$_{1A}$ receptors, 128
    in human β_2-adrenergic receptors, 35, 39
    in luteinizing hormone-chorionic gonadotropin hormone receptor, 214
    in mas/angiotensin receptor, 219
    in muscarinic receptors, 181
    in rhodopsin, 6, 7, 10, 11, 15, 21, 38
    in substance K receptor, 205
    in tachykinin receptors, 208
Cytoskeleton, rhodopsin interaction with,
    23

4-DAMP, binding by muscarinic m1-m5
    receptors, 173
DDT$_1$MF-2 cells
    α_1-adrenergic receptor purification from,
        78–79, 83, 93, 94
    β_2-adrenergic receptor studies on, 48, 52
Deletion mutagenesis studies
    of β-adrenergic receptors, 63, 65
    of G-protein-receptor interaction, 70–71
Dentate gyrus, 5-HT$_{1A}$ receptors in, 119,
    120
Depression, role of 5-HT receptors in, 118
Desensitization, role of receptor gene expression in, 49–55
DGα1 gene, protein encoded by, 258
Diacylglycerol
    effect on calcium conductance, 180

in M-current activity, 178
Diamagnetic anisotropy, of rhodopsins, 5
Diazirine analogue, of retinal, binding to
    rhodopsin glutamic acid analogue, 15
Diazodiiodosulfanilic acid, use in opsin
    structure studies, 4
Dibutyryl-cAMP, effects on receptor
    mRNA, 49, 51, 104
Dictyostelium discoideum, α subunits of,
    239
    comparison with other organisms,
        255–257
Dicylcohexylcarbodiimide reagent, use in
    rhodopsin structure studies, 8
Digitonin, 48
Dihydroretinal derivatives, retinal replacement with, effects on rhodopsin, 15
8-Dihydroxy-2-(di-*n*-propylamino)tetralin.
    See 8-OH-PAT
Disease, dysregulated G-protein expression
    in, 289–290
Disulfide bonds
    in β_2-adrenergic receptors, 39
    in dopamine D1 receptors, 148
    in peptide receptors, 206, 215, 227
    in rhodopsin, 7
    role in G-protein receptors, 224, 226
Dog, dopamine D2 receptors of, 38
Dopamine
    α_1- and α_2-adrenergic receptor binding
        of, 90
    disorders associated with, 142, 160
    as dominant brain catecholamine transmitter, 160
    effect on dopamine D1 and D5 receptor
        binding, 149, 150, 154
    5-HT$_{1A}$ receptor role in release of, 124
    muscarinic receptor-mediated release of,
        186, 187
Dopamine receptor(s), 36
    alternative splicing in, 98
    amino acids in, 37
    G-protein interaction with, 84
    interactions between, 142–143
    ligand binding by, 65, 66, 68
    role in psychomotor diseases, 142, 151
    sequence homology among, 144–146
Dopamine D1 receptors, 142–159
    amino acids in, 146
        sequence homology of, 143, 144–145,
            147
    binding studies on, 150
    chromosomal location of, 153–154
    cloning of, 143–144

expression of, 149–150
function of, 142
glycosylation of, 144, 148
hydrophobic regions in, 144
ligand binding pocket of, 147
mRNA distribution of, 150–151, 155
open reading frame in, 144
phosphorylation of, 148
protein structure of, 144–149
restriction fragment-length polymorphism
    in, 151
role in neuropsychiatric diseases, 142,
    151, 155
selective drugs for, 141
transmembrane topology of, 145
variants of, 154–155
Dopamine D2 receptors, 160–169
    amino acids in, 146
    auto- and postsynpatic subtypes of, 166
    cloning of, 161–162, 167
    comparison with dopamine D3 receptors,
      145–146
    copurification with Go and Gi proteins,
      289
    coupling to second messenger system,
      162–164
    expression of, 152, 162
    function of, 142
    gene localization of, 164–166
    5-HT$_{1A}$ receptor homology with, 127,
      128
    model of, 164, 165
    mRNA distribution of, 150–151
    oligosaccharides of, 38
    open reading frame in, 144
    purification of, 161
    radioligand binding to, 116
    regulation of, 155
    role in dopamine-associated disorders,
      160
    role in psychomotor disorders, 155
    RS1 clone homology with, 144
    similarity to α_2-adrenergic receptors, 167
    structure-function linkages in, 129
Dopamine D3 receptors
    amino acids in, 146
    comparison with dopamine D2 receptors,
      145
    regulation of, 155
Dopamine D4 receptors, regulation of, 155
Dopamine D5 receptors
    binding studies on, 150
    properties of, 155
Dorsal raphe, 5-HT$_{1A}$ receptors in, 121

Dorsal root ganglion, muscarinic receptor
    activity in, 170
Downregulation, of β_2-adrenergic receptors,
    42, 53
DRD1 gene
    as candidate gene for schizophrenia and
      other neuropsychiatric disorders, 154
    as dopamine D1 receptor gene, 153
Drosophila melanogaster
    α subunits of, 238–239
    β subunit of, 274
    frizzled gene product of, 227
    G-proteins in, 284
    opsin structure of, 3, 13, 19
    rhodospin mutant of, 14
Drug addiction, dopamine receptor role in,
    142
DRW sequence, in LH-CG receptor, 215
DRY sequence
    in mas/angiotensin receptor, 218
    in peptide receptors, 205, 206, 214, 215
Dyskinesia, in Huntington's chorea, dopa-
    mine receptor role in, 142

EcoRI restriction enzyme, use in studies of
    RFLP in dopamine D1 receptors,
    151, 152
Ectothiorhodospira halophila, photoactive
    yellow pigment from, 16
E(D)-R-Y(W) motif, of rhodopsin, role in
    G-protein binding, 21
Effectors, of G-protein subunits, 271–272
EF-Tu protein, 234, 251
    point-mutation sites on, 250
    structure of, 235, 243, 274
    X-ray crystal structure of, 248
Elderly persons, oral dyskinesia in, dopa-
    mine receptor role in, 142
ELISA, use in studies of α subunits, 280,
    281, 289
Emotional stability, dopamine system mod-
    ulation of, 160
Endothelin receptor, mechanism of activity
    of, 199
Epinephrine
    α_1-adrenergic receptor binding of, 90,
      104, 114
    α_2-adrenergic receptor binding of, 90
    downregulation promotion by, 49
    effect on receptor desensitization, 43
    as receptor agonist, 65, 70, 73
ERC sequence, in mas/angiotensin receptor,
    218

ERW sequence
  in luteinizing hormone-chorionic gonado-
      tropin receptor, 214
  in thyroid-stimulating hormone receptor,
      215
Erythrocytes, of β_2-adrenergic receptor
    studies on, 42–43, 46, 48
Erythroleukemia cells, α_2-adrenergic
    receptor activation in, 96
Escherichia coli
  α subunit expression in, 262, 278, 280,
      281, 282
  EF-Tu of. *See* EF-Tu protein
  use in muscarinic receptor studies, 184
Ethanol, modulation of G-protein expres-
    sion by, 289
Eticlopride, as dopamine D2 receptor agon-
    ist, 149, 150
"Extra"-MII, induction by G-protein, 20–21

Farnesylation, of ras protooncogenes, 273
Fat cells, G-proteins in, 286, 289
Fatty acylation, of α-adrenergic receptors,
    85, 98
p-F-HHSiD, binding by muscarinic m1–m5
    receptors, 173
Fibroblasts
  of Gi2α-induced neoplastic transforma-
      tion of, 260
  receptor downregulation studies on, 49
α-Flupenthixol, effect on dopamine D1 and
    D5 receptor binding, 150
Fluphenazine, effect on dopamine D1 and
    D2 receptor binding, 150
F-Met-Leu-Phe receptor, mechanism of ac-
    tion of, 199
Forskolin
  adenylyl cyclase stimulation by, 121, 163
  receptor downregulation induction by, 48,
      53
  stimulation of cAMP formation by, 104
Fourier transform infrared spectroscopy, of
    rhodopsin-retinal interaction, 12–13,
    18, 20
FPCα2 cells, adrenergic receptor subtype
    studies on, 103

GABA, muscarinic receptor effect on re-
    lease of, 180
Galanin receptor, mechanism of, 199
Gallamine, binding by muscarinic m1-m5
    receptors, 173, 174

Gα1 and Gα2 proteins
  cloning of, 258
  structure of, 255–257
Gαi3 protein, effects on adenylyl cyclase
    and phospholipase C, 124
Gα subunits. *See* α subunits
γ subunits
  analogy to STE18, 220
  of GP2 protein, 258
  immunochemical studies on, 270, 272–273
  structural studies on, 272–273
GAP. *See* GTPase-activating protein
G-36 clone, of dopamine D1 receptor, 143,
    144, 150, 151
G-21 DNA, transfection of, 114
Genetic analysis, of β-adrenergic receptor,
    62–75
Gepirone, 5-HT$_{1A}$ receptor affinity for, 116
G functional box, of Ha-ras p21 protion
    and α subunits, 251
GH$_4$C$_1$ cells
  dopamine D2 receptor expression in, 163
  5-HT$_{1A}$ receptor expression in, 122, 123
GH$_4$ZR$_7$ cells, dopamine D2 receptor ex-
    pression in, 163
GH3 cells
  Goα subtype expression in, 260
  muscarinic receptor expression in, 180
Gi protein
  antibody-induced uncoupling of opiate
      receptors from, 288
  dopamine D2 receptor copurification
      with, 289
  function of, 233, 277
  immunochemical studies of, 271
  membrane association of, 286
  muscarinic M2 receptor copurification
      with, 289
  peptide receptor linkage with, 199
  subtype antisera specificity of, 279–282
Giα protein
  chimera (with Goα protein)
    conserved region of, 249
    effector binding region of, 249, 251
    mutational studies on, 251
  comparison with yeast α subunits, 254,
      258
  human gene organization for, 247
  human genes for subtypes of, 244–246
  mutation of, 253
  possible oncogenic mutation of, 253
  properties of, 238
  structure of, 254, 260–261
  subtypes of, 234, 277, 280

Gi1α protein, 274, 278
  amino acid sequence of, 240–242, 276,
      277
  antisera specificity of, 279, 281, 282
  chimera (with Gi21α protein), genetic and
      biochemical studies on, 261
  classification and nomenclature of,
      236–237
  cloning of, 236–237, 277
  expression in *E. coli*, 278
  Gtα antibody reaction with, 287–288
  human gene for, 235, 244–246, 247
  membrane association of, 284
  properties of, 234, 236–237, 238
  structure conservation in, 248
Giα1 protein, mutant, transfection studies
    on, 286
Gi2α protein, 274, 277
  ADP-ribosylation of, 262
  altered expression in metastatic
      melanoma, 289–290
  amino acid sequences of, 240–242, 274,
      275, 277
  antisera specificity of, 279, 281, 282
  chimera (with Gsα protein)
      binding site of, 249
      mutational analysis of, 260, 262
  chimera (with Gs1α protein), genetic and
      biochemical studies on, 261
  classification and nomenclature of,
      236–237
  cloning of, 236–237, 277
  comparison with other organisms, 253,
      254, 255–257
  coupling to α2-adrenergic receptors, 288
  expression in *E. coli*, 278
  gene organization of, 246–247
  human gene for, 235, 244–246
  membrane association of, 284, 288
  mutational analysis of, 248–249, 250,
      251, 260
  myristoylation of, 262
  oncogenic mutation of, 261
  phosphorylation of, 284
  properties of, 234, 236–237, 238
  structure conservation in, 248, 255–257,
      274
  structure of, 243, 248, 255–257, 274
Gi3α protein, 234, 274, 277, 278
  amino acid sequences of, 240–242, 274,
      275, 277
  antisera specificity of, 279, 281, 282, 289
  classification and nomenclature of,
      236–237

cloning of, 236–237
expression in *E. coli*, 278
gene organization of, 246–247
human gene for, 235, 244–246
membrane association of, 284, 288
properties of, 234, 236–237, 238
stucture conservation in, 248, 274
gip protein, as oncogenic mutation of Gs1α
    protein, 261
Glands, muscarinic M3 receptors in, 171,
    178
Globus pallidus, dopamine D2 receptor
    mRNA in, 151
Glucagon receptor, mechanism of action of,
    199
Glucocorticoids, modulation of G-protein
    expression by, 289
Glutamate, muscarinic receptor effect on
    release of, 180
Glutamic acid residues
  in β-adrenergic receptors, 67
  in luteinizing hormone-chorionic gonado-
      tropin hormone receptor, 214
  in rhodopsins, 10, 11, 14–15, 18, 20, 23
  in STE2 receptor, 221
N-Glycanase, use in α2-adrenergic receptor
    studies, 92
Glycine residues
  in α2-adrenergic receptors, 44
  in α subunits, 234
  in muscarinic receptors, 181
  in rhodopsins, 10, 13, 19, 23
Glycogen phosphorylase, α1-adrenergic
    receptor stimulation of, 95–96
Glycoprotein hormone receptors,
    198–232
  G-protein interaction with, 84, 125
Glycoprotein hormones, secretion by pitu-
    itary gland, 211
Glycosylation
  of α-adrenergic receptors, 85, 92, 98
  of dopamine D1 receptors, 144, 148
  of G-protein-coupled receptors, 38, 39,
      63
  of 5-HT1A receptors, 125–126, 128
  in luteinizing hormone CG receptor, 206
  in mas/angiotensin receptor, 206, 220
  in muscarinic receptors, 181, 182
  in STE3 and STE4, 206
  in STE2 receptor, 221, 222
  in substance K receptor, 205, 206, 207
  in substance P receptor, 206
  in thyroid stimulating hormone receptor,
      206

Go protein
  antibody reaction with, 288
  antisera against, 282, 288
  discovery of, 276
  dopamine D2 receptor copurification
    with, 289
  function of, 288
  immunochemical studies of, 271
  in muscarinic inhibition of calcium cur-
    rent, 180
  muscarinic M2 receptor copurification
    with, 289
Goα protein, 234, 274
  ADP-ribosylation of, 262
  alternate splicing of transcript of,
    259–260
  amino acid sequences in, 276, 277
  antisera specificity of, 279, 281, 282,
    289
  classification and nomenclature of,
    236–237
  cloning of, 236–237
  comparison with yeast α subunits, 254,
    258
  expression in *E. coli*, 278
  gene organization of, 246–247
  human gene for, 235, 243
  immunochemical studies on, 289
  membrane association of, 284, 285
  mutation of, 253
  myristoylation of, 262, 284
  N-terminal cleavage of, 262
  possible oncogenic mutation of, 253
  properties of, 276–277
  structural model and mutational analysis
    of, 249–250
  structure conservation in, 248
  structure of, 243, 254, 277
  trypsin cleavage of, 262
Go1α protein, properties of, 238
Goα1 protein
  amino acid sequences of, 240–242, 259
  properties of, 236–237
Goα2 protein
  amino acid sequences of, 240–242, 258
  properties of, 236–237, 238
Golfα protein, 274. *See also* Gs2α protein
  cloning of, 234, 274
  expression in olfactory cells, 234, 276
GPA1, as yeast G-protein gene, 223, 253,
    263
GPA2, as yeast G-protein gene, 253
gpa1 gene, of *S. pombe*, 263
G$_{PLC}$. *See* Gqα

Gp protein, vasopressing receptor coupling
    to, 289
GP1α protein
  comparison with other α subunits, 254,
    255–257, 258
  conserved areas of, 253
  properties of, 239
  structure of, 253, 254, 263
GPα1 protein
  properties of, 239
  structure of, 255–257
GP2 protein, β and γ subunits of, 258
GP2α protein
  comparison with other α subunits, 254,
    257, 258
  conserved areas of, 253
  properties of, 239
  structure of, 253, 254
G-protein-coupled receptors
  amino acid sequence similarity among,
    82, 146–148
  antibody studies of coupling in, 286–289
  general structure of, 8, 83–85, 200
  glycosylation of, 38
  hydropathicity plot of, 115
  hydrophilic and hydrophobic domains of,
    63
  light absorption by, 12
  members of, 83–84
  modification of, 38–40
  molecular cloning of, 54
  protein sequences of, 31–38
  retinal isomerization in, 20
  rhodospins as, 1, 7
  schematic model of, 84
  sequence conservation among, 10–11
  structure-function relationships of, 62
G-proteins
  activation of, 19, 20, 21
  binding site for, 20–22
  common binding element of, 148
  conserved sequences of, 244, 284
  dysregulated expression in disease,
    289–290
  expression by transfection, 282, 283
  expression of, immunochemical studies
    of, 289–290
  function of, 233, 286–289
  Gi-like, role in cAMP inhibition by 5-
    HT$_{1A}$ receptors, 123
  immunoblotting of, 271, 283
  immunological studies on, 270–298
  interactions with receptor, 70–72, 142, 170
  as potential proto-oncogenes, 253

protein staining of, 270
receptor coupling to, 40–41, 62
receptor interactions of, α subunit
    C-termini role in, 262
from *Saccharomyces cerevisiae*, 253
structural studies on, 272–284
subunits of. *See* α subunits, β subunits,
    γ subunits
Gq protein
    receptor linkage with, 199
    rhodopsin interaction with, 21
Gqα subunit
    cloning of, 259
    stimulation of phospholipase CV
        β-subtype by, 259
Green pigments
    absence of double cysteine in, 22
    structure of, 3, 11, 13, 16–17
    of tamarins and squirrel monkeys, 18
Growth hormone, secretion by human pitu-
    itary tumors, 252–253
Gs protein
    function of, 233
    immunochemical studies of, 271
    membrane association of, 286
    peptide receptor linkage with, 199
    receptors, activation of, 22
    RM antibody inhibition of, 287
    in thyroid-stimulating hormone receptor,
        215
Gsα protein, 234, 274, 275–276. *See also*
    Gs1α protein
    activating mutation of, 249–250
    antibodies to, blockage of adenylyl cy-
        clase activation by, 287
    antisera specificity of, 279, 281
    chimera (with Gi2α protein), mutational
        analysis of, 260
    comparison among organisms, 248,
        255–257
    effector-recognition sites of, 262
    expression in *E. coli*, 278
    gene organization of, 235, 243, 246–247
    lack of myristoylation membrane of,
        285
    membrane association of, 285, 286
    mutational analysis of, 249–250, 260
    oncogenic mutation of, 252–253
    in platelet membranes, 288
    species of, 235, 243
    structure of, 243, 249–250, 254, 255–257,
        260, 277
    subtypes of, 234
GsαL protein, properties of, 238

Gs1α protein, 234
    ADP-ribosylation of, 262, 275
    amino acid sequences of, 240–242
    classification and nomenclature of,
        236–237
    cloning of, 236–237
    comparison with other organisms,
        255–257, 275
    mutational analysis of, 261
    oncogenic mutations of, 261
    structure of, 243, 255–257
Gsα1 protein
    generation of mRNA for, 238
    properties of, 238
    structure of, 238, 275
Gs2α protein. *See also* Golfα protein
    amino acid sequences of, 240–242
    classification and nomenclature of,
        236–237
    cloning of, 236–237
    properties of, 238
    structure of, 243
Gsα2 protein
    generation of mRNA for, 238
    properties of, 238
    structure of, 238
Gs3α protein
    properties of, 238
    structure of, 243
Gsα3 protein
    generation of mRNA for, 238
    properties of, 238
    structure of, 238
Gs4α protein, structure of, 243
Gsα4 protein
    generation of mRNA for, 238
    properties of, 238
    structure of, 238
GsαS protein, properties of, 238
G site, 234, 251
    conserved sequence of, 244, 248
    mutational analysis of, 260
G′ site, 234, 235, 251
    conserved sequence of, 244, 248
    mutational analysis of, 260
gsp protein, as oncogenic mutation of Gs1α
    protein, 261
G11α subunit, cloning of, 259
G12α subunit, cloning of, 259
G13α subunit, cloning of, 259
G14α subunit, cloning of, 259
G15α subunit, cloning of, 259
G16α subunit, cloning of, 259
Gt protein, antisera against, 282, 284

Gtα protein
  amino acid sequence in, 275, 276
  antisera against, 279, 281, 287, 289
  chimeric (with Gsα subunit), binding site
    of, 249
  cloning of, 275
  human gene organization for, 247
  mutational analysis of, 251
  myristoylation of, 262
  rhodopsin interaction with, 250–251, 286,
    287
Gtβ protein, 273
Gtγ protein, structure of, 272
Gt1 protein, function of, 233
Gt1α protein, 274
  ADP-ribosylation of, 262
  amino acid sequences of, 240–242, 277
  classification and nomenclature of,
    236–237
  cloning of, 236–237, 274
  N-terminal cleavage of, 262
  properties of, 236–237, 238
  structure conservation in, 248
  structure of, 243
  trypsin cleavage of, 262
Gt2 protein, function of, 233
Gt2α protein, 274
  ADP-ribosylation of, 262
  amino acid sequences of, 240–242, 277
  classification and nomenclature of,
    236–237
  cloning of, 236–237, 274
  properties of, 236–237, 238
  structure conservation in, 248
GTPase, α subunit mutation and structural
    effects on, 234–235, 260, 262
GTPase-activating protein, of ras protein,
    252
GTP-binding proteins. See G-proteins
GTP binding site, of transducin, 21
Guanine nucleotide regulatory proteins. See
    G-proteins
Guanine triphosphate hydrolase, receptor
    stimulation of, 43, 44, 97
Guanylate cyclase, muscarinic receptor acti-
    vation of, 175
Guanylate cyclase peptide receptor, of sea
    urchin, 228
GW/6 antiserum, as Gt antiserum, 277
Gxα protein. See also Gzα protein
  amino acid sequences of, 240–242, 279
  cloning of, 236–237
  human gene organization for, 247
  properties of, 238

structure conservation in, 248
structure of, 234, 235, 279
Gzα protein, 274. See also Gxα protein
  amino acid sequences in, 276
  antisera specificity of, 279, 281
  classification and nomenclature of,
    236–237
  cloning of, 236–237
  human, 234
  human gene for, 235
  mutational analysis of, 260
  myristoylation of, 262
  phosphorylation of, 284
  in platelet membranes, 288
  properties of, 235, 236, 238, 279

Hallucinations, in schizophrenia and
    Alzeheimer's disease, dopamine re-
    ceptor role in, 142
Halo-opsin, topography of, 35
Hamster, β₂-adrenergic receptors of, 38
H21a mutant
  of Gsα protein, 250
  receptor downregulation studies on, 50
Ha-ras p21 protein, 234
  effector region of, 251, 252
  functional boxes of, 251
  structure of, 243, 249, 251
Heart, muscarinic receptors in, 170, 171,
    175, 177–178, 179, 184
Heart failure, altered G-protein expression
    in, 289
HEAT. See BE 2254
HEK cells, muscarinic receptor expression
    in, 176
HeLa cells
  β₂-adrenergic receptor uncoupling studies
    on, 42
  5-HT₁ₐ receptor expression in, 122, 123,
    124, 125–126
Helical structures
  of β-adrenergic receptors, 68, 72
  of muscarinic receptors, 181
  of rhodopsins, 5–6, 7, 8, 13, 15, 16, 24
Heparin
  as inhibitor of α₂-adrenergic receptor ki-
    nase, 45
  as inhibitor of receptor desensitization, 46
Heparin-agarose chromatography,
    α₂-adrenergic receptor purification
    by, 79
Hexbutinol, binding by muscarinic m1–m5
    receptors, 173

Hexocyclium, binding by muscarinic m1–m5
    receptors, 173
HHD, binding by muscarinic m1–m5 recep-
    tors, 173
HHSiD, binding by muscarinic m1–m5 re-
    ceptors, 173
Himbacine, binding by muscarinic recep-
    tors, 172, 173, 183
HindIII endonuclease
    in cloning studies of α_2-adrenergic recep-
        tors, 88
    in cloning studies of dopamine D1 recep-
        tor, 144
    in studies of RFLP in dopamine D1 re-
        ceptors, 151
Hippocampus
    5-HT$_{1A}$ receptors in, 119–120, 121, 125,
        126
    muscarinic receptors in, 170, 180, 185, 186
Histidine residues
    in β-adrenergic receptors, 67
    in rhodopsin, 15, 19
        proline replacement in retinitis pigmen-
            tosa, 4
    in tachykinin receptors, 208
HL-60 cells, Gi3 protein expression in, 278
Hormones, stimulation of transmembrane
    signaling by, 41
HT-29 cells, α_2-adrenergic receptor expres-
    sion in, 96, 104, 105
5-HT. See Serotonin
5-HT receptors
    radioligands for, 113
    subtypes of, 113
5-HT$_1$ receptors
    binding site of, 114
    purification of protein similar to, 125
    subtype of, 119
5-HT$_{1A}$ receptors, 91, 105, 113–141
    amino acid sequence of, 127
    amino acids in, comparison with dopa-
        mine D1 receptors, 147
    antagonists for, 118
    binding site of, 115, 128
    biochemistry of, 124–127
    clinical aspects of, 118
    cloning and expression of, 114, 130
    desensitization of, 129
    genes for, 113–114
    glycosylation of, 125–126, 128
    G-proteins linked to, 124
    homology with other G-protein-coupled
        receptors, 127–128, 129
    hydrophilic domains in, 128–130

    hydrophobic transmembrane domains of,
        127–128
    as mediators of vasorelaxation, 119
    molecular biology of, 127–131
    molecular mass of, 125
    pharmacology of, 114–118
    phosphorylation of, 130
    photoaffinity labeling of, 126
    possible role in modulation of killer cell
        cytotoxicity, 119
    recombinant, effect on chloride and po-
        tassium channels, 124
    role in transport processes, 124
    second messenger linkages of, 121, 122
    solubilization of, 125
    structure-function linkages in, 129
    subtypes of, 130
5-HT$_{1A2}$ receptors, pressor response media-
    tion by, 118–119
5-HT$_{1B}$ receptors
    affinity for β_1- and β_2-adrenergic
        blocking agents, 116
    binding site of, 115
    characteristics of, 130
5-HT$_{1C}$ receptors
    cloning of, 130
    5-HT$_{1A}$ receptor comparison with, 127, 128
    phospholipase C coupling to, 129
    structure-function linkages in, 129
5-HT$_{1D}$ receptors, characteristics of, 130
5-HT$_2$ receptors
    binding site of, 114–115
    cloning of, 130
    5-HT$_{1A}$ receptor comparison with, 127,
        128
    possible role in vasoregulation, 119
    spiroxatrine binding to, 116
    structure-function linkages in, 129
5-HT$_{2C}$ receptors, phospholipase C coupling
    to, 129
5-HT$_3$ receptors, activation of, 118
5-HT$_{pre}$ receptors, radioligand binding of,
    116, 118
Humans
    α_2-adrenergic receptor of, 80
    green and red pigments of, 3, 4, 16–17
    rhodopsin structure of, 2, 16–17
Huntington's chorea, dopamine receptor
    role in, 142, 151
HVA calcium current, muscarinic receptor
    effects on, 180
Hybridization histochemistry, use to map
    dopamine D1 receptor mRNA
    expression, 150–151

Hydrogen bonds, in rhodopsin, 7
Hydrogen ions, effect on α_2-adrenergic receptor binding, 105, 106
3-Hydroxy-3-methyl-glutaryl coenzymeA reductase, topography of, 35
5-Hydroxytryptamine, structure of, 117
Hyperplasia, mas role in, 218
Hypothalamus, dopamine receptor mRNA in, 151

IBMX, receptor downregulation induction by, 48, 53
Idazoxan, α_2-adrenergic receptor binding of, 85, 86
Ileum, muscarinic receptors in, 184
i1 loop, in β-adrenergic receptors, 71
i2 loop
   in β-adrenergic receptors, 71
   in dopamine D1 receptors, 145
   in 5-HT$_{1A}$ receptors, 129
   in substance K receptor, 205
i3 loop
   in α-adrenergic receptors, 84
   in α_2-A-adrenergic receptors, 105
   in α_2-C10-adrenergic receptors, 98, 106
   in β-adrenergic receptors, 147–148
      amino acid residues in, 41, 71
      role in G-protein interactions, 71, 72, 100
   in β_2-adrenergic receptors, 148
   in dopamine D1 receptors, 145, 146, 147, 148
   in dopamine D2 receptors, 148, 164
   in dopamine D3 receptors, 148
   in G-protein-coupled receptors, 129
      phosphorylation at, 40, 148
   in 5-HT$_{1A}$ receptors, 120, 129
   in human α_2-adrenergic receptor, 77
   in muscarinic receptors, 181, 183
   in peptide receptors, 224, 227
   in rhodopsin, 71, 149
   in STE2 receptor, 222
   in thyroid-stimulating hormone receptor, 215
i4 loop
   in β_2-adrenergic receptors, 148
   in muscarinic receptors, 181
   in rhodopsin, 148
Immune function, possible role of 5-HT$_{1A}$ receptors in, 119
Immunoblotting, use in α subunit studies, 276, 277, 278, 279, 281, 282, 283, 289
Immunochemical studies, of G-protein subunits, 270–298

Immunoprecipition, of α subunits, 285, 286
Infrared dichrosism, of rhodopsins, 5
Inositol phosphates
   dopamine D1 receptor-stimulated production of, 154
   5-HT$_{1A}$ receptor-induced increase of, 123
   production by tachykinin receptors
Inositol phospholipid pathway, α_1-adrenergic receptor function in, 95
Insulin receptor tyrosine kinase, α subunits as substrates for, 284
Invertebrates, G-proteins of, 284
Iodine-125, use as receptor label, 82
Iodocyanopindolol, 5-HT$_{1A}$ receptor affinity for, 116
Ion channels, peptide receptor regulation of, 198
α-Ionone ring, role in opsin shift, 13
IP3
   effect on M-current, 178
   effects on calcium release, 176, 179
Ipsapirone
   α_1-adrenergic receptor affinity for, 116.118
   as 5-HT$_{1A}$ receptor radioligand, 116
   structure of, 117
Isoleucine residues, in rhodopsins, 11
Isoproterenol
   adenylyl cyclase stimulation studies using, 45, 122
   downregulation promotion by, 49
   receptor binding of, 65, 69

JWR21 antiserum, use in 5-HT$_{1A}$ receptor sequencing, 125

Kidney, dopamine D1-like receptors in, 154
Kidney 293 cells, luteinizing hormone-chorionic gonadotropin hormone receptor expression in, 212
Killer cell, 5-HT$_{1A}$ receptor in modulation of cytotoxicity of, 119

λ EMBL 3 SP6/T7 human genomic library, G-36 use to probe, 144
λHD2G1 clone
   use in dopamine D2 receptor studies, 165
   use in linkage of dopamine D2 receptor to genetic disorders, 166
Lamprey rhodopsin, structure of, 2, 10, 11, 14, 19
Lectin chromatography
   α_1-adrenergic receptor purification by, 79
   α_2-adrenergic receptor purification by, 79

Leucine residues
  in luteinizing hormone-chorionic gonado-
    tropin hormone receptor, 212
  in rhodopsin, 7, 10, 16, 21
Leucine-rich glycoproteins, PG40 and
    luteinizing hormone- chorionic
    gonadotropin hormone receptor as,
    212
Ligand binding sites
  in G-protein-coupled receptors, 36, 62,
    64–70, 73
  in peptide receptors, 224, 227
  in rhodospins, 8–13, 15–16, 73
"Light" membrane vesicles, adenylyl cyclase
    absence in, 46
Linear dichroism, of rhodopsins, 6, 8
Lipid membrane, G-protein segment spans
    of, 31–34
Liposaccharides, N-linked, in adrenergic
    receptors, 38
Lisuride, 5-HT_{1A} receptor affinity for, 116
LLC-PK1 cells, α_2-A-adrenergic receptor
    expression in, 104
Lock-and-key model, of peptide binding,
    198–199
Loop sizes, in peptide receptor structures,
    205
Low-molecular-weight GTP-binding pro-
    teins, 233
L-RGB2Zem-1 clone, of dopamine D2 re-
    ceptor, 162
d-LSD, 5-HT_{1A} receptor affinity for,
    116
Ltk$^-$ cells, dopamine D2 receptor expres-
    sion in, 162
Lung, muscarinic "M4" receptor in, 185
Luteinizing hormone-chorionic
    gonadotropin hormone receptor
  clone of, 201, 202, 211–212
  comparison with thyroid-stimulating
    hormone receptor, 215
  glycosylation in, 206
  ligand binding in, 225
  mechanism of action of, 199
  palmitoylation in, 206
  peptide-receptor interaction in, 201
  structure of, 213
  truncated forms of, 212
Luteinizing hormone releasing hormone
    receptor
  loop sizes of, 205
  mechanism of action of, 199
LY-171555, as dopamine D2 receptor
    agonist, 149, 150

Lysine residues
  in β-adrenergic receptors, 67
  in rhodopsins, 6, 10, 11, 13, 14, 20, 65
  in STE2 receptor, 221
Lysosomes, receptor degradation in, 49

Mamillary bodies, dopamine D2 receptor
    mRNA in, 151
Manic depression, dopamine D2 receptors
    and, 166
Mannose oligosaccarides, in adrenergic re-
    ceptors, 38
MARCKS, as indicator of protein kinase C
    activation, 123
mas/angiotensin receptor, 217–220
  cloning of, 201, 217
  comparison with other pituitary glyco-
    protein hormone receptors, 219–220
  glycosylation in, 206, 220
  luteinizing hormone-chorionic gonado-
    tropin hormone receptor compared
    to, 214
  mitogenic potential of, 253
  peptide-receptor interaction in, 201, 204
  pharmacology and regulation of,
    217–218
  structure of, 218–220
mas gene, location on chromosome 6, 218
mas oncogene product, as angiotensin
    receptor, 35
MASS-NMR technique, use in studies of
    rhodopsin-retinal interaction, 12–13
Mating factor receptors, from
    Saccharomyces spp., 35, 258, 263
Mating hormone response, in yeast, 220
Matrix factors, altered G-protein effects on
    tumor cell response to, 290
McN-A-343, as muscarinic receptor agonist,
    174
M-current, mechanism of action of, 178,
    180
MDL 72823, 5-HT_{1A} receptor affinity for,
    116
MDL 73005, as 5-HT_{1A} receptor partial
    antagonist, 118
Melittin, as initiator of arachidonic acid
    release, 123
Membrane proteins
  models of, 1, 4
  rhodopsin structure and, 1
Membrane-spanning helices, of rhodopsins,
    13–14
Mental disorders, drugs used for, affinity
    for dopamine D2 receptors, 160

5-MeO-DPAC
  as 5-HT$_{1A}$ receptor radioligand, 116
  structure of, 117
Mesolimbic tract, dopamine in, 160
Metabotropic glutamate receptor,
    N-terminus of, 227.228
Meta-rhodopsin, equilibrium of, 18–20
Meta-II rhodopsin
  decay of, 22
  interaction with G-protein, 20
  as meta-stable bleaching intermediate of
    rhodopsin, 18, 19–20
  phosphorylated, arresting binding to, 22–23
Metastatic melanoma, altered Gi2α protein
    in, 289–290
Metergoline, 5-HT$_{1A}$ receptor affinity for,
    116
Methionine residues
  in β_2-adrenergic receptors, 128
  in dopamine D1 receptors, 144
  in 5-HT$_{1A}$ receptor, 128
  in rhodopsins, 11
Methoctramine, binding by muscarinic m1-
    m5 receptors, 172, 173
Methoxamine
  as α_1-adrenergic receptor agonist, 77
  α_1-adrenergic receptor binding of, 90, 95
  α_2-adrenergic receptor binding of, 90
5-Methoxy-3-(di-*n*-propylamino)chroman.
    See 5-MeO-DPAC
8-Methoxy-2-[*N-n*-propyl-*N*-3-(2-nitro-4-
    azidophenyl)- aminopropyl]tetralin,
    as photoaffinity label, 125
9-Methyl group
  removal from retinal, effects on
    rhodopsin activation, 20
  of retinal, 24
N-Methylscopolamine. *See* NMS
Methylsergide, 5-HT$_{1A}$ receptor affinity for,
    116
Metitepine, 5-HT$_{1A}$ receptor affinity for,
    116, 118
Migraine, role of 5-HT receptors in, 118
MII. *See* Meta-II rhodopsin
Mitogenesis, muscarinic receptor inhibition
    of, 176
Mitogenic potential, of serotonin 1c and
    angiotensin/mas oncogene receptor,
    253
m1 loop, in receptor hybrid, 72
Movement, dopamine system modulation
    of, 160
mRNA
  cAMP effects on levels of, 49, 52

of dopamine D1 and D2 receptors, distri-
    bution of, 150–151, 155
MspI restriction enzyme, use in studies of
    RFLP in dopamine D1 receptors, 151
Muscarine, as muscarinic receptor agonist,
    174
Muscarinic m1–m5 receptors
  agonist action on, 174
  correlation with muscarinic M1–M3 re-
    ceptors, 184
  functional differences among, 175–176
  as molecular subtypes, 171
  muscarinic antagonists for, 171, 172, 173
  stimulation of, 176
  structure/function of, 181–183
  tissue distribution of, 184, 185, 186
Muscarinic M1 receptors
  agonists for, 174, 175
  biochemical responses of, 175
  as cerebral cortex subtype, 171
  correlation with muscarinic m1–m3
    subtypes, 184
  5-HT$_{1A}$ receptor homology with, 127, 128
  phospholipase C coupling to, 129
  structure-function linkages in, 129
  tissue distribution of, 184, 185, 186
Muscarinic m1 receptors
  amino acids in, comparison with dopa-
    mine D1 receptors, 147
  physiological responses mediated by, 178,
    179, 180
Muscarinic M2 receptors
  agonists for, 174
  biochemical responses of, 175
  copurification with Go and Gi proteins,
    289
  correlation with muscarinic m1–m3
    subtypes, 184
  gallamine effects on, 225
  as heart subtype, 171
  physiological responses mediated by,
    178–179, 180
  structure-function linkages in, 129
  tissue distribution of, 184, 185
Muscarinic m2 receptors
  physiological responses mediated by, 178,
    180, 181
  structure of, 181
Muscarinic M3 receptors
  correlation with muscarinic m1–m3
    subtypes, 184
  as gland subtype, 171
  structure-function linkages in, 129
  tissue distribution of, 184, 185

Muscarinic m3 receptors, physiological responses mediated by, 178, 179, 180
Muscarinic M4 receptors, 185
   structure-function linkages in, 129
Muscarinic m4 receptors, physiological responses mediated by, 178, 179, 181
Muscarinic m5 receptors, physiological responses mediated by, 178, 179, 180, 187
Muscarinic receptors, 36, 170–197
   agonist action on, 174, 178
   amino acids in, 37
   binding site of, 127, 174, 182
   biochemical responses to, 175–181
   chimeras of, 183
   cloning of, 80, 171
   coupling to functional responses of, 175–181
   glycosylation in, 181, 182
   G-protein coupling with, 40, 84
   ligand binding in, 65, 66, 68, 182–183, 209
   model for, 181, 182
   nomenclature for, 171
   palmitoylation in, 181
   pharamcological differentiation and properties of, 171–175
   phosphorylation of, 37, 40
   physiological responses to, 176–181
   purification of, 171
   structure/function of, 179, 181–183, 222
   tissue localization of, 183–187
Mutagenesis (site-directed)
   in α_2-adrenergic receptors, 38, 44, 148
   of β-adrenergic receptors, 64, 65, 70, 100–101
   of muscarinic receptors, 181
   in rhodopsin, 7, 8, 13, 21, 22
Myristoylation, effects on α subunits, 262, 284–286

Na^+/H^+ exchanger, as second messenger, 96
Na^+/K^+-ATPase, 5-HT$_{1A}$ receptor stimulation of, 123
NAN-190, as 5-HT$_{1A}$ receptor partial antagonist, 118
N_3-NAPS(N-p-azido-m-iodophenethyl) spiperone, as photoaffinity label for 5-HT$_{1A}$ receptor, 125
Neocortex
   dopamine receptor mRNA in, 150–151
   muscarinic receptor activity in, 180
Neuroblastoma, chromsome 6 region associated with, 218

Neuromedin K receptor
   clone of, 201, 202, 208
   comparison with substance K receptor, 208–209
   desensitization of, 211
   glycosylation in, 206
   loop sizes of, 205
   palmitoylation in, 206
   serine and threonine residues in, 211
   structure of, 208, 210
Neurons, mas/angiotensin receptor expression in, 218
Neuropeptide receptors, G-protein interaction with, 84
Neuropeptide Y receptor, mechanism of, 199
Neurotensin receptor, mechanism of, 199
Neurotransmitters, muscarinic receptor-mediated release of, 180–181
NG108 cells, opiate-stimulated GTPase in, 288
NG108-15 cells
   M-current studies on, 178
   muscarinic receptor expression in, 175, 179, 180, 185
NG 115–40IL cells, mas/angiotensin receptor expression in, 217
Nicotinic receptors, discovery and function of, 170
Nifedipine, 87
Nigrostriatal tract, dopamine in, 160
NIH 3T3 cells
   mas/angiotensin expression in, 217
   muscarinic receptor expression in, 172, 176
Nitrogen starvation, signal transmission for, in yeast, 263
NK-1 tachykinin receptor. See Substance P receptor
NK-2 tachykinin receptor. See Substance K receptor
NK-3 tachykinin receptor. See Neuromedin K receptor
NMS, binding by muscarinic m1–m5 receptors, 173–174
N132N1 cells, muscarinic receptor expression in, 175
Noradrenaline, effect on dopamine D1 and D5 receptor binding, 149, 150
Norepinephrine
   α_1- and α_2-adrenergic receptor binding of, 90
   as receptor agonist, 65, 70, 82–83

Northern blot analysis
  of α_2-adrenergic receptor subtypes, 92,
    103, 104
  of dopamine D1 receptor mRNA,
    143
  of human fetal RNA, 119, 120
  of luteinizing hormone-chorionic
    gonadotropin hormone receptor,
    212
  of pituitary glycoprotein hormone recep-
    tors, 212, 214
  of RGB-2 mRNA sequences, 162
N-termini
  of dopamine D1 receptors, 144
  of dopamine D2 receptors, 164
  extended, in metabotropic glutamate re-
    ceptor, 227–228
  of G-protein-coupled receptors, 18, 63,
    71, 72, 84
  of luteinizing hormone-chorionic gonado-
    tropin hormone receptors, 212,
    213–214, 215, 224–225
  of muscarinic receptors, 181
  of neuromedin K, 225–226
  of pituitary glycoprotein hormone recep-
    tors, 211, 215, 217, 224–225
  of STE2 receptor, 221
  of substance K receptor, 205
Nuclear magnetic resonance techniques
  bovine rhodopsin studies using, 38
  magic angle sample spinning method,
    MASS-NMR technique
Nucleus accumbens, dopamine receptor
  mRNA in, 150

8-OH-DPAT
  as 5-HT$_{1A}$ receptor radioligand of choice,
    114, 116, 118
  lack of effect on arachidonic acid release,
    123
  structure of, 117
OK cells, adrenergic receptor subtype
  studies on, 103
Olfactory cells, Golfα expression in, 234
Olfactory tubercle, dopamine receptor
  mRNA in, 150
o1 loop, of dopamine D1 receptor, 148
o2 loop, of dopamine D1 receptor, 146,
    148
o3 loop
  of β-adrenergic receptors, 205
  of muscarinic receptors, 205
  of substance K receptor, 205

Oncogene
  mas/angiotensin as possible, 217
  muscarinic receptor cellular transforma-
    tion similar to, 176
Oncogenic mutations, of α proteins,
    251–253, 261
μ-Opiate receptor
  mechanism of, 199
  spiroxatrine binding to, 116
δ-Opioid receptor, mechanism of, 199
Opsins
  alignment of sequences of, 2–3
  cloning of, 198
  constituent peptides of, 4
  functions of, 8–23
  G-protein interaction with, 84
  sequence conservation among, 10–11
  sequence resemblance to other receptors,
    82, 127
Opsin shift, 12, 13
Oral dyskinesia, in elderly persons, dopa-
    mine receptor role in, 142
Ovarian sex cord stromal tumors, α subunit
    activation in, 261
Ovine rhodopsin, structure of, 1, 2, 5, 6,
    11, 16
Oxotremorine, as muscarinic receptor agon-
    ist, 174
Oxotremorine M, as muscarinic receptor
    agonist, 174
Oxymetazoline
  as α_2-adrenergic receptor agonist, 77, 89,
    92, 93
  α_1-adrenergic receptor binding of, 90
  α_2-adrenergic receptor binding of, 85, 90
Oxytocin receptor, mechanism of, 199

Palmitoylation
  of G-protein-coupled receptors, 38–39,
    98, 181
  in luteinizing hormone-chorionic gonado-
    tropin hormone receptor, 214
  in peptide receptors, 206, 208
  of rhodopsin, 4, 7, 38
Pancrease, muscarinic receptor activity in,
    178
PAPP
  as 5-HT$_{1A}$ receptor radioligand, 116
  structure of, 117
Parathyroid gland, dopamine D1 receptors
    in, 142, 143, 144, 149, 154
Parathyroid hormone receptor, mechanism
    of, 199

Parkinson's disease
dopamine receptor role in, 142, 155, 160,
165
muscarinic drug action on, 186
pBC12B, dopamine D1 receptors expression
into, 149
pBMT3X plasmid, α_2-A-adrenergic receptor
expression in, 104
PC12 cells, transfection of Gsα cDNA mutations into, 253
pCMV4 plasmid, α_2-A-adrenergic receptor
subcloned into, 104
Peptide receptors, 198–232
characteristic sequence elements of,
227–228
cloned, 201, 202–224
C-terminal tails of, 227
general properties of, 227–228
intracellular loops of, 226
loop sizes in, 205
mechanisms of, 199
signaling of adenylyl cyclase, 198
Peripheral nervous system, muscarinic receptor activity in, 180
Pertussis toxin, 123
effects on α subunits, 262, 271
G-protein sensitivity to, 96, 97, 121, 124,
163, 175, 176, 177, 180
PG40, luteinizing hormone-chorionic gonadotropin hormone receptor
homology with, 212
Phenoxybenzamine, use in α-adrenergic receptor subclassification, 77, 78
Phentolamine, α_1- and α_2-adrenergic receptor binding of, 90
Phenylalanine residues
in α_2-A-adrenergic receptor subtypes, 105
in β-adrenergic receptors, 67, 69
in rhodopsins, 11, 15, 16, 18
Phenylephrine
α_1-adrenergic receptor binding of, 77, 90,
95
α_2-adrenergic receptor binding of, 90
Phorbol esters, as protein kinase C activators, 218
Phosphate functional box, of Ha-ras p21
protion and α subunits, 251
Phosphate group, attachment point in opsin, 4
Phosphatidylinositol
norepinephrine-stimulated turnover of,
82–83
receptor stimulation of, 87, 95, 96, 170,
174, 175, 176, 180, 181–182
Phosphatidylinositol 4,5-bisphosphate
mas/angiotensin receptor effect on, 217,
219–220
peptide receptor effect on, 226
Phosphatidylinositol kinase, 218
Phosphoenolpyruvate carboxykinase, sequence responsible for cAMP regulation in, 53
Phospholipase A$_2$, receptor activation of,
233, 261
Phospholipase C
activation of, 72, 180, 233, 277, 279
α_1-adrenergic receptor subtype coupling
to, 87
5-HT$_{1A}$ receptor coupling to, 121, 123,
129
mas/angiotensin receptor coupling to, 220
muscarinic receptor stimulation of, 179
peptide receptor stimulation of, 198
β-subtype of, stimulation by Gqα, 258
tachykinin receptor coupling to, 210
Phospholipid vesicles, G-proteins in, 187
Phosphorhodopsin, amino acid sequencing
of, 44
Phosphorus-82, use as receptor label, 82
Phosphorylation
of α-adrenergic receptors, 85
of β_2-adrenergic receptors, 43, 54–55
of dopamine D1 receptors, 148
of rhodopsin, 18, 19, 22, 23, 40
Phosphoryl-binding region, in α subunits,
261
Photoaffinity labeling
of 5-HT receptors, 113
use in studies of mutant receptors, 64
Pig, D2-dopaminergic receptors of, 38
Pilocarpine, as muscarinic receptor agonist,
174
Pindolol, 5-HT$_{1A}$ receptor affinity for, 116
Pirenzepine
binding by muscarinic receptors, 173, 175
as muscarinic receptor antagonist, 171
Pituitary gland, dopamine D2 receptors in,
38, 162
Pituitary glycoprotein hormones, 211–217
Pituitary tumors
altered G-protein expression in, 289
α subunit activation in, 252–253, 260, 261
Plants, G-proteins in, 284
Plaque hybridization analysis, of
α_2-adrenergic receptor, 80
Plasma membrane
β_2-adrenergic receptor topography in, 35
receptor translocation in, 48–49

320   Index

Platelet-derived growth factor receptor, phosphatidylinositol kinase activation by, 218
Platelet membranes, α subunits in, 288
"Point-charge" hypothesis, of rhodopsin wavelength shift, 14
Polar residues, in rhodopsin, 7
Polymerase chain reaction
  cloning of α subunit by, 259, 274, 279
  cloning of dopamine D1 receptor by, 143
  cloning of G-protein-linked receptors by, 35–36, 38, 91
  cloning of mas/angiotensin receptor by, 217
  isolation of dopamine D5 receptor genes using, 155
  pituitary glycoprotein receptor cloning by, 212, 214
  in substance P receptor structure studies, 208
Polyphosphoinositide, stimulation of metabolism by chimeric adrenergic receptors, 100
Potassium channels
  activation of, by muscarinic stimulation, 170, 178–179
  calcium-dependent, second messenger effect on, 170
  G-protein gating of, 233, 286–287
  inhibition of, by muscarinic receptors, 179
  muscarinic receptor modulation of, 176–178
  as second messengers, 96
Potassium ions, 5-HT$_{1A}$ receptor mediation of, 121
Prazosin
  as α_1-adrenergic receptor anatagonist, 77
  α_1-adrenergic receptor binding by, 87, 90
  α_2-adrenergic receptor binding by, 85, 86, 90, 92
Preprotachykinin, peptide-receptor interaction of, 203
γ-Preprotachykinin-peptide amide, in brain, 207
Proenkephalin, sequence responsible for cAMP regulation in, 53
Prolactin, dopamine system regulation of, 160, 162–163
Proline residues
  in β-adrenergic receptors, 205
  in luteinizing hormone-chorionic gonadotropin hormone receptor, 214
  in mas/angiotensin receptors, 219

  in peptide receptors, 206
  in rhodopsin, 10, 11, 13, 20, 23, 24
    histidine replacement in retinitis pigmentosa, 4
  in STE3 receptor, 223, 224
  in tachykinin receptors, 208
  in TM 2 and TM 4, 37
Prolyl cis-trans isomerase, as possible *Drosophila* rhodopsin defect, 14
Propranolol, 5-HT$_{1A}$ receptor affinity for, 116
N-Propylbenzilylcholine mustard, as muscarinic receptor antagonist, 182–183
Protein immunoblotting, use in studies of mutant receptors, 64
Protein kinase
  cAMP-dependent, dopamine D1 receptor activation of, 142
  receptor phosphorylation by, 40, 43, 45, 46, 53, 55, 148
  role in receptor downregulation, 48
  thyroid-stimulating hormone receptor compared to, 217
Protein kinase A, 129
  muscarinic receptor activation of, 170, 182
  phosphorylation sites for, in dopamine D2 receptors, 164
Protein kinase C
  α subunits as substrates for, 284
  5-HT$_{1A}$ receptor activation of, 129
  MARCKS indication of activation of, 123
  in M-current activity, 178
  muscarinic receptor activation of, 170, 180, 182
  receptor phosphorylation by, 40, 55, 148
Protein sequences, of G-protein-coupled receptors, 31–38
Proto-oncogene(s)
  G-proteins as, 253
  mas gene as, 218
PS120 cells, α_2-C10-adrenergic receptor activation in, 97–98
P site, 234, 235, 251
  conserved sequence of, 244, 248, 249
  mutational analysis of, 260
pSP73, dopamine D1 receptors clone fragments subcloned into, 144
PstI restriction endonuclease
  use in cloning studies of α_2-adrenergic receptors, 88
  use in studies of RFLP in dopamine D1 receptors, 151

PvuII restriction enzyme, use in studies of
RFLP in dopamine D1 receptors, 151
Pyramidal cells, 5-HT$_{1A}$ receptor effect on,
121
PZP, binding by muscarinic m1–m5 recep-
tors, 172, 173

QNB, gallamine effects on muscarinic acid
binding of, 174
Quinidine, binding by muscarinic m1–m5
receptors, 173
Quinpirole, as dopamine D2 receptor agonist, 163

Rab protein, 233
RA42 clone, of α_1-adrenergic receptor, 95
Ral protein, 233
Rap protein, 233
ras1 gene, function in yeast, 263
RAS2 gene, 253
Ras p21 proteins
oncogenic mutation of, 251
structure of, 262
Ras protein, 233, 252
mutational analysis of, 260
RAS2 protein, structure of, 243
Rat, α_2-adrenergic receptors of, 38
Rauwolfia serpentina, yohimbine from, 77
Rauwolscine
as α-adrenergic receptor antagonist, 77
α_1-adrenergic receptor binding of, 90,
103, 114
α_2-adrenergic receptor binding of, 85, 86,
89, 90, 93
RBL 2H3 cells, muscarinic receptor expression in, 178, 181
R179C mutation, of Gi2α, neoplastic transformation from, 260
RDC1, as peripheral peptide receptor, 227
RDC4, as 5-HT$_{1D}$ receptor, 130
RDC7 and RDC8 receptors, as clones from
thyroid cDNA library, 36, 38, 39
R* derivative, of light-activated rhodopsin,
18, 22
comparison with MII, 18
Recombinant DNA techniques, for
α-adrenergic receptor purification, 79
Red pigments
absence of double cysteine in, 22
structure of, 2, 11, 13, 161–17
of tamarins and squirrel monkeys, 18
Refractoriness, receptor desensitization as,
41

RESACT/SPERACT peptide, of sea urchin, 228
Resonance Raman spectrosocopy, of
rhodopsin-retinal interaction, 12–13
Restriction fragment-length polymorphism,
in dopamine D1 receptors, 151, 155
Retina
α subunits in, 275
dopamine D1 and D2 receptors in, 142,
151
11-*cis*-Retinal
attachment point to opsin, 4, 13
in binding pocket of rhodopsin, 65
light absorption by, 12
9-methyl group removal effect on activity, 12
photoisomerization of, 12
role in rhodopsin structure, 8, 12
Retinitis pigmentosa
altered G-protein expression in, 289
altered rhodopsin structure in, 4
RG10-adrenergic receptor clone, classification of, 102
RG20-adrenergic receptor clone, amino acid
sequence of, 101
RGB-2 clone, of dopamine D2 receptor,
161–162, 164
Rhodopsin(s), 1–30
activation of, 18–20
binding pocket of, 65, 127
dark-adapted, rhodopsin kinase effect on,
22
deactivation of, 22–23
3-D modeling of, 4–7, 24
functions of, 8–23
glycosylation of, 63, 126
G-protein binding site of, 20–22, 24,
70–71
Gtα interaction with, 250–251, 286
helical wheel representation of, 16–17
hydrophilic and hydrophobic domains of,
63
i3 loop of, role in transducin interaction,
71
ligand binding sites in, 8–13, 15–16, 226
light absorption by, 12
palmitoylation of, 38, 39
phosphorylation of, 18, 19, 22, 23, 37, 40
retinal binding to, 4, 13, 19, 65
site-directed mutagenesis of, 7, 8, 22
structure of, 1–8, 12, 18, 38, 114, 161
thermodynamic stability of, 8
Rhodopsin kinase, phosphorylation by, 22,
40, 44, 149

rho protein, 233

RMB 2H3 cells, muscarinic receptor expression in, 172

RNGα_2-adrenergic receptor clone, classification of, 102

Rods and cones, Gt1α, 234

RS 86, as muscarinic receptor agonist, 174

RS1 clone of dopamine receptor, 144

RS2 clone of dopamine receptor, 144

RU 24969, 5-HT$_{1A}$ receptor affinity for, 116

RV/3 antiserum, specificity for Goα protein, 275

Saccharomyces cerevisiae
  α subunits of, 239, 253
  mating factor receptors of, 35, 198, 258
  STE2 and STE3 receptor clones from, 201
  STE gene family of, 220–223

Saccharomyces kluyveri
  STE2 and STE3 receptors of, 222, 226
  STE3 receptor clone from, 201

Salivary glands, muscarinic receptors in, 171, 178

S49 cells
  gα chimera construct expression in, 251
  Gsα protein expression in, 249

SCG1, α subunit relationship to, 220

SCH-23388, effect on dopamine D1 and D5 receptor binding, 149, 150

SCH-23390, binding to dopamine D1 receptor, 147, 148, 149, 154

Schiff base linkage, in rhodopsin-retinal binding, 13, 14, 15, 19, 20, 65

Schizophrenia
  dopamine receptor role in, 142, 151, 155, 160, 164, 165
  gene for susceptibility to, on chromosome 5, 153
  muscarinic receptors as drug targets for, 186

Schizosaccharomyces pombe
  α subunits of, 255–257, 263
  G-proteins from, 239, 253

s-cis conformation, in bovine rhodopsin, 12

SDS-PAGE
  of 5-HT$_{1A}$ receptors, 125
  use in α-adrenergic receptor purification, 79, 80
  use in studies of mutant receptors, 64

Second messengers
  coupled to muscarinic receptors, 170, 175

dopamine D2 receptor coupling to, 162–164

peptide receptor mechanisms with, 198

Secretin receptor, mechanism of, 199

Septum
  dopamine D2 receptor mRNA in, 151
  5-HT$_{1A}$ receptors in, 119, 121

Sequestration, of β_2-adrenergic receptors, 46–47

Serine residues
  in α_2-adrenergic receptors, 105
  in β-adrenergic receptors, 67, 68, 69, 72, 73
  in β_2-adrenergic receptors, 40, 44, 101
  in dopamine D1 receptor clone, 143
  in dopamine D1 receptors, 145, 146–147, 149
  in luteinizing hormone-chorionic gonadotropin hormone receptor, 214
  in rhodopsin, 6, 7, 18, 23, 44
  in STE2 receptor, 221–222
  in substance K receptor, 205
  in tachykinin receptors, 211

Serotonergic receptors, 36. *See also* 5-HT$_{1A}$ receptor
  amino acids in, 37
  G-protein interaction with, 84
  ligand binding in, 65, 66, 68

Serotonin
  effect on dopamine D1 and D5 receptor binding, 149
  5-HT$_{1A}$ receptor affinity for, 116, 127
  lack of effect on arachidonic acid release, 123
  as vasoregulatory agent, 118

Serotonin 1c receptor, mitogenic potential of, 253

Serum-derived growth factors, possible muscarinic receptor interaction with, 176

S functional site, of Ha-ras p21 protion and α subunits, 251

Signal transduction
  mechanisms, of 5-HT$_{1A}$ receptors, 121–124
  via G-protein-coupled receptors, 62, 142, 233–269

Sila-hexocyclium, binding by muscarinic m1–m5 receptors, 173

Size-exclusion HPLC, α_1-adrenergic receptor purification using, 79

SKF-82526, effect on dopamine D1 and D5 receptor binding, 149, 150

SKF 104078
  α_1-adrenergic receptor binding of, 90
  α_2-adrenergic receptor binding of, 86, 90, 93
  chemical name for, 77
SK-N-SH cells, muscarinic receptor expression in, 175
Sleep, role of 5-HT receptors in, 118
Slime mold, G-proteins in, 284
S49 lymphoma cells, β_2-adrenergic receptor uncoupling studies on, 42, 43, 46
Smooth muscle, muscarinic receptor activity in, 179
Sodium ions, effect on α_2-adrenergic receptor binding, 105, 106
SOMAP program, use in studies of rhodopsin sequences, 2
Somatostatin, sequence responsible for cAMP regulation in, 53
Somatostatin receptors, mechanism of, 199
Southern blot analysis
  of α_2-adrenergic receptors, 88, 89
  of dopamine D1 receptor clones, 144
  of Giα subtypes, 244
Spectroscopy
  of opsins, 5–6
  of rhodopsin ligand binding sites, 8, 12–13
  of rhodopsin-retinal interaction, 12–13
Spinal cord, 5-HT$_{1A}$ receptors in, 119, 121
Spiperone
  dopamine D2 receptor binding of, 162, 163
  as 5-HT binding inhibitor, 115, 118
Spiroxatrine
  5-HT$_{1A}$ receptor affinity for, 116
  receptor binding of, 116
  structure of, 117
Squirrel monkeys, red and green pigments of, 18
STE18, gene coding for, 258
STE gene family, 220–223
  β and γ subunit coding by, 254, 258
STE2 receptor
  cloning of, 201, 221
  disulfide bond lack in, 226
  genes for, 222
  glycosylation of, 221, 222
  growth arrest from, 204
  loop sizes of, 205
  mutations affecting, 222, 226
  structure of, 221, 224, 225, 226
STE3 receptor
  cloning of, 201, 223

glycosylation in, 206
growth arrest from, 204
loop sizes of, 205
structure of, 223, 225, 258
STE4 receptor, β subunit relationship to, 220
STE18 receptor, γ subunit relationship to, 220
Striatum
  dopamine D1 receptors in, 143, 144
  dopamine D2 receptors in, 162
    in schizophrenics, 164
  muscarinic receptor activity in, 180
Substance K receptor
  clone of, 198, 201, 204
  comparison with neuromedin K receptor, 208–209
  desensitization of, 211
  function and structure of, 204–208
  glycosylation in, 204–208
  loop sizes of, 205
  mas/angiotensin receptor similarity to, 217, 219, 220
  palmitoylation in, 206
  peptide-receptor interaction in, 201, 203
  resemblance to substance P and neuromedin K receptors, 208
  serine and threonine residues in, 211
Substance P receptor
  clone of, 201, 202, 208
  comparison with substance K receptors, 209
  desensitization of, 211
  glycosylation in, 206
  loop sizes of, 205
  palmitoylation in, 206
  serine and threonine residues in, 211
  structure of, 208, 209
Substantia nigra
  dopamine D1 receptor absence in, 143, 151
  muscarinic receptors in, 186
Succinic anhydride, use in opsin structure studies, 4
Sulfhydryl reagents, use in opsin structure studies, 4
Sulpiride, effect on cAMP inhibition, 163, 164
Superior and inferior colliculli, dopamine D2 receptor mRNA in, 151
SV40 early promoter, cAMP stimulation of, 50
Synaptosomes, muscarinic receptor effects on acetylcholine release from, 180

Tachykinin receptors, 204–207
  clones of, 202
  comparison among, 208–211
  mas/angiotensin receptors compared to,
    219
Tachykinins, cell-proliferation regulation
    by, 204
Tamarins, red and green pigments of, 18
TaqI restriction enzyme, use in studies of
    RFLP in dopamine D1 receptors,
    151
Tardive dyskinesis, role of cholinergic
    systems in, 151
Tetrahydroaminoacridine, binding by
    muscarinic m1–m5 receptors,
    173
TGACGTCAA sequence motif, in cAMP
    response elements, 52, 53
Thalamus, muscarinic receptors in, 186
Three-dimensional modeling, of rhodopsin,
    4–7, 24
Threonine residues
  in β_2-adrenergic receptors, 40, 44
  in dopamine D1 receptors, 149
  in 5-HT$_{1A}$ receptors, 129–130
  in luteinizing hormone-chorionic
    gonadotropin hormone receptor,
    214
  in muscarinic receptors, 183
  in rhodopsin, 7, 10, 18, 22, 44
  in STE2 receptor, 221–222
  in substance K receptor, 205
  in tachykinin receptors, 211
Thrombin, as initiator of arachidonic acid
    release, 123
Thrombin receptor, mechanism of, 199
Thyroid cDNA library, G-protein-coupled
    receptor clones form, 36
Thyroid hormone, modulation of G-protein
    expression by, 289
Thyroid-stimulating hormone receptors,
    214–217, 227
  clone of, 201, 202, 214
  comparison with luteinizing hormone-
    chorionic gonadotropin hormone
    receptor, 215
  glycosylation in, 206
  human, 215
  ligand binding in, 224
  loop sizes of, 205
  mechanism of, 199
  palmitoylation in, 206
  peptide-receptor interaction in, 202
  structure of, 216

Thyrotropin-releasing hormone
  enhancement of prolactin release by,
    163–164
  mechanism of, 199
TM 1
  of α_2-adrenergic receptor subtypes, 88
  amino acid residues in, 10, 19
  of dopamine D1 receptors, 145
  of βedit $_2$-adrenergic receptor, 70
  of 5-HT$_{1A}$ receptors, 128
  of mas/angiotensin receptor, 219
  of STE3 receptor, 223, 224
  of thyroid-stimulating hormone receptor,
    215
TM 2
  of α_2-adrenergic receptor, 70
  amino acid residues in, 10, 37
  of dopamine D2 receptor model, 164
  of 5-HT$_{1A}$ receptors, 127, 128
  of luteinizing hormone CG receptor, 206
  of mas/angiotensin receptor, 206, 219
  of neuromedin K receptor, 206
  of peptide receptors, 224
  in STE3 receptor, 223
  of substance K receptor, 205, 206
  of substance P receptor, 206
  of tachykinin receptors, 209
  of thyroid-stimulating hormone receptor,
    206
TM 3
  of α-adrenergic receptor, 70
  of β_2-adrenergic receptors, 67, 68, 69,
    101, 205
  of α_2-adrenergic receptor subtypes, 88
  amino acid residues in, 10, 35, 36, 67,
    73.226
  of dopamine D1 receptors, 145, 146
  of dopamine D2 receptors, 143, 147, 164
  of 5-HT$_{1A}$ receptors, 127, 128
  of luteinizing hormone CG receptor, 206
  of mas/angiotensin receptor, 218
  of muscarinic receptors, 182–183
  of peptide receptors, 224, 226, 227
  of rhodopsin, 7, 21
  of STE2 and STE3, 206
  of STE2 receptor, 221
  of tachykinin receptors, 209
  of thyroid-stimulating hormone receptor,
    206, 215
TM 4
  of α_2-adrenergic receptor, 100
  of β_2-adrenergic receptor, 70, 101
  amino acid residues in, 10, 27, 37, 39
  of 5-HT$_{1A}$ receptors, 128

of mas/angiotensin receptor, 219
of rhodopsin, 7
TM 5
of β-adrenergic receptors, 68, 69, 71, 72, 73, 101
of β_2-adrenergic receptors, 40, 70
of dopamine D1 receptors, 145, 146–147
of G-36 clone of dopamine D1 receptor, 143
of 5-HT$_{1A}$ receptors, 128
of mas/angiotensin receptor, 206, 219
of muscarinic receptors, 182–183
of neuromedin K receptor, 206
of rhodopsin, 7, 21
of substance K receptor, 206
of substance P receptor, 206
TM 6
of β-adrenergic receptors, 69, 71, 72
of β_2-adrenergic receptors, 40, 161
of α_2-C10-adrenergic receptors, 100
of 5-HT$_{1A}$ receptors, 128
of luteinizing hormone-CG receptor, 206
of mas/angiotensin receptor, 206, 219
of muscarinic receptors, 182, 183
of neuromedin K receptor, 206
of peptide receptors, 224
of rhodopsin, 7, 21
of STE2 and STE3, 206
of substance K receptor, 206
of substance P receptor, 206
of tachykinin receptors, 209
of thyroid stimulating hormone receptor, 206
TM 7
of adrenergic β_2/α_2 chimeric receptors, 225
of β-adrenergic receptors, 69, 105
of β_2-adrenergic receptors, 100, 161
of α_2-adrenergic receptor subtypes, 88, 106
of α_2-C10-adrenergic receptors, 100
of dopamine D2 receptor model, 164
of 5-HT$_{1A}$ receptors, 128
of luteinizing hormone CG receptor, 206
of mas/angiotensin receptor, 206, 219
of muscarinic receptors, 181, 182
of neuromedin K receptor, 206
of peptide receptors, 224
of rhodospin, 6, 19, 20, 21, 65
of substance K receptor, 206
of substance P receptor, 206
of tachykinin receptors, 209, 210, 225
of thyroid stimulating hormone receptor, 206

Tolerance, receptor desensitization as, 41
Tourette's syndrome
dopamine D2 receptors in etiology of, 160, 165, 166
muscarinic receptors as drug targets for, 186
Transducins
activation of, 13, 18, 20
GTP binding site of, 21
rhodopsin binding of, 22–23, 71
subtypes of, 234
Transfection, G-protein expression by, 282
Translational elongation factors, in signal transmission, 233
Transmembrane segments. *See also* TM
of α-adrenergic receptors, 84
of bacteriorhodopsin, 5, 6
of dopamine D1 receptors, 145
of dopamine receptor family, 146
of G-coupled receptors, 31–34, 36, 198
helical wheel representation of, 16–17
of 5-HT$_{1A}$ receptors, 127–128
of muscarinic receptors, 181, 182
of rhodopsins, 4
amino acids in, 8
in substance K receptor, 205
Transmembrane signaling
G-protein role in, 233
receptor phosphrylation role in, 55
Trihexyphenidyl
as anti-parkinsonian drug, 186
binding by muscarinic m1–m5 receptors, 172, 173
Trypsin, α cleavage by, 262
Tryptophan residues, in rhodopsins, 6, 10, 11, 16
Tuberoinfundibular tract, dopamine in, 160
d-Tubocurarine, binding by muscarinic m1–m5 receptors, 173
TVX Q-7821. *See* Ipsapirone
$(-)21009$, 5-HT$_{1A}$ receptor affinity for, 116
Two-photon absorption, of rhodopsin-retinal interaction, 12–13
Tyrosine hydroxylase, sequence responsible for cAMP regulation in, 53
Tyrosine kinases, effect on thyroid-stimulating hormone receptor, 217
Tyrosine residues
in α_2-adrenergic receptors, 101
of β-adrenergic receptors, 69
in muscarinic receptors, 183
in rhodopsin, 6, 10, 11, 16, 18